Critique as Uncertainty

A Volume in:
Montana Mathematics Enthusiast: Monographs in Mathematics Education

Series Editor
Bharath Sriraman

Montana Mathematics Enthusiast: Monographs in Mathematics Education

Series Editor

Bharath Sriraman
The University of Montana

Selected writings from the Journal of the Mathematics Council of the Alberta Teachers' Association: Celebrating 50 years (1962–2012) of delta-K (2014)
Edited by Egan J. Chernoff and Gladys Sterenberg

The Development of Teaching Expertise from an International Perspective (2013)
By Su Liang

Crossroads in the History of Mathematics and Mathematics Education (2012)
Edited by Bharath Sriraman

Interdisciplinarity for the 21st Century: Proceedings of the 3rd International Symposium on Mathematics and its connections to the Arts and Sciences, Moncton 2009 (2010)
Edited by Bharath Sriraman and Viktor Freiman

The Role of Mathematics Discourse in Producing Leaders of Discourse (2010)
By Libby Knott

Relatively and Philosophically Earnest: Festschrift in honor of Paul Ernest's 65th Birthday (2009)
Edited by Bharath Sriraman and Simon Goodchild

Critical Issues in Mathematics Education (2009)
Edited by Paul Ernest, Brian Greer, and Bharath Sriraman

Interdisciplinarity, Creativity, and Learning: Mathematics with Literature, Paradoxes, History, Technology, and Modeling (2009)
Edited by Bharath Sriraman, Viktor Freiman, and Nicole Lirette-Pitre

Creativity, Giftedness, and Talent Development in Mathematics (2008)
Edited by Bharath Sriraman

Interdisciplinary Educational Research In Mathematics and Its Connections to The Arts and Sciences (2008)
Edited by Bharath Sriraman, Claus Michelsen, Astrid Beckmann, and Viktor Freiman

Mathematics Education and the Legacy of Zoltan Paul Dienes (2008)
Edited by Bharath Sriraman

Beliefs and Mathematics: Festschrift in honor of Guenter Toerner's 60th Birthday (2007)
Edited by Bharath Sriraman

International Perspectives on Social Justice in Mathematics Education (2007)
Edited by Bharath Sriraman

Critique as Uncertainty

Ole Skovsmose

IAP

INFORMATION AGE PUBLISHING, INC.
Charlotte, NC • www.infoagepub.com

Library of Congress Cataloging-in-Publication Data

The CIP data for this book can be found on the Library of Congress website (loc.gov).

Paperback: 978-1-62396-753-6
Hardcover: 978-1-62396-754-3
eBook: 978-1-62396-755-0

Cover design by Rodrigo Carriel.
Cover picture by Ole Skovsmose.

Printed in the United States of America

CONTENTS

PART 3

DEMOCRACY AS A CHALLENGE

PART 4

MATHEMATICS AND POWER

PART 5

CRITICAL MATHEMATICS EDUCATION

ACKNOWLEDGMENTS

I want to thank Bolema, Educational Studies in Mathematics, For the Learning of Mathematics, Inter-American Conference on Mathematics Education, International Journal of Mathematical Education in Science and Technology, Journal for Mathematics Teacher Education, Nova Science Publishers, Peter Lang International Academic Publishers, Philosophy of Mathematics Education Journal, Springer, and Zentralblatt für Didaktik der Mathematik (ZDM) for kind permission to include already published material in this book. In a note following each previously published chapter I provide the reference to the original.

All these original publications have been accompanied by acknowledgments. In this present book I do not repeat these acknowledgments, but again I do want to thank everybody for their help and support: Helle Alrø, Bill Barton, Richard Barwell, Alan Bishop, Morten Blomhøj, Jo Boaler, Gunnar Bomann, Arindam Bose, Laurinda Brown, Henning Bødtkjer, Iben Maj Christiansen, Paul Dowling, Lisser Rye Ejersbo, Kathrine Krageskov Eriksen, Paul Ernest, Peter Gates, Núria Gorgorió, Brian Greer, Halfdan Grage, Susan Hart, Arne Astrup Juul, Anne Kepple, Steve Lerman, Ruth Loshak, Kristina Brun Madsen, Klaus E. Mortensen, Marilyn Nickson, Danny Martin, Pernille Pind, Núria Planas, Norma Presmeg, Ole Ravn, Kenneth Ruthven, Jeppe Skott, Mikael Skånstrøm, Shikha Takker, Allan Tarp, Paola Valero, Renuka Vithal, Tom Webb, Tine Wedege, and Keiko Yasukawa.

Critique As Uncertainty, pages vii–viii.

I want to thank Lucas Penteado for assisting in ensuring the many different parts of the manuscript have a uniform format; Patricia Perry for the very careful way she prepared the manuscript for publication; the Department of Learning and Philosophy, Aalborg University for financial support for completing this publication; and Bharath Sriraman, the Montana Mathematics Enthusiast and Information Age Publishing for making it all happen.

Finally I want to thank Denival Biotto Filho, Renato Marcone, Raquel Milani, and Miriam Godoy Penteado for writing some chapters together with me and for their collaboration in bringing this volume together. And many more thanks to Miriam, my wife, for her support throughout this whole process.

Ole Skovsmose
—*June, 2014*

INTRODUCTION

The title of the book is *Critique as Uncertainty*, and it is a central part of my concern here to relate these two notions. The concept of *critique* has a long history in philosophy. It forms a part of the whole development of the Modern outlook, and to illustrate that, let me just mention René Descartes, Immanuel Kant, and Karl Marx. As a philosophical concept, *uncertainty* is still in need of some exploration.

Often Descartes has been presented as the first Modern philosopher. As part of his critical endeavour, he introduced the universal doubt: what was possible to doubt could not be true *by necessity*. Furthermore, since Descartes assumed knowledge should only be made up of necessary truths, he let the universal doubt sweep away all forms of assumed knowledge and beliefs. The universal doubt is a powerful critical activity, and one has to remember that this doubt was activated by Descartes within a historical context dominated by the celebration of faith within the Christian religion. When the proper ground for knowledge had been established, eliminating whatever could possibly be doubted, only what must be true with certainty would remain. Upon this foundation the deductive construction of knowledge could be built. Thus, according to Descartes, a strict rational deduction could confront doubt.

Through his philosophy, Kant addressed the questions: What is knowledge? What can be known? What can be known with certainty? Kant wanted to look behind all particular examples of knowledge and to identify the universal conditions

Critique as Uncertainty, pages ix–xv.

for obtaining knowledge. This is the defining element in his critical philosophy. He wanted to provide a foundation for knowledge as well as to identify the possible limits of all possible kind of knowledge. He wanted to study knowledge *a priori* to any specific examples of knowledge. Following Kant, critique turns into an examination of our universal conditions for obtaining knowledge.

Marx, however, added a new dimension to critique. He addressed political economy, both in its theoretical formulations and in its real-life manifestations. In this way, he elaborated critique not only as a logical and epistemological endeavour, but also as a political one. From being the elements of reflection and theory, critique also became a real-life activity. Marx was a firm believer in progress, and he even found it possible to identify the logic of progress. For Marx, this logic is composed of economic and material changes. This crushing dialectical logic had brought about the destruction of feudalism and made space for capitalism which, according to Marx, represented tremendous social progress compared to feudalism. However, Marx's logic of progress also includes the destruction of capitalism, making space for true socialism. According to Marx, a critical activity obtains its essence from this deep logic of progress, which ultimately takes the form of class struggle. While Kant established critique as an investigation of the general epistemic human condition, Marx elaborated critique as an engagement with the general human condition in its material, economic, and political forms.

There are many differences between the critical enterprises of Descartes, Kant, and Marx. There is also, however, an important similarity. They all represent preoccupations with obtaining certainty, either with respect to what one knows, or with respect to what one is doing.

I want, however, to develop critique outside any alliance with certainty. I am not trying to establish any solid foundation for knowledge. Nor am I trying to identify any solid foundation for political actions. Yes, critique includes a political dimension, but critique does not include guidelines or well-defined directions. A critique cannot be positioned on solid foundations. Instead, I want to associate critique with uncertainty, also in the educational domain. In fact I see critique as an expression of uncertainty.

This book is composed of some already published papers and some new texts and is organised in five parts. The chapters in each part are not placed in any strict chronological order of threir original publication. Sometimes I start with the oldest; sometimes with the most recent; sometimes I mix them up. In deciding the final order, I tried to provide the best possible flow for the reader.

I like to write. I like to work on a new manuscript. I like to let ideas develop as I am writing. I am curious to see where the writing might take me. Sure, I have some ideas in advance, but writing is for me a powerful process. It seems to create its own dynamic. I also like to carry out the final editing: changing a word here and a word there, shortening a sentence, and deleting a repetition. Then, finally, the manuscript goes to the printer.

I like to see the printed text. However, I do not read the text again. I had a friend, who published many articles in journals and newspapers. He told that when the publication arrived, he always read it again, from start to finish. Sometimes he read it twice. I cannot image myself doing this. In fact, I do not think that I ever had read one of my published articles or books when it was first published. I admit that I have looked at some, for instance when I have to prepare a lecture. But I have never done a complete rereading.

So, it was some of an experience to bring together this book of previously published work. I have read it all again—and more than once. I have done some minor editing, basically by deleting. Thus I have deleted some references and footnotes and modified some sentences, first of all by shortening them. In some cases I have deleted longer pieces of texts. When one publishes a paper in one journal, and then a related paper in a different journal, the papers might, with good reasons, include some overlapping material. But when these articles appear next to each other in the same book, such repetitions appear disturbing.

Part 1, "Working with Mathematics," includes the paper "Landscapes of Investigation," and let me tell a bit about the context for writing it. In 1996, I was about to give a lecture to about 100 teachers. I was a bit delayed in my preparations, but I knew that I was going to talk about project work in mathematics education. And, sure enough, I would emphasise the importance of project work; the importance of addressing real-life problems; and the importance of students getting ownership of the whole process. I had many examples to draw on and I had given several such lectures, so I was somehow prepared.

I then imagined the teachers' faces when I was about to start; they were also prepared. So many times they must have heard a presenter telling them what they had to do: activating the students; organising group work; and also doing project work. But, they also knew about real-life classrooms, and about the obstructions that tend to turn well-intended ideas into wordy illusions. I imagined their preparation and decided to try to give the lecture a different slant.

First, I did not make any particular use of the notion of project work. Instead I talked more freely about different possibilities for creating investigations. The notion of landscapes of investigation emerged through this lecture. I talked about different possible landscapes, also of landscapes located within mathematics. I did not talk about what a teacher should do, but about possibilities.

My presentation was supported by a figure (Figure 4 in Chapter 1) in which I presented six different teaching-learning milieus. This figure was not assumed to provide any principal categorisation of teaching-learning milieus—there could be many other milieus—it was just used to provide a simple overview of possibilities that I addressed in the lecture. The figure turned out, however, to be very useful and I have used it in many presentations subsequently.

I came to realise how this figure facilitated discussions of educational possibilities and difficulties. It provided invitations for many different discussions; thus the notion of comfort zone and risk zone—introduced by Miriam Godoy

Penteado and elaborated further by Denival Biotto Filho—can be related back to this figure. Certainly it can be elaborated much further, I agree completely. It is a simplistic diagram, but still it works as an invitation for discussing possibilities and challenges in mathematics education and for developing shared pedagogical imaginations.

Part 2, "Foregrounds and Possibilities," addresses a notion I have been struggling with for a long time, and which I am continuing to try to develop. This is the notion of *foreground*, which has to do with the interpretation of learning.

I see learning as a form of action. One can naturally ask if all kinds of learning can be seen as a form of action. This issue of generality I have not found too productive, and I am ready to assume that some forms of learning cannot be called actions. Certainly soldiers are learning something at the drill ground. They are forced to do so, but one can consider to what extent they are acting as learners. One can also consider the almost subconscious assimilations of values within a cultural context. Is this an example of or a counter example to learning interpreted as action?

I leave such questions aside here, and just emphasise that in many cases I find it relevant to consider learning as action. This invites questions such as: What *motives* could students possibly have for engaging in particular learning activities? What *meaning* could they associate to an educational task? What kind of *intentions* could they establish in the learning process?

I characterise a foreground as a combination of two features: First, the possibilities which the social, political, economic, and cultural context provide for the person. Second, the person's experience and interpretation of these possibilities. The notion of foreground provides a way of capturing the space of possibilities and obstructions that constitutes motives, meanings, and intentions.

Part 3, "Democracy as a Challenge," includes my paper "Mathematics Education and Democracy." It has been much referred to, which is something I am very happy about. It also includes a notion which has troubled me: "democratic competence." What kind of competence are we in fact talking about? Does it make any sense to talk about democratic incompetence?

The problem has to do with the use of concepts, like democracy, social justice, inclusion, equality, equity, not to forget critique and critical. Such notions are drifting around in a sky of positive associations. Who would explicitly work for social injustice? Or for more inequality through education? In many contexts it is taken for granted that such concepts demonstrate universally attractive values. One really needs a philosophical anarchism of Nietzschean brutality to be able to point out democracy as an instantiation of a most appalling slave morality. Without mercy, Nietzsche punctured the too-cozy floating blue conceptual balloons.

I do not refrain from using blue balloon concepts. In fact I think it is important to use notions like democracy, social justice, inclusion, and such. It is important, also in education, to try to express visions, imaginations, and aspirations. It is important to try to formulate a mathematics education for social justice; to establish democratic classroom practices; and to work for social inclusion through education. However, it is important to remember that any such visions are not substantiated through the very notions of democracy, social justice, and equity. Such concepts do not contain much firmness or gravity. Nevertheless, I think it is important to use such notions, acknowledging that imaginations can be as fragile as any other blue balloons.

Part 4, "Mathematics and Power," addresses an issue that is important for the whole conception of mathematics. Long ago I started writing about the *formatting power of mathematics*. I considered this an important extension of the social interpretation of mathematics. Many studies have emphasised that mathematics is socially structured: that the apparent pure logical structures of mathematics do not represent any universal rationality, but, instead, include a historicity. The logic of mathematics is a sociological phenomenon, a position with which I am very comfortable.

By talking about the formatting power of mathematics, I wanted, however, to add a second part to this story, namely that social structures are also formed through mathematics. I saw the possibility of a complex looping between mathematics and the social. I got to this idea through discussions of mathematical modelling. Through a range of examples, I pointed out how mathematical models form procedures in production, economy, managements, politics, medicine, warfare, and daily life. It is such observations that I have tried to condense by the construct of the formatting power of mathematics.

It was pointed out to me, however, that the very construct "formatting power of mathematics" invites misunderstandings. By talking about the formatting power of mathematics it sounds like it was *mathematics* as such that did the formatting. However, it is crucial to address the complexity of the agency of the formatting. This agency includes the people who constructed the model; the institutions that asked for the model building; and the routines established through the model. A model could serve a multitude of interests. All this could hardly be assigned to mathematics.

In order to prevent any simplification of this issue, I started talking about *mathematics in action*. This expression did not invite any simple conclusions about agency. In fact I felt that this change of terminology provided a more adequate way of addressing the relationship between mathematics and power. This is a relationship in need of being further explored. It was, however, not addressed in any classic philosophy of mathematics, which appears blind to power issues. Foucault provided a grand opening of the science-power issue. He concentrated on psychology, psychiatry, medicine, and related issues. He demonstrated how apparent science-based standards and procedures incorporate particular interests and perspectives of the particular historical period. He did not address, however, the so-called exact sciences including mathematics. Within this domain, we find a principal interaction between knowledge and power.

I have found it helpful to think of mathematics as discourse. Naturally, I do not want to claim that mathematics is *only* a discourse. However, the discourse perspective helps to broaden the interpretation of mathematics and power. To address mathematics in action becomes similar to address discourses in action. Certainly discourses are operating through people; still they can form ideologies, structures, and routines, just as in the case of mathematics.

Part 5, "Critical Mathematics Education," concludes with my paper "Mathematics Education versus Critical Education." This was my first publication in English. In 1982 I had finished my PhD and I had published books and papers, all in Danish. I had got a job at Aalborg University and was working in mathematics education. One day one of my colleagues asked whether I had considered publishing something in English. In fact, I had not really thought of that. I saw myself as a Danish mathematics educator, engaged in the Danish context; English was a bit out of my reach.

Anyway, I came to be attracted to challenge, and I started writing in English. It was a paper-and-pencil exercise. When I had drafted some pages, I sent them to a friend of mine, who was very well versed in English. Like a good teacher, she corrected the texts. I rewrote the pages again, and then added some more pages, and so it continued until I got to the end of the paper. There then followed several revisions of the whole text. Eventually I found myself unable to imagine any more improvements, and I sent the paper to *Educational Studies in Mathematics*. It was well received, although the English was still problematic. But the editor, Alan Bishop, supported me tremendously in bringing everything into a proper shape such that in 1985, it was published.

The title, "Mathematics Education versus Critical Education," reflects some of my principal concerns of the time. Let me explain. From the mid-1970s, I began trying to formulate a critical mathematics education. However, there was one issue, which at the time I experienced as a challenge. Critical education in general was far from expressing any interest in mathematics education. If not simply ignored, mathematics education was considered as being almost an obstruction to critical education. This position found some justification in Critical Theory and in particular in Habermas' formulations. He claimed that the interest in emancipation would establish the social sciences in their proper critical format, and many formulations of critical education emphasised that education had to be guided by this interest. As a consequence, it appeared contradictory to talk about a critical mathematics education, as mathematics was claimed to be constituted by a technical interest.

Certainly I was ready to acknowledge that mathematics education, as acted out in the school mathematics tradition, did not demonstrate any affinity with the visions of critical education. Nor did mathematics education, in the form suggested by the Modern Mathematics Movement, demonstrate any such affinity. One had to search for new directions for establishing a critical mathematics education. In other words, critical mathematics education had to be worked out, not through any theoretical import, nor even from Critical Theory, but through its own theoretical and educational creativity.

Critical mathematics education has been on its way for a long time. However, it is not anything specific and stable; it is a process, a challenge, a preoccupation. I do not assume the existence of any blueprints for social and political improvements. I do not think that any theoretical structures can provide any solid foundation for a critical activity. For me critique is an open activity, and an uncertain activity. This applies just as much to critical mathematics education.

PART 1

WORKING WITH MATHEMATICS

CHAPTER 1

LANDSCAPES OF INVESTIGATION

In his observations in English classrooms, Cotton (1998) noticed that a mathematics lesson is divided into two parts: first, the teacher presents some mathematical ideas and techniques; then the students work with selected exercises. He has also noticed that there are variations of the same pattern, reaching from a full-lesson teacher presentation to a full-lesson student occupation with exercises. According to this and many other observations, traditional mathematics education falls within the *exercise paradigm*. Most often, the mathematical textbook represents a "given" for the classroom practice. Exercises are formulated by an authority external to the classroom. This means that the justification of the relevance of the exercises is not part of the mathematics lesson itself. Furthermore, a central premise of the exercise paradigm is that one and only one answer is correct.

The exercise paradigm can be contrasted with an *investigative approach*. Such an approach can take many forms, one example being project work, as described for primary and secondary school education in Nielsen, Patronis, and Skovsmose (1999) and in Skovsmose (1994), and for university studies in Vithal, Christiansen, and Skovsmose (1995). In general, project work is located in a "landscape"

Critique as Uncertainty, pages 3–20.

which provides resources for making investigations. Project work represents a learning milieu, different from the exercise paradigm.

My interest in the investigative approach is related to critical mathematics education, which can be characterized in terms of different concerns,[1] one of which is the development of *mathemacy*, seen as a competence similar to literacy, as characterized by Freire. Mathemacy refers not only to mathematical skills but also to a competence in interpreting and acting in a social and political situation structured by mathematics. Critical mathematics education includes a concern for developing mathematics education in support of democracy, implying that the microsociety of the mathematics classroom must also show aspects of democracy. Critical mathematics education emphasizes that mathematics as such is not simply a subject to be taught and learnt (no matter whether the processes of learning are organized according to a constructivist or a sociocultural approach). Mathematics itself is a topic which needs to be reflected upon, as mathematics is part of our technology-based culture, and it exercises many functions, which may best be characterized by a slight reformulation of "Kranzberg's First Law": What mathematics is doing is neither good nor bad, nor is it neutral (see Kranzberg, 1997). D'Ambrosio (1994) has used a more harsh formulation emphasizing that mathematics makes part of our technological, military, economic and political structures, and as such it becomes a resource for wonders as well as for horrors.[2] Making a critique of mathematics as part of mathematics education is a concern of critical mathematics education. Such concerns seem better taken care of outside the exercise paradigm.

The following presentation is partly based on my work on project-based mathematics education, and it is related to my work with teachers, with whom I have discussed these ideas—teachers working in very different political, economic, and cultural contexts. I always start with an example.

An Example

A landscape which can support investigative work, I call a *landscape of investigation*.[3] We take a look at the good old table of numbers, which has certainly decorated the walls of many mathematics classrooms and served as basis for a variety of exercises. We concentrate on a rectangle drawn on the table. If the numbers in the corners of the rectangle are labelled a, b, c, and d, and it is possible to calculate the value of determined by $F = ac - bd$.

The rectangle can then be translated to another position, and the value of $F = ac - bd$ can be calculated again.

[1] See Skovsmose and Nielsen (1996).

[2] See also D'Ambrosio (1998).

[3] The following example is inspired by Ole Einar Torkildsen's lecture at the NOMUS Conference in Aalborg (Denmark) in 1996.

1	2	3	4	5	6	7	8	9	10
11	12	13	14	15	16	17	18	19	20
21	22	23	24	25	26	27	28	29	30
31	32	33	34	35	36	37	38	39	40
41	42	43	44	45	46	47	48	49	50
51	52	53	54	55	56	57	58	59	60
61	62	63	...						

FIGURE 1. The good old table of numbers.

For instance (see Figure 1), we observe that $22 \times 34 - 24 \times 32 = -20$, and that $37 \times 49 - 39 \times 47 = -20$. Let us try to translate the rectangle to a different position and again calculate the value of F. By the way, what will happen if we rotate the rectangle 90° and make the same calculation? Well, . . . What is going to happen, if we choose a bigger rectangle and make a similar translation? What will now be the value of How does the value of $F = ac - bd$ depend on the size of the rectangle?

Naturally, it is possible to investigate translations of other figures. What will happen if we calculate the values $F = ac - bd$ and a, b, c, and d refer to the numbers determined by the corners of the shapes shown in Figure 2? Which of these figures can be "translated" within the table of numbers without the value of F being changed?

Why not investigate a function different from F? For instance, what will happen if we permute the operations subtraction and multiplication and instead of $F = ac - bd$ calculate $G = (a - c)(b - d)$ (a, b, c, and d and still refer to the corners of a rectangle)? Would G be constant under translation? What about the other figures shown at Figure 2? Do other functions exist which are rectangle-translatable (meaning that the value of the function is kept constant during a translation)? Yes, of course a function H defined as $H = 0a + 0b - 0c + 0d$. But do more "in-

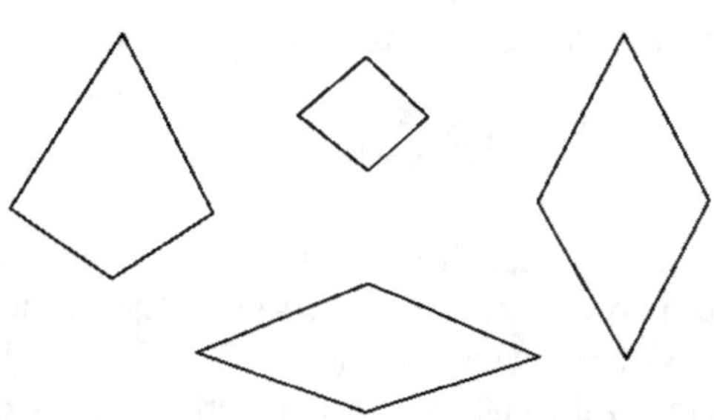

FIGURE 2. Other figures to be translated.

1	2	3	4	5	6	7
8	9	10	11	12	13	14
15	16	17	18	19	20	21
22	23	24	25	26	27	28
29	30	31	32	33	34	35
36	37	38	39	40	41	42
43	44	45	...			

FIGURE 3. A different set up for the table of numbers.

teresting" rectangle-translatable functions exist? If we succeed in finding such a function, would it, then, also be rhombus-translatable? Would, in fact, any rectangle-translatable function be rhombus-translatable? In more general terms, what functions make which figures translatable?

What if we consider negative numbers? Thus, the number table from Figure 1 could be extended adding numbers to the left and to the right of each line, so that we have to do with number lines placed on top of each other. We could then consider translations which bring the figures into areas with negative numbers. Incidentally, what would happen if the table was set up as shown at the Figure 3?

It must also be possible to carry out the calculation in a different number base. Would the quality of "translatability" depend on which number base we are considering?

Naturally, we need not concentrate on configurations of numbers determined by the corners of a figure with four corners. We could consider any configuration of numbers, a_1, ..., a_n, and any function, $F = F(a_1, ..., a_n)$. The question would then be, what functions defined on a configuration of numbers are constant with respect to translation of the configuration? And why not consider rotation as well? Or any other movement of the figure? Furthermore, up to now we have concentrated on a particular property of the function F, being constant or not, but we could observe many other properties of the function F. This leads to the question, What functions defined on a configuration of numbers exhibit "nice" properties under translation?

What if . . . ?

We imagine that this example has occupied some students and a teacher for a while. We have been observing their conversation. The teacher has asked "What if . . . ?" and later we hear again his or her "What if . . . ?" The students might be surprised by some of the mathematical properties indicated by the questions. Mumbling is heard all around. Later it becomes possible to hear students' voices more clearly "What if . . . ?" "Yeah, what if . . . ?" Maybe the teacher asks "Why is it that . . . ?" which leads to more mumbling and, maybe longer periods of silence. Later on some of the students voices can be heard "Yes, why is it that . . . ?"

A landscape of investigation invites students to formulate questions and to look for explanations. The invitation is symbolized by the teacher's "What if . . . ?" The students' acceptance of the invitation is symbolized by their "Yes, what if . . . ?" In this way the students become involved in a process of *exploration*. The teacher's "Why is it that . . . ?" provides a challenge, and the students' "Yes, why is it that . . . ?" illustrates that they are facing the challenge and that they are searching for *explanations*. When the students in this way take over the process of exploration and explanation, the landscape of investigation comes to constitute a new learning milieu. In a landscape of investigation the students are in charge.

Does the example about the translation of figures then, in fact, function as a landscape of investigation? Maybe, maybe not, because a landscape only functions as a landscape of investigation if the students *do* accept the invitation. Serving as a landscape of investigation is a relational property. Acceptance of the invitation depends on the nature of the invitation (the possibility of exploring and explaining pure mathematical properties of a number table might not appear so attractive to many students). It depends on the teacher (an invitation can be presented in many ways, and to some students an invitation from a teacher might sound like a command). And it depends certainly on the students (they might have other priorities for the time being). What might serve perfectly well as a landscape of investigation for one group of students in one particular situation might not provide any invitation to another group of students. The question whether a certain landscape might support an investigative approach or not is an empirical question which has to be answered through an experimental educational practice by the teacher and students involved.

Milieus of Learning

Classroom practices based on landscapes of investigation contrast strikingly with the exercise paradigm. The distinction between the two can be combined with a different distinction which has to do with the "references" which might provide the mathematical concepts and the classroom activities with some meaning to the children.

In philosophy, many attempts have been made to clarify the notion of meaning in terms of reference. Such attempts have inspired mathematics educators to discuss meaning in terms of possible references of mathematical concepts. For instance, the notion of fraction can be introduced by referring to the division of pizzas, while, later, the meaning of "fraction" can be developed further by introducing different sets of references. However, meaning can also be seen, first of all, as a characteristic of actions, and not just as a characteristic of concepts. In my interpretation, references also include motives for actions; in other words, it includes the context for locating an aim of an action (performed by the student in a mathematics classroom). When, in what follows, I talk about different types

	Tradition of exercises	Landscapes of investigation
References to pure mathematics	(1)	(2)
References to a semireality	(3)	(4)
Real-life references	(5)	(6)

FIGURE 4. Milieus of learning.

of references, I will generally be alluding to meaning production in mathematics education.[4]

Different types of reference are possible. First, mathematical questions and activities can refer to mathematics and to mathematics only. Second, it is possible to refer to a semireality—not a reality that we actually observe, but a reality constructed by, for instance, an author of a mathematical textbook.[5] Finally, students and teachers can work with tasks referring to real-life situations.

Combining the distinction between the three types of reference and the distinction between two paradigms of classroom practices, one gets a matrix showing six different types of *learning milieus* (Figure 4). What I mean by learning milieu, I shall try to clarify further by commenting on the different types of milieus suggested by the matrix. Type (1) is positioned in a context of "pure mathematics" as well as in the tradition of exercises. This learning milieu is dominated by exercises, which can be of the form:

$$(27a - 14b) + (23a + 5b) - 11\text{a} =$$

$$(16 \times 25) - (18 \times 23) =$$

$$(32 \times 41) - (34 \times 39) =$$

Type (2) is characterized as a landscape of investigation located in numbers and geometric figures. The introductory example about the translation of geometric figures in a number table illustrates this type of milieu.

The type (3) milieu is located in the paradigm of exercises with references to a semireality. The nature of such a semireality can be illustrated by the following example:

> Shopkeeper A sells dates for 85p per kilogram. B sells them at 1.2 kg for £1. (a) Which shop is cheaper? (b) What is the difference between the prices charged by the two shopkeepers for 15 kg of dates?

---

[4] For an analysis of meaning production in mathematics education, see Lins (2001).

[5] Christiansen (1997) refers to a "virtual reality" as a reality which is established by the mathematical exercise itself. I use the notion "semireality" in a similar way.

Certainly there is talk about dates, shops, and prices. But I do not suppose that the person who constructed this exercise made any empirical investigation of how dates are sold or interviewed a person in order to find out under what circumstances it would be relevant to buy 15 kg of dates. The situation is artificial. The exercise is located in a semireality. This example is presented in Dowling's book, *The Sociology of Mathematics Education: Mathematical Myths/Pedagogical Texts* (1998), in which he describes the "myth of references." It is certainly a myth that such an exercise refers to any reality. But, as I see it, it has a reference: a semireality imagined by the author of the problem.

It might be a reference which can support some students in solving the problem. However, the practice of mathematics education has established specific standards for how to operate in such a semireality. If, for instance, a student asks the teacher about the distance between the shops and the home of the person who is going to buy the dates, and if the student wants to figure out how long distance it is possible to carry a bag of 15 kg. by making an experiment in the school yard, and if the student asks whether both shops can be expected to deliver the dates or not, and whether it can be assumed that the qualities of the dates from the two shops are the same, then the teacher would most likely regard the student as trying to obstruct the whole mathematics lesson.

Certainly, such questions generate obstruction considering the general "agreement" between teacher and students operating in the exercise paradigm. Solving exercises with reference to a semireality is an elaborated competence in mathematics education, based on a well-specified contract between teacher and students. Some of the principles from the agreement are the following: The semireality is fully described by the text of the exercise. No other information is relevant in order to solve the exercise. Further information is totally irrelevant; the sole purpose of presenting the exercise being to solve it. A semireality is a world without sense impressions (to ask for the taste of the dates is out of the question), only the measured quantities are relevant. Furthermore, all quantitative information is exact, as the semireality is *defined* in terms of these measures. Thus, the question whether it is relevant to negotiate the price or to buy, say, a little less than15 kg of dates is devoid of meaning. The exactitude of the measurement combined with the assumption that the semireality is fully described by the information provided, makes it possible to maintain the one-and-only-one-answer-is-correct assumption. The metaphysics of the semireality makes sure that this assumption can be maintained, not only when references are made exclusively to numbers and geometric figures, but also when references in exercises are made to "shops," "dates," "kilograms," "prices," "distances," as well as to other seemingly empirical entities.[6] In particular, this metaphysics has structured the communication between teacher and students.

---

[6] If it is not realized that the way mathematics fits the semireality has nothing to do with the relationship between mathematics and reality, then the ideology of certainty has found a habitat. For a discussion of the ideology of certainty, see Borba and Skovsmose (1997).

				x						
				x	x			x		
	x	x		x	x	x	x	x		x
2	3	4	5	6	7	8	9	10	11	12

FIGURE 5. The terrain of the horse race.

It has been considered irrelevant to make actual observations of how mathematics is operating in real-life situations to being able to construct exercises of type (3). But recently, much more careful studies of mathematical practices in different work situations have been carried out.[7] Real-life based exercises provide a learning milieu of type (5). For instance, figures concerning unemployment can be presented as part of the exercise, and based on such figures questions can be asked about the decrease or the increase of the employment, comparisons can be made between different periods of time, different countries, and so on.[8] All figures referred to are real-life figures, and this provides a different condition for the communication between teacher and students, as it now makes sense to question and to supplement the information given by the exercise. Still, the activities are settled in the exercise paradigm.

Like milieu (3), milieu (4) also contains references to a semireality, but now this semireality is not used as a resource for a production of exercises but as an invitation for the students to make explorations and explanations. The "big horse race" can serve as an example. The racecourse is drawn on the blackboard, and eleven horses: 2, 3, 4, . . . , 12, are ready for start. Two dice are thrown, the sum of the number of spots shown is calculated and a cross is made at the diagram. As Figure 5 shows, the sum 6 came out three times before any of the other sums. Horse 6, therefore, became the lucky winner, followed by horse 7 and horse 10.

This horse race can be developed into a greater classroom activity. Imagine that we have to do with children, about 11 years old. Two bookmaker agencies are set up behind desks in corners of the classroom. A small group of students run each agency. Independent of each other, the agencies announce their odds. The rest of the class, the very wealthy gamblers, make their bets: "Look, agency A pays back 8 times the money on horse number 9. But look at agency B! They pay back 40 times for horse number 10!" Placing the bets has to be done in a hurry, as the next race is soon going to start. Another group of children are in charge of the race, they ring the bell, and (a kind of) silence enters the classroom. The dice are thrown, the sums are calculated, crosses are made, the horses race towards the goal line. Some of the gamblers show big smiles.

[7] See, for instance, Wedege (1999).

[8] See, for instance, Frankenstein (1989) for exercises of this type.

Agency A has only few customers. Their odds seem far less favorable than those provided by agency B. However, a new race is going to start. New odds are suggested. The gamblers become surprised: "What marvellous odds this agency A is now offering!" New bets, new races, new winners, new losers. The horses are not anonymous anymore, and horse number 2 is called the turtle. Suddenly, one agency loses its whole fortune. Anyway, a new millionaire sets up a new agency.

The teacher suggests that it is time for a derby. Up to now the races have had the length of 3 units, but a derby must be of 5 units, at least. Odds are produced by the agencies. Some of the gamblers are wearing paper hats. After the second derby, some of the gamblers start wondering: Could horse number 7 possible be particularly fit for derbies?

Even after several races, there is no smell of horses in the classroom. The great horse race takes place in a semireality, but not in the educational paradigm defined by exercises. And the many remarks about the abilities of the different horses ("Horse number 11 needs some vitamin pills") are not perceived as obstructions. The strict logic governing the semireality of learning milieu number (3) is no longer in operation. The whole activity is located in a landscape of investigation. Many discoveries are waiting for the children. Strategies are to be produced and improved. And, as I have chosen to describe the activity, the children certainly accepted the invitation to participate in the big horse race.[9]

Another Example

Naturally, it is possible to develop landscapes of investigation with a greater degree of reality involved than the big horse race. In *Towards a Philosophy of Critical Mathematics Education* (1994), I have discussed some examples organized as project work, and such work can illustrate the learning milieu (6).

The project "Energy" concentrated on input-output models for energy. As an introduction, the students calculated how much energy certain types of breakfast contained (energy was measured in kJ). Then it was calculated, using formulas from sports research, how much energy was used during a certain trip on a bike. Formulas expressed the use of energy as a function of different parameters like speed, length of the trip, type of bike, and the "frontal area" of the cyclist. How to measure this area? The students found a method and completed their calculations. In this way they were introduced to the idea of making an input-output model for energy.

Then the project concentrated on input-output models for farming. The students investigated a specific farm, not far away from the school. First, it was calculated how much energy in terms of, for instance, gas was used in preparing a certain field during a year. Sitting in the barn, the students listened to the farmer explaining the methods of preparing the field. The students then measured the

9 I made this description while I was a visiting scholar in England. Had I stayed in Denmark, I would most likely have described the big cycling race.

breadth of the different tools: the plough, the harvest, and such. This made it possible for them to estimate how many kilometers the farmer, on an annual basis, had to drive the tractor in preparing the field. On this field barley was grown, and it was calculated how much energy the harvested barley contains. In these calculations, research statistics about farming were used. According to the students' calculations, the input-output figures were very profitable: the harvested barley contained about 6 times the energy that was "supplied" to the field, the reason, of course, being that the sun is a great supplier of energy. This result could be compared to the official statistics in Denmark, which reveals that the factor is somewhat smaller, one reason being that the students did not consider all the relevant types of transports which are necessary in order to manage the farming.

In this particular farm the barley was used as food for pigs, and the students could set up a new input-output model. They collected information about how much pigs were eating depending on their weight, and about the time needed before they were brought to the bacon factory. It was then possible to set up a new input-output model, and the factor was calculated to be about 0.2. Only one fifth of the energy contained in the food supplied to the pigs is contained in the meat. Meat production, thus, seems to be a bad "economy" in terms of energy.

Were these findings characteristic of the chosen farm only? The official statistics about Danish farming could report that also in this case the students' results were similar to official results. From an energy point of view, turning barley into meat costs a lot of energy. In this sense, the students' investigations became exemplary, and this is an essential element in project work. The discussion can be carried on. Is Danish farming doing things in a particular bad way, from an energy point of view? Not particularly bad. As statistics can tell, not as bad as, for instance, the U.S. farming which demonstrates a most problematic energy account.

This project illustrates different aspects of learning milieu (6). The references are real, and they provide the activities (and not only the concepts) with meaning. The students are making calculations related to real farming. This means that "authorities" which were exercising their power in the exercise paradigm are eliminated; thus, the assumption one-and-only-one-answer-is-correct does not make sense any longer. When such a project is running, textbooks can rest safely in a corner of the classroom. The teacher comes to serve as a supervisor, and new inquiry-oriented discussions emerge: How in fact to calculate the front area of a cyclist? The problems now become setting up models for the input-output calculations, and it becomes important to reflect on the result of the calculations. Are the results reliable? Did we consider the relevant factors? Well, we can compare to official statistics. But are these results correct? Critical reflection on mathematics and on mathematical modeling gets a new significance.

In Denmark the official curriculum is no hindrance for students and teacher to operate in learning milieu (6). No exams after each school year, where the students can fail, determine specific classroom activities. Only after the 9th year will the students face a national exam in mathematics. Marks will be given, but everybody will pass. Furthermore, this exam supports an investigative approach since, in its written part, it does not presuppose any memorized knowledge and, in its oral part, it concentrates on groups of students making mathematical investigations. Nevertheless, the exercise paradigm also finds a strong support in this corner of the world.

Moving between Different Learning Milieus

Naturally, the matrix in Figure 4 represents a strong simplification. The vertical line separating the exercise paradigm from landscapes of investigation is certainly a very "broad" line, representing a huge terrain of possibilities. Some exercises can provoke problem-solving activities, which might turn into genuine mathematical investigations. Problem posing means a further step into landscapes of investigation, although problem-posing activities can be very different from project work. No doubt the horizontal lines are also "fluffy." My point, of course, is not to try to provide any clear-cut classification but to elaborate the notion of milieus of learning in order to facilitate discussions about making changes in mathematics education.

A good deal of mathematics education is switching between the milieus (1) and (3). In this sense the exercise paradigm provides a foundation for "tradition" in mathematics education. Many studies in mathematics education have been carried out providing a desolate picture of what is going on in a traditional classroom. Some of these studies, however, do not acknowledge that other possible learning milieus do exist in mathematics education, and that the observations are linked to a particular organization of the mathematics classroom, although a most typical one.[10] A differentiation between "the school mathematics tradition" and "the inquiry mathematics tradition" has been suggested by Richards (1991). This differentiation certainly also fits into the matrix. Exercises are a defining element of the school mathematics tradition.

In Denmark a challenge to the school mathematics tradition has been presented by the type (6) learning milieu. However, I find it important that the challenges are organized in terms of learning milieus of types (2) and (4) as well as of (6). I do not want to make the claim that milieu (6) is the only essential alternative to the exercise paradigm. In fact, I do not want to suggest that a particular learning milieu can become designated to represent the ultimate goal for mathematics education, critical or not.

I support a mathematics education moving between the different milieus as presented in the matrix. In particular, I do not regard it as an aim to abandon exercises from mathematics education altogether. It might make good sense after, say, the big horse race to use a period for "consolidation" in which the students work with exercises related to the notion of probability. It is important that students and teacher together find their route among the different milieus of learning. The

---

[10] See, for instance, Walkerdine (1988).

"optimal" route cannot be determined in advance but has to be decided upon by students and teacher. The matrix of learning milieus can also be used as an analytic tool. For instance, it is possible for the students and the teacher to reconsider last year's route: Which learning milieus did we experience? Have we spent all the time in one or two milieus? In which milieu did we experience a particular success? Did some moves from one milieu to another cause difficulties? Many considerations of planning can be referred to the matrix.

Long ago, I was engaged in a mathematical project involving young children, about 7 years old. The main aim of the project was to plan and to construct a playing ground outside the windows of the classroom where there was a small piece of ground available for the class. Certainly, this activity took place in a learning milieu of type (6), and, as a result of the project, a small playground was in fact set up outside the windows of the classroom with the active help of parents during a few weekends. Before that, however, much activity had taken place. As the first thing, the children visited other playing grounds in order to test which one was a "good" one. Children, 7 years old, are experts in carrying out this kind of test. More difficult, however, was specifying the exact quality of the good playing ground. How tall are the swings? How much sand is needed? And so on. Many things have to be measured, and in order not to forget such measures it becomes important to make notes about the observations. Not an easy task!

Such periods of intense activity are very fruitful and important, but other more relaxed types of activities are important as well, both for the teacher and for the children. As part of the project about the playground (which lasted for a few months), there were organized periods of "office work," which actually looked like an excursion into the learning milieu of type (1). The children were organized in small groups working in their "offices." As in any public office, voices were low. The children had juice or lemonade in plastic cups standing on their desks which, by some magic, now looked like real office desks. Sometimes the office workers nibbled at a cookie while they added up numbers. Sometimes the radio poured out low, soft music. Sometimes the teacher played the guitar. The papers scattered around the desks contained first of all exercises in adding and subtraction. The point is that the children during the more intensive periods of project work had recognized the importance of being able to add numbers, and to add them correctly. During office hours, these kinds of skills could then be consolidated, and reasons for doing such office work were found in the previous periods of the project work. The actual setup of "office work" broke the pattern of the normal exercise paradigm, although the activity as such was of type (1). This illustrates that the route between the different milieus might help to provide the students' activities with new meaning. The office work did not take place in an atmosphere of the school mathematics tradition, although it took place in the exercise paradigm. In particular, the communication between teacher and students in the office was not governed by the same logic as the communication between teacher and students adjusted to the school mathematics tradition.

The consolidation provided by office work also serves as a preparation for being engaged in a new project. To create harmony between project work and course work has been a big challenge to the project-based mathematics education—no matter whether we have to do with project based university studies in mathematics or with elementary school mathematics.

Sometimes, in discussions with teachers, it has been suggested to me that before trying to investigate any landscape the students should be equipped with some understanding and techniques which, most efficiently, can be produced within an exercise paradigm. The big horse race illustrates why, in my opinion, this is not generally the case. Had the children, before the race, been introduced to some basic notions of probability illustrated by the canonical diagram: the number of eyes of the red dice is shown at the -axes, the number of eyes of the blue dice is shown at the -axes, and the sum . . . , then the fascination of the game could be lost. However, an opposite route, from (4) to (3), is relevant in many cases. When the game has been tried out and the children have become familiar with the strengths and weaknesses of the different horses and they have got an idea about the reliability of odds, then the children and the teacher can start making particular observations and finding explanations. And exercises can be used as a means of stabilizing some experiences.

Generalisation: Culture of the Mathematics Classroom

The six types of learning milieus have been specified in terms of references (to pure mathematics, to a semireality or to real-life situation) and forms of organizing tasks (exercises or investigations), but many other elements can be considered in specifying learning milieus.

A learning milieu is certainly also determined by the nature of the stratification of the students which may take place. By a stratification, I mean a way of providing an order of the students according to ability. A stratification can be made in the most brutal way, as for instance in many schools in England, by a public streaming of the students into A, B, C, and D levels. An evaluation of a student can be communicated to the student, and to the student only. But explaining to a student where he or she is "positioned" in terms of performance is very different from making a stratification public. Such stratification is a most powerful way of disciplining students. And when a stratification is integrated in exam systems of passing and failure, then the stratification takes the form of school violence towards students. Public stratification contradicts the concern of developing the mathematics classroom as a microsociety where democratic values can be experienced. The nature of the communication between teacher and students definitely depends on the stratification in operation.

Many other aspects are important for characterizing a learning milieu: forms of communication which can vary from being fixed by the exercise-discourse to being a dialogue; use of information and communication technologies; economic resources of the school; the students' background; political conflicts represented in the classroom; the students future possibilities in life; and so on.[11] Taken together all such aspects influence the learning milieu, and they establish the *culture of the classroom*. But in order to give this notion a full value we have to consider all such aspects.[12]

The Risk Zone

French research in mathematics education has paid much attention to the notion of *didactical contract*.[13] With reference to the notion of learning milieu, a didactical contract can be defined in terms of "balance in a learning milieu." Thus, a didactical contract refers to an established harmony between parameters of the learning milieu: the way meaning is produced, the tasks are organized, the textbook is structured, the communication are carried out, and so on. And, furthermore, this harmony must be recognized and accepted by both teacher and students. That a didactical contract is established does not, however, reveal anything about the quality of the learning milieu. It first of all indicates that teacher and students have a shared understanding and acceptance of the priorities of the learning milieu. Their interaction is not problematic as long as both parties recognize the contract.

A didactical contract can be broken in many ways, for instance when students start asking about details of a semireality, as described previously. The contract can be broken if the evaluation is drastically changed. In general, improvement of mathematics education is closely linked to breaking the contract. And when, initially, I suggested a challenge to the paradigm of exercise, it can also be seen as a suggestion for breaking the contract of the school mathematics tradition.

From the teacher's perspective this might appear as moving from a comfort zone into a *risk zone*. This notion has been introduced by Penteado (2001) in her study of teacher's experiences in a new learning environment where computers play a crucial role. Moving between the different possible learning milieus, and paying special attention to landscapes of investigation, will cause a great deal of uncertainty. My point is that uncertainty is not to be eliminated. The challenge is to face uncertainty.

Computers in the mathematics classroom have helped to establish new landscapes of investigation, thus the computer will easily come to challenge the authority of the (traditional) mathematics teacher. Students working with, say, dynamic geometry will easily come to face situations and experience possibilities not foreseen by the teacher as part of his or her planning of the lesson. A student's eager clicking on the mouse might quickly lead to an unknown corner of the

[11] For a discussion of communication in the mathematics classroom see, for instance, Alrø and Skovsmose (1996a, 1996b, 1998).

[12] See, for instance, Lerman (1994); Nickson (1992); Powell and Frankenstein (1997); Valero (1999); Vithal (1999, 2003); and Volmink (1994).

[13] See, for instance, Brousseau (1997).

program: What to do now? How to get out of here? The teacher must always be ready to face questions which cannot easily be answered. The traditional teacher-authority can be broken within seconds, and nobody knows which time next. Certainly not the teacher. The degree of unpredictability is high.

When students are exploring a landscape of investigation, the teacher cannot predict what questions may come next. One way of eliminating this risk is, for the teacher, to try to guide everybody back into the exercise paradigm and into the comfort zone. Thus, the whole exploration of translatability of geometric figures in the number table could be reorganized as a sequence of exercises. And instead of letting the students play around with the program of a dynamic geometry, the teacher could specify each step to be taken: "First you select a point. Yes, all of you! This point we call A. Then you select another point. This other point we call B . . . " By organizing the activities by means of such orders, the teacher can bring about (almost) the same picture on all the screens in the classroom. When students in this way are moving slowly forward like soldiers in columns and rows, the teacher can prevent the occurrence of the unpredictably events and challenges. By doing so, however, many learning opportunities are lost as well.

Any landscape of investigation raises challenges to a teacher. A solution is not to rush back into the comfort zone of the exercises paradigm, but to be able to operate in the new environment. The task is to make it possible for the teacher and students to operate in cooperation within a risk zone, and to make this operation a productive activity and not a threatening experience. This means, for instance, accepting that "what if . . . " questions can lead the investigation into unknown territory. According to the research of Penteado (2001), an important condition for teachers to be able to operate in a risk zone is the establishment of new forms of cooperative work in particular among teachers, but also along the line of students-parents-teachers-researcher.

However, why bother about operating in the risk zone? Why not simply accept the didactical contract of the school mathematics tradition which has been so carefully elaborated? Cobb and Yackel refer to "intellectual autonomy" as an explicitly stated goal for their efforts to establish an inquiry mathematics tradition in contrast to a school mathematics tradition. Intellectual autonomy is characterized "in terms of students' awareness of and willingness to draw on their own intellectual capabilities when making mathematical decisions and judgements" (Cobb & Yackel, 1998, p. 170). Intellectual autonomy can be associated to the activities of exploration and explanation as facilitated by landscapes of investigation. It is difficult to see this autonomy rooted in those rules which constitute the adequate behavior when operating in a semireality in milieu (3). In particular, leaving the "risk zone" in search for a comfort zone also means eliminating learning opportunities linked to computers-as-reorganizers.

Making a move in the matrix of Figure 4 away from the exercise paradigm and into the direction of landscapes of investigation may help to abandon the authorities of the traditional mathematics classroom and make students the acting subject in their learning process.

Making a move in the matrix of Figure 4 away from references to pure mathematics and to real-life references may help to provide resources for reflections on mathematics.[14] Thus, studies of classrooms where real-world problems were the starting point for mathematical considerations made Voigt state the following hope:

> As future citizens, students will have to cope with many real-world problems that seem to be mathematically intransparent . . . Is the citizen competent to distinguish between necessary mathematical inferences and the suppositions of modelling that depends on interests? It could be hoped that paying more attention to the quality of the negotiation of mathematical meaning in the classroom could improve the education of the "competent layman." (Voigt, 1998, p. 195)

I certainly share this hope, and real-life references seem necessary in order to establish a detailed reflection on the way mathematics may be operating as part of our society. A critical subject is also a reflecting subject.

How can we develop a mathematics education as part of our concern for democracy in a society structured by technologies that include mathematics as a constituting element?[15] How can we develop a mathematics education which does not operate as a blind introduction of students to mathematical thinking but makes students recognize their own mathematical capabilities and makes them aware of the way mathematics may operate in certain technological, military, economic, and political structures? I would never dare to claim that leaving the exercises paradigm in order to explore landscapes of investigation would provide an answer to these questions. Nor would I claim that it is sufficient to build mathematics education solely on real-life references. My only hope is that finding a route among the different milieus of learning may offer new resources for making the students both acting and reflecting and in this way providing mathematics education with a critical dimension.

NOTE

This paper was first published in 2001 in *Zentralblatt für Didaktik der Mathematik, ZDM, 33*(4), 123–132.

REFERENCES

Alrø, H., & Skovsmose, O. (1996a). On the right track. *For the Learning of Mathematics, 16*(1), 2–9, 22.

Alrø, H., & Skovsmose, O. (1996b). The students' good reasons. *For the Learning of Mathematics, 16*(3), 31–38.

[14] See also Cobb, Boufi, McClain and Whitenack (1997).

[15] See, for instance, Chapter 11.

Alrø, H., & Skovsmose, O. (1998). That was not the intention! Communication in mathematics education. *For the Learning of Mathematics, 18*(2), 42–51.

Borba, M., & Skovsmose, O. (1997). The ideology of certainty. *For the Learning of Mathematics, 17*(3), 17–23.

Brousseau, G. (1997). *Theory of didactical situations in mathematics: Didactique des Mathématiques, 1970–1990* (N. Balacheff, M. Cooper, R. Sutherland & V. Warfield Trans. & Ed.). Dordrecht, The Netherlands: Kluwer Academic.

Christiansen, I. M. (1997). When negotiation of meaning is also negotiation of task. *Educational Studies in Mathematics, 34*(1), 1–25.

Cobb, P., Boufi, A., McClain, K., & Whitenack, J. (1997). Reflective discourse and collective reflection. *Journal for Research in Mathematics Education, 8*(3), 258–277.

Cobb, P., & Yackel, E. (1998). A constructivist perspective on the culture of the mathematics classroom. In F. Seeger, J. Voigt, & U. Waschescio (Eds.), *The culture of the mathematics classroom* (pp. 158–190). Cambridge, UK: Cambridge University Press.

Cotton, T. (1998). *Towards a mathematics education for social justice* (Unpublished doctoral dissertation). Nottingham, UK: Nottingham University.

D'Ambrosio, U. (1994). Cultural framing of mathematics teaching and learning. In R. Biehler, R. W. Scholz, R. Sträßer, & B. Winkelmann (Eds.), *Didactics of mathematics as a scientific discipline* (pp. 443–455). Dordrecht, The Netherlands: Kluwer Academic.

D'Ambrosio, U. (1998). Mathematics and peace: Our responsibilities. *Zentralblatt für Didaktik der Mathematik, 98*(3), 67–73.

Dowling, P. (1998). *The sociology of mathematics education: Mathematical myths/ pedagogic texts.* London, UK: Falmer.

Frankenstein, M. (1989). *Relearning mathematics: A different third r— Radical maths.* London, UK: Free Association.

Kranzberg, M. (1997). Technology and history: "Kranzberg's Laws." In T. S. Reynolds & S. H. Cutcliffe (Eds.), *Technology and the West: A historical anthology from technology and culture* (pp. 5–20). Chicago, IL: University of Chicago Press.

Lerman, S. (Ed.). (1994). *Cultural perspectives on the mathematics classroom.* Dordrecht, The Netherlands: Kluwer Academic.

Lins, R. (2001). The production of meaning for algebra: A perspective based on a theoretical model of semantic fields. In R. Sutherland, T. Rojano, A. Bell, & R. Lins (Eds.), *Perspectives on school algebra* (pp. 37–60). Dordrecht, The Netherlands: Kluwer Academic.

Nickson, M. (1992). The culture of the mathematics classroom. In D. A. Grouws (Ed.), *Handbook of research on mathematics teaching and learning* (pp. 101–114). New York, NY: Macmillan.

Nielsen, L., Patronis, T., & Skovsmose, O. (1999). *Connecting corners of Europe: A Greek Danish project in mathematics education.* Århus, Denmark: Systime.

Penteado, M. G. (2001). Computer-based learning environments: Risks and uncertainties for teachers. *Ways of Knowing, 1*(2), 23–35.

Powell, A. B., & Frankenstein, M. (Eds.). (1997). *Ethnomathematics: Challenging Eurocentrism in mathematics education.* Albany: State University of New York Press.

Richards, J. (1991). Mathematical discussion. In E. von Glasersfeld (Ed.), *Radical constructivism in mathematics education* (pp. 13–52). Dordrecht, The Netherlands: Kluwer Academic Publishers.

Skovsmose, O. (1994). *Towards a philosophy of critical mathematics education.* Dordrecht, The Netherlands: Kluwer Academic.

Skovsmose, O., & Nielsen, L. (1996). Critical mathematics education. In A. Bishop, M. A. Clements, C. Keitel, J. Kilpatrick, & C. Laborde (Eds.), *International handbook of mathematics education* (pp. 1257–1288). Dordrecht, The Netherlands: Kluwer Academic.

Valero, P. (1999). Deliberative mathematics education for social democratization in Latin America. *Zentralblatt für Didaktik der Mathematik, 98*(6), 20–26.

Vithal, R. (1999). Democracy and authority: A complementarity in mathematics education? *Zentralblatt für Didaktik der Mathematik, 98*(6), 27–36.

Vithal, R. (2003). *In search of a pedagogy of conflict and dialogue for mathematics education.* Dordrecht, The Netherlands: Kluwer Academic.

Vithal, R., Christiansen, I. M., & Skovsmose, O. (1995). Project work in university mathematics education—A Danish experience: Aalborg University. *Educational Studies in Mathematics, 29*(2), 199–223.

Voigt, J. (1998). The culture of the mathematics classroom: Negotiating the mathematical meaning of empirical phenomena. In F. Seeger, J. Voigt, & U. Waschescio (Eds.), *The culture of the mathematics classroom* (pp. 191–220). Cambridge, UK: Cambridge University Press.

Volmink, J. (1994). Mathematics by all. In S. Lerman (Ed.), *Cultural perspectives on the mathematics classroom* (pp. 51–68). Dordrecht, The Netherlands: Kluwer Academic.

Walkerdine, V. (1988). *The mastery of reason: Cognitive development and the production of rationality.* London, UK: Routledge and Kegan Paul.

Wedege, T. (1999). *Matematikviden og teknologiske kompetencer hos kortuddannede voksne* [Mathematical knowledge and technological competencies at adults with shorter education] (Unpublished doctoral dissertation). Roskilde University Centre, Roskilde, Denmark.

CHAPTER 2

HOW TO DRAG WITH A WORN-OUT MOUSE?

Searching for Social Justice through Collaboration

Miriam Godoy Penteado and Ole Skovsmose

Education for social justice can refer to many things—and in reality, one could hardly consider promoting education for social *injustice*. Consequently, one could claim that the expression "education for social justice" is therefore empty. However, if we consider "education for social injustice," not to be referring to educational *aspirations* but rather to features of educational *practices*, then the notion appears to be far from meaningless. There is a myriad of ways in which we can see *injustice* in education (see, for instance, Gates, 2006).

This injustice can take the form of social exclusion. A most direct example is the way in which some groups of people are prevented from having access to (certain forms of) education. For far too long such restrictions have been in effect in terms of racism: Blacks have been prevented from certain forms of education in the United States; during the apartheid regime in South Africa this type of injustice was exercised to an extreme; Jewish people were banned from educational institutions during a gruesome part of German history. Furthermore, education

Critique as Uncertainty, pages 21–36.

has been socially unjust in terms of sexism; one only has to consider when women got the opportunity to attend universities in many countries. These are all very diverse examples but reflect the depths to which injustice has reached.

Fortunately, such extreme forms of explicit exclusion are difficult to find now. In fact, we find more and more legislation trying to ensure social inclusion through education. In Brazil, for example, great effort was put into providing a "school for everybody." This motto has been expounded with particular reference to children between 6 and 14 years, and inclusion through education has been encouraged at other levels of the educational system, including access to universities. However, even though some considerable effort goes into providing social inclusion via education, there are still many obstacles.

Social inclusion is an attractive motto, but it could easily remain no more than a motto if it is not considered how inclusion can be worked on in practice. Here, we explore what a concern for social justice in terms of social inclusion might mean for teacher education. Naturally, social justice refers to many more issues than social inclusion, but we are going to concentrate on this one aspect. Furthermore, we address the issue with a particular reference to the use of information and communication technology (ICT) in mathematics education for students in a *borderland position* (described below).

Our discussion proceeds along the following lines: (1) We explore what a borderland position might be in order to address what social inclusion might mean. (2) We consider the significance of mathematics education and the use of ICT for processes of social inclusion. (3) We briefly sketch the Interlink Network, as many of our observations refer to this project of collaboration. (4) We present different issues of particular importance to teacher education when aiming to establish a mathematics education for social inclusion. This brings us to (5), our final considerations, where we return to the notion of social justice.

The Borderland Position

In order to clarify the notion of a borderland position, we refer to the notions of the *Fourth World* and the *Network Society* as used by Castells (1998). The Fourth World refers to part of society that falls outside the Network Society. Castells' point is that this Fourth World stretches across many areas of the world; it is not a clearly defined geographic region as, say, the Third World is normally described to be. The Fourth World is made up of regions representing the dropout of the Network Society. Large parts of Western cities—New York, London, and Paris, for instance—make up part of the Fourth World. This world includes not only slums and squatter settlements but also regions where traditions and trades are separated from the global network.

By a *borderland school,* we mean a school positioned close to the Fourth World but still with a good view of the Network Society.[1] It is a school where the inequalities of society are clearly visible and where the diversity of students is reflected in both their poverty and their affluence. Some students come from poor neighborhoods, and for them the school is a guarantee of at least one appropriate meal a day. For many students the school might provide the only possibility for working with ICT.

The Network Society and the Fourth World are connected through different processes of inclusion and exclusion, which are acted out in borderland schools. The poorest students in such a school can "see" what is going on in the well-to-do neighborhoods. They watch TV and they are familiar with all forms of wealthy lifestyles, which could be reality just one block away. They know about the Internet, online games, chat and e-mail, but such things are not easily accessible for them. Their parents do not have credit cards, there is no cable TV at home, and certainly no computer or Internet.

It is not realistic to claim that education can solve all such problems caused by poverty. But poverty easily turns into forms of social *exclusion,* and to promote education for social *inclusion,* requires that we address poverty as an educational challenge. For example, what could it mean to establish a prosperous learning environment within a school in a borderland position? Furthermore, it is necessary to think of food for the students as well as providing the opportunity for them to experience ICT. Social inclusion is not only a set of philosophical concerns, but also a very practical concept.

Mathematics Education, Inclusion, and ICT

A most severe form of social exclusion occurs when a particular group of people never have the opportunity to learn how to read and write. It was with reference to this observation that Freire (1972) embarked on his reading-writing program. However, at the same time he provided a broad interpretation of literacy as not only referring to competencies in reading and writing, but also referring to sociopolitical competencies such as *reading the world* and *reading the world as open to change.* Thus, Freire gave the notion of literacy a profound sociopolitical dimension, and in this way he sought to make social inclusion available through the development of literacy. Mathematical literacy can be narrowly defined as a capacity in reading and writing mathematics, for example a capacity to deal with mathematical calculation, formulae, diagrams, and so on. But as Eric Gutstein suggests, we may alternatively imbue mathematical literacy with a more profound meaning as referring also to a competency in "reading the world" and as "reading it as open to change":

---

[1] For a fuller discussion of the notion of borderland, see Skovsmose, Scandiuzzi, Valero, and Alrø (2008).

> Students themselves are key participants in the social movements of the day. They need tools with which to join the local, and ultimately the global, struggles for justice, peace, and true equality. Among these tools is the capacity to read the mathematical world and to read and write the world with mathematics. (Gutstein, 2006, p. 221)

We use *mathemacy* as referring to a mathematical literacy in the deeper sense of the term. Mathemacy can be related to interpretations of mathematics in action (see Skovsmose, 2005), which emphasizes that mathematics is part of many different social, economic and technological phenomena. Mathematics in action is an integral part of a variety of working processes, of economic management, of processes for design and construction, of strategies for automatization, and of systems for control and security for marketing. We live in an environment built up of mathematics in action to a considerable extent. To make sense of this environment, and to be able to operate within it, one needs to be able to "read" it and to "read it as open to change."

Today competencies in operating with ICT form an integral part of both literacy and mathemacy. Social inclusion (as well as exclusion) concerns access to building such competencies, whether they have to do with mathemacy or literacy. We are claiming that how these competencies are developed by students in a borderland position is of particular significance because of the starkness of the divisions within such schools.

Research finds that ICT may bring huge contributions to the teaching and learning of several school subjects (see, for instance, Tabach, Hershkowitz, Arcavi, & Dreyfus, 2002). In the case of mathematics, the power and speed of computers—in particular for making graphs, tables, and drawings—facilitate the exploration of many mathematical ideas and so open possibilities for new methodologies based on an investigative approach, such as project work (Alrø & Skovsmose, 2002). Forms of project work have increased due to the use of ICT, which makes new forms of group dynamics possible, and new resources become accessible; for example, project work could take the form of telecollaboration (Penteado, Biotto Filho, & Silva, 2006) where new learning environments become established and a rich variety of programs can be applied. Reports of such ICT-based learning possibilities, however positive, bring us to ask the question: What are the implications of the observed development for those many students who have no or only limited access to ICT and are they somehow left behind? To a large extent such questions have been neglected in the many enthusiastic investigations about learning gains associated with the use of ICT. However, social injustice is linked directly to the limitation of access to such learning facilities, so the question is critical.

Social inclusion not only concerns overall social and political issues about formal access for everybody to schooling and further education, nor is it just a matter of ensuring that school buses reach each and every neighborhood. Social inclusion includes a great deal of content-matter issues such as concerns over literacy and mathemacy, two broad domains of competencies with respect to which processes of social inclusion can be established, but where social exclusion may be feared as an outcome.

The Interlink Network

Acknowledging the value of ICT for establishing a "school for everybody," the Brazilian government subsidizes several programs for the implementation of computers in public schools. In general, they include an effort to provide schools with computers as well as to provide some professional development for practicing teachers. Although many more things need to be done, we cannot deny that there has been considerable progress in the last years in several states in Brazil. The schools are getting computers as well as financial investment for educating those who are going to work with this equipment.

However, distributing the educational strategy by means of professional development courses did not result in actions that reached the classroom level. Professional development courses have a value in dissemination, but their efficiency appears almost zero with respect to bringing about new practices (Bovo, 2003). These courses last 30 hours on average, and here the teachers become acquainted with software and possible ways of using it with their students. This, however, does not relate to questions that arise when the teachers return to their schools and try to put into practice what they studied during the course. This of course is a widespread concern in education. More often than not, there is nobody to talk with about these issues, as the colleagues on the course will have been from other schools, and there is no simple way of keeping in contact. Hence, the question remains: How can we put all those suggestions provided by the course into practice?

In the light of this acknowledgement, it was decided to create the Interlink Network project established in 2000, which has run since then with several ramifications (see www.rede-interlink.blogspot.com). The project, directed by the first author, includes collaboration between practicing mathematics teachers, prospective mathematics teachers, and researchers in mathematics education. A particular emphasis of the project is to address the use of ICT within mathematics education. However, we are not going to present particular research results from the project (such results are presented in Penteado, 2004). In this article, we present instead some more general issues for teacher education inspired by the Interlink Network. But before we do that, a few more remarks about the project.

The aim of the Interlink Network is to bring together those who research the use of ICT in educational areas and those who are undertaking to use it in the classroom. It is a locally based initiative and the schools that participated are all located in a borderland position. Many of the students come from homes in poor areas of the city and travel by bus to the school (paid for by the City Council) as there is no school in their neighborhood because it appeared cheaper to provide such transport than to build a school in a position more convenient for the students.

The interaction among the participants in the Interlink Network is established in two basic forms: face-to-face and virtual meetings. The face-to-face interaction takes place in weekly meetings during the Horário de Trabalho Pedagógico Coletivo (Collective Pedagogical Working Time), teachers' time for collective pedagogical work, in their respective schools. Every week, or every second week, researchers from the São Paulo State University together with some prospective teachers are present at these meetings. During this time, the type of activities that may be adequate for the students is discussed. It is important to highlight that the teachers are not obliged to participate in the "Interlink meeting." It is a voluntary activity and only those who want to know more about the educational use of ICT are expected to join in. The second way of interaction is the virtual meetings by means of the Internet.

More than promoting the use of ICT in the classroom, the Interlink Network is aimed at promoting critical discussion about such use and about the implications for teacher practice and teacher education. We have teachers investigating their own practice and prospective teachers investigating their coming practice.

Issues for Teacher Education

Social inclusion does not arise from any simple educational scheme, like for instance, ensuring the presence of computers in school. Establishing social inclusion is a huge challenge. It is not simply a question of personal style or attitude. It concerns teacher education, both preservice and in-service; it concerns teacher professionalism. We illustrate this with a discussion of some of the themes related to mathematics and ICT that have emerged from working with students in a borderland position, as it was experienced through the Interlink Network.

Moving away from the Comfort Zone

Various researchers argue that in order to use ICT in schools, teachers need to be open to changes which range from the physical organization of the classroom to more complex educational change, including the production of new meanings for mathematical ideas (Wilson, 2008; Zbiek & Hollebrands, 2008). To move in this direction implies leaving a *comfort zone* in order to enter a *risk zone* (Penteado, 2001). One must give away the control of a predictable state of affairs and come to operate within uncertain conditions. To follow this path is not easy. It requires, among many other things, familiarity with computers and software, a fresh view on the mathematical content, and much attention to the students' thoughts while they are working at the computer.

A risk zone is characterized as an unpredictable situation where teachers have to be open to negotiate with students on issues they are not sure about, maybe not even familiar with. They have to plan activities for using ICT that they did not learn about in their initial teacher education. Many questions about the math-

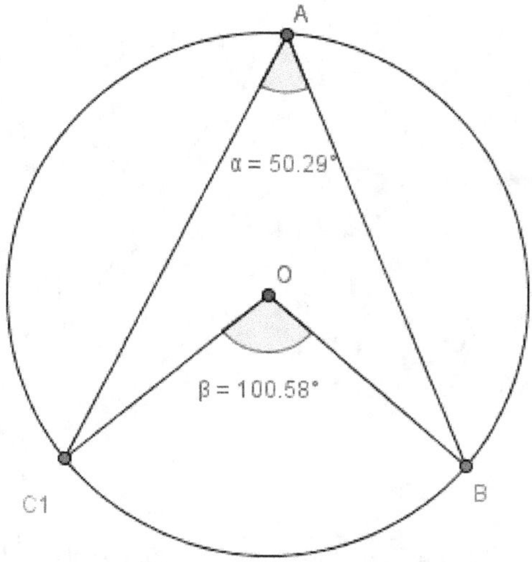

FIGURE 1. Relationship between the circumferential angle and the peripheral angle.

ematical content that the teacher never thought of before are raised by the students when they are working with computers.

Dynamic geometry software makes it possible to provide rich illustrations of geometric properties, and dragging is a powerful tool for doing this. We can, for

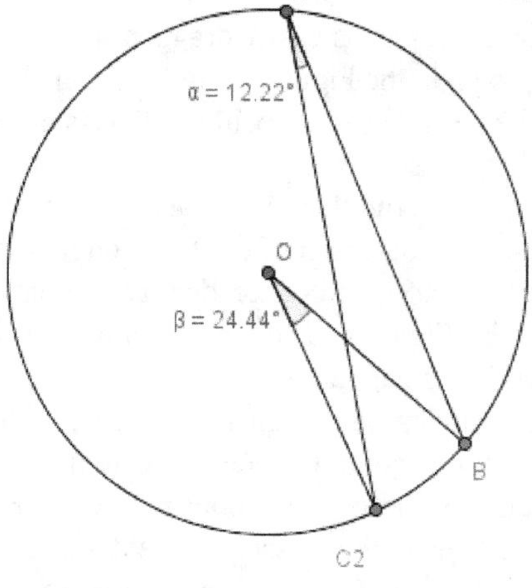

FIGURE 2. Relationship between the circumferential angle and the peripheral angle.

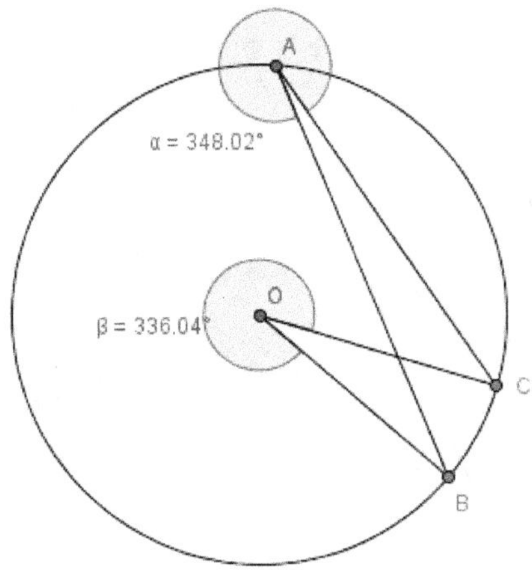

FIGURE 3. Relationship between the circumferential angle and the peripheral angle.

instance, look at the property: The size of the inscribed angle in a circumference is half the size of the correspondent central angle. This property can be illustrated as in Figure 1.

If we are dealing with a printed figure (as here), the illustration works perfectly well. If, instead, the figure is on a computer screen, produced by dynamic geometry software, we can start dragging the point C_1 along the circumference. We could move it to a new position C_2, as shown in the Figure 2, and the figure shows that the property is upheld.

The point C_1 is, however, just as easily dragged in the opposite direction, say, to the position C_3, as shown in the Figure 3. This dragging brings about an illustration which appears strange: What is happening? This is one small example showing that working with such software can bring about surprising results, which need particular explanations. And there are several examples of this kind.

Students may investigate the graph of the function $f(x) = x(2/3)$, using the software *Winplot* (Parris, 2008), and conclude that its domain is all positive, real numbers, as illustrated by Figure 4. The teacher, however, may know that this is not correct and ask, "What is happening?"

Using ICT in mathematics education brings teachers outside their comfort zone. Many surprising things could happen, challenging the traditional authority of the teacher, so characteristic of the school mathematics tradition. According to this tradition the activities in the classroom are defined through preestablished exercises, and it is assumed that the teacher is able to solve all exercises and to check if the students' solutions are correct or not. Operating within this assumption provides a kind of comfort zone in the sense that nothing happens that

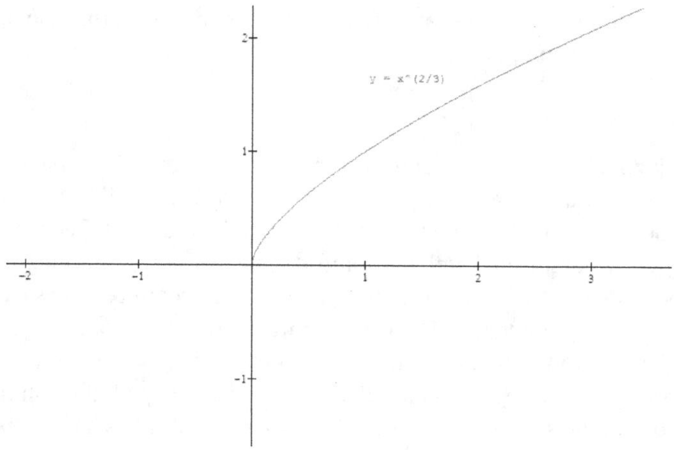

FIGURE 4.

could challenge the teacher's mathematical authority. However, operating with an investigative approach, the teacher and the students will enter landscapes of investigation with many more questions than known answers; and this may happen immediately when ICT becomes applied. This observation concerns all possible situations whether we have to deal with borderland schools or not.

However, many of the uncertainties which the teachers from the Interlink Network communicated were of a different nature. In borderland schools some immediate risks have to do with security. It is not easy for teachers and staff to operate in a situation where computers are very costly compared to the rest of the equipment in the school. A computer provides new learning opportunities but also new opportunities for crime. The computer room has to be secured in such a way that it becomes difficult to access. The teachers found that much of their attention had to be directed towards the controlling of the students and conversely, the teachers themselves experienced being controlled. Such measures are not addressed in the prototypical mathematics education discourse celebrating the educational learning gains provided by the computer. The point, however, is not to return to the comfort zone by, for instance, closing down the computer room. A retreat to the comfort zone does not solve the problem of digital inclusion but exacerbates it.

Furthermore, difficulties emerge due to the lack of quality of the equipment. Few and often old computers have to serve too many students. Many apparently small technical issues have to be addressed, such as for instance, how do you drag with a worn-out mouse? This is a most common problem that faces students working with dynamic geometry and worn-out technical equipment. Certainly some bad-mouse dragging techniques can be developed, like making rapid and repeated small movements with the mouse. Nevertheless, in many cases it is impossible to measure an angle because it requires a good control of the mouse. This has to be taken into consideration when the teachers are preparing the tasks to be worked with at the computer.

Establishing Networks

The risk zone, though, is not simply a "problematic zone"; it also contains possibilities. We could naturally choose to talk about a risk/possibility zone, but we use the word risk to include the meaning of possibility as well. Security and predictability can be associated with the comfort zone, while new learning opportunities can be associated with the risk zone. When a classroom becomes experimental, new things might happen. The teacher might lose some control, but students may also be able to become experimental and make discoveries. If a mathemacy of the form referred to by Gutstein (2006) is to be developed, the solution is not to try to control the situation. Instead, educational possibilities have to be opened up to the students. Social inclusion presupposes risk-taking.

However, it might be impossible for a teacher acting alone to deal with all these risks. It is necessary to find support for operating in a risk zone, and teacher engagement in a network could provide such support. This support is important in ensuring that risks become not destructive but rather entrances to new educational possibilities. In fact we consider that collaboration and the establishment of networks make up part of a teacher's professionalism. It is important that this networking comes to be part of both preservice and in-service teacher education. The potentiality of such networking is discussed by Reid, Brain, and Boyes (2007) addressing examples from a network operating in the UK.

The Interlink Network provides a collective forum where teachers can reflect on their own practices, formulate problems, and search for actions to be taken. It is important that prospective teachers are part of the networking, as they get an opportunity to address the diversity present in a borderland school and reflect on the forms of inequality with which they have to cope. In this way they get close to both teachers and researchers while planning tasks and discussing the problems faced during the development of these tasks, including suggestions for how to cope with insufficient equipment. Meetings among teachers, prospective teachers, and researchers are essential in any teacher education, and particularly in a teacher education guided by a concern for establishing social justice. When we consider a borderland school, a concern for social justice could easily remain a well-intended motto, in the case where there is no profound professional development related to it. Such a development has to be accompanied by an adequate networking forming part of the teacher education itself.

Identifying New Approaches

The Interlink Network brought about new approaches to some of the problems experienced and, most important, many kinds of "disturbances" could be addressed, not as personal problems for the teachers, but as general problems associated with working within the conditions set by a borderland school. It was a principal result of the established network that problems experienced by the teachers could be addressed in a public format. This changed the status of the problems. They were general problems and realizing this is important if social inclusion is to be part of a general education approach.

The Interlink Network is also a space where what the teachers learn is disseminated, particularly in the courses offered to them by the State Secretary of Education. For example, a teacher attended such a course about dynamic geometry software at the Regional Centre for Educational Technology in her city, and she used several Interlink meetings in her school to carry out the activities she learned in collaboration with colleagues who did not have the possibility to attend the course. In another case, a teacher attended a long-distance course about the use of spreadsheets, and, via the Internet, she shared what she learned with the Interlink members. She sent the activities to the e-list and distributed papers to the other schools. In this case the teacher acted as a *multiplier teacher*. However, we find that it is only possible for a teacher to function as a multiplier if he or she can draw on the funds of knowledge of a network such as Interlink (Moll & Greenberg, 1990; Sichieri, 2003). For instance, one mathematics teacher, Sueli (pseudonym), wanted to produce a set of tasks using calculators, but she had no clear idea about how to do it and no time to complete the preparation. She announced her intention, and two prospective teachers who had some expertise with this topic joined her in the school during the planning and the activities in the classroom. After this experience, Sueli felt more secure in applying the same approach on her own with other groups of students. This is an example that knowledge produced by the students from the university can promote educational innovations in the school.

Moving beyond Prototypical Research

Any teacher education should draw on literature, and there is much research which appears relevant. However, when we are dealing with the context of a borderland school, much of that research appears irrelevant.

If we consider research literature in mathematics education as presented in international journals and books, we might realize that a dominant part of this includes, as its principal empirical foundation, observations from what can be referred to as a "prototypical mathematics classroom" (see Skovsmose, 2005, 2006). Such a classroom may be characterized in the following way: it is not crowded with students; the students are not hungry; relevant school books and computers are available; students seem ready to learn, and there are few disturbances. A main part of the theorizing about learning and teaching processes draws on empirical material of such a prototypical format. If, however, we consider the global scene for learning and teaching mathematics, the prototypical mathematics classroom only represents a minority of the sites for teaching and learning, and certainly it does not represent the situation in the borderland school. It is important to be aware that the theorizing in mathematics education might include devastating bias.

When addressing schools in a borderland position, it is imperative that a research-practice interaction is not trapped within the prototypical theoretical discourse. It is important that a different discourse is developed, where all experienced problems are treated not as exceptions to an (idealised) norm, but as principal parts of the common conditions. Furthermore, it does not make sense to think of the communication between researchers and teachers, including prospective teachers, as one-way—from the researcher to the teachers—as if the researcher were the only one who knows what should be done in the classroom. We have, instead, to establish a multidirectional communication. It is necessary that teachers also find their way to expressing what in their opinion needs to be done, and in this way to set up priorities for the research to be carried out. Teachers have to assume their role in educational research programs. This is a complex task, and it depends much on the form of communication established between teachers and researchers. The Interlink Network was intended to create conditions for facilitating this task.

Beyond practical activities, at the Interlink meetings we also discuss theoretical interpretations of the use of ICT in education and its integration in the school environment. In Brazil a large number of master's dissertations and PhD theses in mathematics education are widely available, and some of the prospective teachers engaged in the Interlink Network select part of this material that might be read by the teachers. We discuss articles on how the students produce knowledge in geometry, in particular with dynamic geometry software. The study of literature is always related to the practice we are trying to implement at that moment. We also study literature which addresses problems of violence in the school and what this means for the process of inclusion and at the same time, teachers report on their experiences with respect to such issues. Many times, we discuss the articles at the school but we have also had a productive experience discussing them in an online chat. The important point, however, is to combine the discussion of content and the discussion of contexts, in this case the conditions set by a borderland school.

As part of the Interlink, new ideas for research emerged during this activity. For instance, we have dedicated 2 years to investigating the possibilities of telecollaborative project work where students from different schools could work at the same investigation and share their results through the Internet. Articles produced by Interlink members were published as a result of this research (Penteado et al., 2006). The collaboration made both teachers and prospective teachers more aware of the research that in fact addresses nonprototypical context and borderland situations. Much research from the ethnomathematical approaches addresses such situations (see, for instance, D'Ambrosio, 2006; Ribeiro, Domite, & Ferreira, 2004). Furthermore, much research is now emerging that does not just refer to the ethnomathematics approach, but which addresses nonprototypical educational settings more widely (for an overview, see Skovsmose, 2006).

Getting in Contact

Even though research and insight is being developed with reference to nonprototypical contexts, it is important to be open to the many experiences that working in a borderland school might provide.

Miriam was working with a school student, and they got to talking about having access to computers. The conversation turned to the student's situation at home, and in an off-remark the student mentioned that his father had just gone to jail. He had been painting a car when the police came (apparently his father had the job of changing the appearance of stolen cars). No acquisition of a computer could be expected in this home for the time being. Such an observation does not signify anything out of the ordinary. Usually, this student has some problematic behavior in the classroom, but he likes working at the computer with dynamic geometry software. Computers have a special appeal to him and maybe the reason is that he wants one so much and the school is the only place where he has access to one. The borderland position raises new educational challenges, but it also affords school activities with new forms of significance for the students. The borderland school represents a huge diversity, and the teacher has to consider this when planning the activities for the classroom.

Many students were fascinated by the conjecturing about mathematical properties emerging from an investigation with dynamic geometry software. One of the students, 12 years old, however, stayed quiet. She looked at what the others were doing, but did not want to do anything herself. After some time we convinced her to put her hand on a mouse. This was the first time in her life that she had done so. It was the first time she experienced that moving her right hand holding the mouse established a movement at the screen. In the beginning she was not able to coordinate these two movements. She appeared ashamed, but nobody paid any attention to her. After some time she could smile. She came to experience that *she* was the one who moved the cursor on the screen. She was in fact getting ready for the next step in taking control: how to drag a corner of a triangle.

Motives for learning can have very different formats. The construction of meaning by students in a borderland position might not follow those patterns that have become commonsense through prototypical research. For example it is often emphasized that the teacher has to find situations from the daily or "real" life of students and relate the activities in the mathematics classroom to these. One cannot however, assume that meaning construction takes place with a direct reference to the background of the students. Instead, meaning construction might be related to the foregrounds of the students—and what this could mean specifically for students in a borderland position needs to be explored carefully. Here prototypical research is not of much help. Becoming able to master ICT might have particular significance for students in a borderland position.

"Getting in contact" could be thought of as an important feature of a dialogue (see Alrø & Skovsmose, 2002), and it was a specific feature of the whole scheme of cooperation between teachers, researchers, and prospective teachers within the Interlink Network to get in contact with the students. For the researchers and the prospective teachers this means getting in contact with students whom they did not know in advance; for the teachers, who certainly know the students, the cooperation means, rather, that they come to know their students in a different way. In our case, this new contact occurred through the many possibilities for informal talk that emerges due to fact that there were more adults around. To us this "getting in contact" is important for understanding the contexts in the borderlands, and for establishing activities in the classroom that might become meaningful to the students. "Getting in contact" is crucial in any teacher education (for prospective as well as practicing teachers) that is going to address a borderland context.

Final Considerations

No school can guarantee social inclusion for its students, but if we take education for social justice as being more than a motto, we have argued here that schools need to provide new, different and more socially relevant opportunities for their students. In our view, social inclusion represents a human right that education must take seriously. As a consequence it becomes paramount to address issues that can help teachers work towards inclusion in a professional way.

We have shown through the context of the Interlink Network, that in mathematics teacher education, it is important to consider borderland schools as a potential yet real professional environment. If not, teacher education could easily settle within the boundaries of a prototypical discourse. We find it essential in any mathematics teacher education to address how inequalities are operating in educational contexts not least with respect to content matter issues, and to discuss how to organize learning milieus that acknowledge a concern for social inclusion.

An interaction between teachers, prospective teachers, and researchers may help to establish a discourse which, on the one hand, draws on the experience of the teachers and, on the other hand, includes concepts, ideas, and notions from a more theoretical framework. Through the Interlink Network, we tried to establish such a discourse. Furthermore, it is important to get in contact with the students. They are dialogical partners in an education for social justice, and through such a dialogical relationship we might find some of the resources with which, in practice, to establish an education for social justice.

Talking about an education for social justice may suggest romanticism on our part, but there are no reasons to be romantic. We should not expect that mathematics education will in fact change the position of a borderland school; nor should we expect that the presence of ICT will change the social and economic situation of the students or the inequalities that are engraved in the neighborhood. If their parents have no jobs, if the city does not offer anything much to the students, it is difficult to believe that a changed educational approach could make a profound difference. However, according to Freire (1972) this limit-situation should not paralyze us. Mathematics teacher education programs should be intentionally planned in a way that teachers get the opportunity to learn how to operate within such situations. We find that this should make part of teacher professionalism. Freire claims that teacher education could offer an opportunity to conceive the "*inédito viável*," which we may interpret as the "viable unheard-of," as something that is completely new. In the words of Ana Lúcia Souza de Freitas, "The creation of the viable unheard-of is an exercise of boldness, of taking the risk of facing the new and the pleasure in being the author of what still does not exist" (2005, p. 227; our translation). This is a collective project and we hope that this article may contribute to this—to the construction of more equality in society and toward greater social justice.

NOTE

This paper was first published in 2009 in *Journal of Mathematics Teacher Education, 12*(3), 217–230.

REFERENCES

Alrø, H., & Skovsmose, O. (2002). *Dialogue and learning in mathematics education: Intention, reflection, critique*. Dordrecht, The Netherlands: Kluwer Academic.

Bovo, A. A. (2003). *A informática na escola e os obstáculos ao trabalho do professor de matemática: O que fazer em termos de sua formação?* Paper presented at the VII EBRAPEM. UNESP, Rio Claro, Brazil.

Castells, M. (1998). *The information age: Economy, society and culture. Volume III, End of millennium*. Oxford, UK: Blackwell.

D'Ambrosio, U. (2006). *Ethnomathematics: Link between traditions and modernity*. Rotterdam, The Netherlands: Sense.

de Freitas, A. L. S. (2005). *Pedagogia da conscientização—Um legado de Paulo Freire a formação de professores* (3rd ed.). Porto Alegre, Brazil: EDIPUCRS.

Freire, P. (1972). *Pedagogy of the oppressed*. Harmondsworth, UK: Penguin.

Gates, P. (2006). The place of equity and social justice in the history of PME. In A. Gutiérrez & P. Boero (Eds.), *Handbook of research on the psychology of mathematics education: Past, present and future* (pp. 367–402). Rotterdam, The Netherlands: Sense.

Gutstein, E. (2006). *Reading and writing the world with wathematics: Toward a pedagogy for social justice*. New York, NY; London, UK: Routledge. Retrieved from http://www.unesp.br/prograd/PDFNE2004/artigos/eixo10/possibilidadeselimitacoes.pdf

Moll, L. C., & Greenberg, J. B. (1990). Creating zones of possibilities: Combining social context for instruction. In L. C. Moll (Ed.), *Vygotsky and education: Instructional implications of sociohistorical psychology* (pp. 319–348). Cambridge, UK: Cambridge University Press.

Parris, R. (2008). *Winplot* [software]. Exeter, UK: Phillips Exeter Academy.

Penteado, M. G. (2001). Computer-based learning environments: Risks and uncertainties for teachers. *Ways of Knowing, 1*(2), 23–35.

Penteado, M. G. (2004). Redes de trabalho: Expansão das possibilidades da Informática na educação matemática da escola básica. In M. A. V. Bicudo & M. Borba (Eds.), *Educação Matemática: Pesquisa em movimento* (pp. 283–295). São Paulo, Brazil: Cortez.

Penteado, M. G., Biotto Filho, D., & Silva, R. M. R. (2006). Possibilidades e limitações no desenvolvimento de projetos telecolaborativos na educação matemática escolar. In S. Z. Pinho & J. R. C. Saglietti (Eds.), *Núcleos de ensino*. Retrieved from http://www.unesp.br/prograd/PDFNE2004/artigos/eixo10/possibilidadeselimitacoes.pdf

Reid, I., Brain, K., & Boyes, L. C. (2007). Networked learning communities: Joined up working? In T. Townsend & R. Bates (Eds.), *Handbook of teacher education: Globalization, standards and professionalism in times of change* (pp. 433–444). Dordrecht, The Netherlands: Springer.

Ribeiro, J., Domite, M. do C., & Ferreira, R. (Eds.). (2004). *Etnomatemática: Papel, valor e significado*. São Paulo, Brazil: Zouk.

Sichieri, R. M. (2003). *Formação de professores em informática educativa: sobre a experiência de ser professor multiplicador* (Paper presented at the VII EBRAPEM). UNESP, Rio Claro, Brazil.

Skovsmose, O. (2005). *Travelling through education: Uncertainty, mathematics, responsibility*. Rotterdam, The Netherlands: Sense Publishers.

Skovsmose, O. (2006). Research, practice, uncertainty and responsibility. *Journal of Mathematical Behavior, 25*(4), 267–284.

Skovsmose, O., Scandiuzzi, P. P., Valero, P., & Alrø, H. (2008). Learning mathematics in a borderland position: Students' foregrounds and intentionality in a Brazilian favela. *Journal of Urban Mathematics Education, 1*(1), 35–59.

Tabach, M., Hershkowitz, R., Arcavi, A., & Dreyfus, T. (2002). Computerized environments in mathematics classrooms. In L. English (Ed.), *Handbook of international research in mathematics education* (pp. 784–805). Mahwah, NJ: Lawrence Erlbaum.

Wilson, P. S. (2008). Teacher education: A conduit to the classroom. In G. W. Blume & M. K. Heid (Eds.), *Research on technology and the teaching and learning of mathematics: Cases and perspectives* (Vol. 2, pp. 415–426). Charlotte, NC: Information Age.

Zbiek, R. M., & Hollebrands, K. (2008). A research-informed view of the process of incorporating mathematics technology into classroom practice by in-service and prospective teachers. In M. K. Heid & G. W. Blume (Eds.), *Research on technology and the teaching and learning of mathematics: Research syntheses* (Vol. 1, pp. 287–344). Charlotte, NC: Information Age.

CHAPTER 3

PROJECT WORK IN MATHEMATICS

In Denmark, Roskilde University was established in 1972 and Aalborg University in 1974 with project work as an organizing principle for studies in all subjects. At Aalborg University it is possible to study, for instance, engineering, computer science, land surveying, languages, psychology, communication, philosophy, physics, and also mathematics (see Vithal, Christiansen, & Skovsmose, 1995, for a presentation of project work in mathematics at Aalborg University).

That project work was institutionalized at two universities had a huge impact on education at other levels of the Danish educational system. Project work was introduced in the Danish Folkeskole (the primary and secondary school). The upper secondary school (the Danish "gymnasium") was much slower to embrace the project approach, but now we also in this part of the educational system find such experimental work. Many other institutions in Denmark have come to consider the idea of project work, which now is included also in some study programs at traditional universities, like the University of Copenhagen and the University of Aarhus.

I will take departure in an example of project work, "Run for your Life," and on this basis I address some features of this educational approach.

Critique as Uncertainty, pages 37–44.

An Example of Project Work

The class of about 15-year-old students is imagined to be a division of an American sports factory, "Run for your Life." The following description of "Run for your Life" is based on Alrø and Skovsmose (2002). Mikael Skånstrøm has been planning and guiding this project.

The classroom has become a factory hall: paper, glue, scissors, handballs, footballs, and such are scattered around. Every day new orders and new information arrive to the Danish division. For instance, one day the following fax arrives: "Our sports centre has burnt down. We have rented a bubble hall of meters. We need grounds for handball, basketball, badminton and volleyball. Please help us!"

Another day a particular request arrives from the company Batman & Co., which needs bats for table tennis, it being understood that the price must be no more than 89 Danish Kroner. As the Danish division of "Run for your Life" has no bats available at that price, they have to try to import some. They receive the information that a Swedish supplier is able to sell the bats for 70 Swedish Kroner. This appears a reasonable price. However, the students are also informed that freight charges are estimated to be 1.5%; that the exchange rate between Danish and Swedish Kroner is 82.14; that duty is 8%; that profit is expected to be 25%; and that VAT in Denmark is 25%. One task for the "Run for your Life" project is to handle this particular request.

For a while, a couple of students, Mary and Adam, concentrate on this task. They are trying to construct a spreadsheet, but they face many difficulties. Thus they have to add the "profit." But what could this mean in terms of mathematical operations? It cannot simply mean to add 0.25 or 25%. Mary and Adam consider this question:

Mary:	Plus profit, right?
Adam:	What?
Mary:	That one plus profit, right? That one . . .
Adam:	No, not plus.
Mary:	Yes . . . oh no, it's, oh no, it's per cent. How the hell does one do that?
Adam:	That one . . . times . . . oops . . . times . . .
Mary:	. . . that box on top . . . and then you have to write it in . . .
Adam:	No wait a minute, multiplied by that one
Mary:	. . . and then divided by . . .
Adam:	. . . di . . . divided by . . . 100 . . . will it do that, then?
Mary:	Should we try and find out?

In the end Mary and Adam construct a spreadsheet with the following structure:

C_1 (the original price)
$C_2 = C_1 + 0.015C_1$ (insurance and freight is added)
$C_3 = 0.8129C_2$ (transaction into Danish Kroner)
$C_4 = C_3 + 0.08C_3$ (duty is added)
$C_5 = C_4 + 0.25C_4$ (profit is added)
$C_6 = C_5 + 0.25C_5$ (VAT is added)

It appears that when 70 Swedish Kroner is paid, the price to be offered to Batman & Co. becomes much more that 89 Danish Kroner. In terms of the spreadsheet:

$C_1 = 70.00$
$C_2 = 71.05$
$C_3 = 57.756545$
$C_4 = 62.377069$
$C_5 = 77.971336$
$C_6 = 97.46417$

They would have to charge 97.46 Danish Kroner. As this would be unacceptable to Batman & Co., the students have to revise the whole situation. They are ready to make further experiments with the spreadsheet. They can, for instance, try to figure out if they could reduce the profit so much that they in fact could sell the bats to Batman & Co. at the price the company will accept. They can also try to make business with the Swedish supplier. But how much could they offer to the supplier?

In this situation the idea is brought to the students' attention that they can construct an "inverse spreadsheet." This could calculate the offer to be given to a supplier, using the maximum price as an input. Mary and Adam start out from the idea that they can go backwards by, first, subtracting the VAT by an operation of the form .

It takes a good deal of consideration and discussion, also involving the teacher, before the students become aware of the fact that when subtracting the VAT, a different percentage has to be used. The construction of the inverse spreadsheet turns out to become a huge challenge, and we leave Mary and Adam working on this, while we turn to some more general features of doing project work in mathematics.

Features of Project Work in Mathematics

Traditional mathematics education falls within the *exercise paradigm*. Most often, the mathematical textbook represents a "given" for the classroom practice. When an authority, external to the classroom, formulates exercises, the justification of the relevance of the exercises is not part of the mathematics lesson itself.

The exercise paradigm can be contrasted with *landscapes of investigation*. I have found it useful to introduce this notion in order to develop further some of the ideas of doing project work in mathematics. It is possible for the teacher to invite the student into a landscape of investigation (see Chapter 1). This is not made up of preformulated exercises and problems. The landscape is open for investigations. Such a landscape can be exemplified by the factory hall of the Danish division of "Run for your Life." The students are not given exercises. They are not to be engaged in activities in which they can assume that one and only one correct answer is waiting for being discovered. For instance, one can consider very many different solutions to the Batman & Co. problem.

The Students as Active

One basic idea of project work is that students should become active in a way quite different from that of doing exercises. Different theories of learning, for instance as proposed by radical or social constructivism, have emphasized that it is essential that students construct their own knowledge.

With special reference to the project work at university level, the original idea of the 1970s was that the students should take over responsibility of their own education and that teachers dedicated to traditional academic standards should not set the agenda for university studies. The active and politicized university student was an ideal, which suggested that students must assume responsibility for the educational planning process. This was also a basic principle when project work proliferated at other levels in the educational system. Naturally, it became less obvious what "politicized" could mean with respect to students in the Folkeskole. Still, the idea of "taking responsibility" implied that learning and curriculum issues should be discussed with students. The curriculum could not stay as an authority outside the classroom. The curriculum should be clarified in cooperation with students. In practice, that meant that the specific topics for project work should be identifued by teachers and students in cooperation.

Since the mid-1980s, the notion of the "politicized student" tended to diluted, and the constructivist conception of the active students can be interpreted as including a step towards a depoliticizing of the students. Nevertheless, it remains a basic idea that students cannot be ordered to take responsibility. It is essential that they stay as "owners" of the projects work they are supposed to be engaged in.

Which Problems to Address?

Since the beginning of the 1970s, this question has been answered in different ways. First it was emphasized that project work had to take departure in real-life problems. The basic claim was that a general and profound insight could be obtained by addressing a particular issue or problem. The validity of this claim was argued with reference to the exemplary principle, which had been advocated by Martin Wagenschein (1965, 1970) and Oscar Negt (1964), although from two quite different perspectives.

Wagenschein argued that the whole domain of geometric knowledge (Euclidean as well as non-Euclidian) could be developed by taking departure in the proof of the Pythagorean theorem. Negt's (1964) point was that general economic, political, and social structures could be addressed through episodes taking place, for instance at the labor market. Thus a particular event in a particular company could reveal principal supressing structures of capitalism. It was, in particular, Negt's interpretation of exemplarity that inspired the initial formulations of doing project work based on real-life problems.

At the beginning the identification of real-life problems caused a huge challenge to mathematics. However, soon a broadening up of the notion of real-life problems took place. This kind of problem came to include the mathematical way of addressing technical problems of any kind, and later mathematical issues as such came to constitute problems for project work. The tendency moved from a politicised Negt-interpretation towards a depoliticized Wagenschein interpretation of exemplarity.

In the first phase of project it was emphasized that the students should made the decision about which issue to address. This idea was soon to be modified. At Aalborg University, and not only in mathematics, each semester was addressing a particular theme (and the same theme, year after year). And only within this frame, the students can choose which problem they want to address. In many cases, however, the choice for the students to make become even more specific: they have to choose their project within a given limited set of predefined problems.

I have suggested a new elaboration of the conception of project work in mathematics by presenting the notion of landscapes of investigations. What this could mean is illustrated by "Run for your Life." A scenario is prepared and presented for the students, and within this scenario the students are working with what they find relevant and challenging. Possible landscapes can naturally be discussed in advance by students and teachers.

The Variety of Reflections

Most common types of reflections, supported by the exercise paradigm, address the correctnes of some calculations: Has the exercise been copied correctly from the text book? Has the given procedures for doing the calculation been observed? Is the result correct? However, when the exercise paradigm has been left, the scope of reflections becomes broadened.

Mary and Adam became engaged in reflections addressing their whole approach to the problem. They had to figure out how the request from Batman & Co. could be interpreted in mathematical terms. They have to reconsider the preconditions for constructing the spreadsheet. Is the profit reasonable? What to think of the VAT? How could they make experiments with the spreadsheet? Why should they in fact construct an inverse spreadsheet?

It is important not only to reflect on mathematical issues, but also to consider the contexts where mathematics is put in operation, and project work makes this possible. In the case of selling bats to Batman & Co., the reflection could address the possibilities of changing some of the parameters that determines the (too high) price. Such reflections are reaching beyond the strict mathematical issues related to the set up of the spreadsheet. This "reaching beyond" is crucial in order to broaden the perspective of what is done and what could be done through mathematics. Mathematics in general cannot be assumed to play unproblematic roles when applied.

The Teacher as Supervisor

Many studies are made of the communicative patterns dominating the traditional mathematics classroom. However, such patterns cannot be expected to work particularly well in project work. Here the teacher must also serve as a supervisor, and this is a different role to assume. A classic mistake, often done by teachers unfamiliar with project work, is that he or she continues "lecturing" also when "supervising" is more appropriate.

Helle Alrø and I have presented an Inquiry Cooperation Model which points out dialogic aspects of teacher-students communication (see Alrø & Skovsmose, 2002). This model names some dialogic acts which are important for establishing and maintaining a shared inquiry process.

The dialogic acts can be described in the following way: *Getting in contact* refers to the act of tuning in at each other, which is an important initial step for establishing a dialogical relationship. *Locating* and *identifying* refer to ways of grasping and exploring perspectives, ideas, and arguments of the other. This could, for instance, mean a teacher's attempt to grasp Mary and Adam's preliminary conception of "adding a percentage." *Advocating* means providing arguments for a certain claim, although not necessarily one's own. Thus, an advocating could mean outlining a possible mathematical argument or procedures in order to consider if it might be working. *Thinking aloud* means making details of one's thinking public. Thinking aloud can be understood literally, but also metaphorically as when one represents a mathematical insight by a diagram and when one combines verbal arguments with gestures. This took place many times when Mary and Adam were trying "to add 25%" into an operation at the spreadsheet. *Reformulating* refers to attempts in grasping other ideas by rephrasing them. Thus reformulating can be a crucial strategy for both teachers and students in order to rethink and rework ideas presented by others. *Challenging* is a most important way of opening up new perspectives. Thus the teacher provided a particular challenge by suggesting that it might be possible to construct an inverse spreadsheet. *Evaluating* refers to reflexive questioning, like, did we in fact solve the problem? What insight might we have reached? What new questions have we encountered?

The different dialogue acts are found in Mary and Adam's and in the way the teacher interacted with them. The general point is that in project work one should be expected to find a range of the dialogic acts as presented in the Inquiry Cooperation Model. This form of dialogic interaction is crucial for the teacher operating as a supervisor.

Project Work as Research

One idea of doing project work is to approximate, on the one hand, processes of learning and, on the other hand, processes of researching.

Even if the knowledge obtained through project work is not new in any objective sense, it can be new in a psychological sense. Researching can take place at all different levels in the educational system. When Mary and Adam were setting up the spreadsheet, they were both engaged in a process of researching.

University students who are engaged in project work somehow get used to making investigations, and for them the step from *studying* mathematics to *doing* mathematics is not a big one. This is one reason why mathematical studies at Aalborg University are considered so successful. When the students leave university they are able to do mathematics. This need not mean that they become researcher-mathematicians. But they easily get jobs in industry, where the capacity in problem solving is much more appreciated than an exercise-solving capacity.

The possibility of establishing a close relationship between learning and researching has been pointed out in different contexts. Thus, John Dewey based his progressive education on the idea that the scientific method is universal in the sense that it also applies outside science. In case children are invited to apply a scientific methodology, their learning will be largely improved. Many considerations of doing project work are based on similar ideas.

Limitations of Project Work

During the 1970s and 1980s, many discussions of project work generally assumed that project work constituted an absolute improvement of mathematics education. Certainly there can be observed many improvements when project work is compared to the exercise paradigm. However, it is also possible to observe very problematic examples of project work. There is no point of pursuing project work in all situations and at all costs. Project work can function as an educational possibility, but not as any decisive solution to educational problems.

I do not want to claim that operating in landscapes of investigation always have to be preferred. I would think it more relevant to consider the teaching-learning of mathematics as a route among different learning milieus as presented in Figure 4 in Chapter 1. I would not suggest that a class that has just been engaged in "Run for your Life" should jump from this project directly to a new project work. It makes good sense to spend some time doing exercising and trying to routinise some of the knowledge and understanding, which have been established through the project. Thus it might be relevant after the project "Run for your Life" has been concluded to include a period working with exercises and spreadsheets.

In university studies in mathematics at Aalborg University, the students will also have lectures and do exercises. Only half of their study time is used for project work. During their studies the students will follow a route which brings them to different learning milieus. They are sometimes settled in the exercise paradigm and sometime in landscapes of investigation. A mathematics education should not be parked in any of the possible learning milieus.

NOTE

A preliminary version of this paper was published in 2002 in the *Proceedings of ProfMat, Associação de Professores de Matemática*, Viseu, Portugal (on disk).

REFERENCES

Alrø, H., & Skovsmose, O. (2002). *Dialogue and learning in mathematics education: Intention, reflection, critique*. Dordrecht, The Netherlands: Kluwer Academic.

Negt, O. (1964). *Soziologische phantasie und exemparisches lernen* [Sociological imagination and exemplary learning]. Frankfurt am Main, Germany: Europäische Verlagsanstalt.

Vithal, R., Christiansen, I. M., & Skovsmose, O. (1995). Project work in university mathematics education—A Danish experience: Aalborg University. *Educational Studies in Mathematics, 29*(2), 199–223.

Wagenschein, M. (1965, 1970). *Ursprüngliches Verstehen und exaktes Denken, I-II* [Original understanding and exact thinking, I-II]. Stuttgart, Germany: Ernst Klett Verlag.

CHAPTER 4

INQUIRY GESTURES

Raquel Milani and Ole Skovsmose

The notion of inquiry has a long history in education. A principal contribution was provided by John Dewey who found that turning learning into inquiry processes would provide education with many important qualities: in particular a resonance between educational aims and democratic aspirations would become established.

As soon as it is claimed that inquiry processes include attractive educational qualities, many different kind of educational processes tend to be labeled as inquiry. As a consequence, it becomes relevant to characterize what one could understand by inquiry, and in this chapter we try to characterize such processes in mathematics education.

We keep in mind a characteristic of dialogue as presented by Helle Alrø and Ole Skovsmose (2002). They characterized the following 8 dialogic acts: *Getting in contact, locating, identifying, advocating, thinking aloud, reformulating, challenging,* and *evaluating*. They described dialogic learning as a process rich with such acts. Furthermore, we keep in mind a set of notions presented in a lecture by Helle Alrø and Marit Johnsen-Høines, October 30, 2012 at UNESP (State University of São Paulo) at Rio Claro. They used the following notions in order to characterize inquiry processes in an educational context: *Experimenting, (re) searching, wondering, listening, trying out, dwelling, keeping open, challenging,*

Critique as Uncertainty, pages 45–56.

and *criticising.*[1] We can observe much overlap between dialogic acts and these inquiry acts. This is not surprising as dialogue and inquiry belong to the same family.

Principally, we will draw on some empirical observations, and on this basis try to identify a set of *inquiry gestures* through which one can characterize inquiry processes within mathematics education.

About Dewey and Lakatos

Before we turn to our empirical material, let us just make a couple of remarks about John Dewey and Imre Lakatos.

Dewey (1963, 1964) explored widely the notion of inquiry. He related the process of inquiry to the natural sciences, in particular as characterized, simplified, and celebrated by positivist and pragmatic philosophy at the time.

Dewey found that a scientific inquiry process could be characterized in the following way: It starts with experiences and observations. These can be registered, summarized, and condensed into theories. Thus, it is through induction that theories become formulated on the basis of observations. From formulated theories it is possible to deduce consequences, which in turn can be confronted with new observations. In this way it is possible to provide a verification of theories, although not in any absolute sense. This is the assumed basic pattern of scientific inquiry.

According to Dewey, this pattern is, however, not particular to science, but has validity far beyond the range of science. This brings Dewey to claim that education can be radically improved if learning processes accommodate to this pattern.

Furthermore, Dewey finds that an inquiry approach not only supports a better learning of different subjects; it can also serve a much more profound development. Dewey claims that scientific methodology is closely connected to democratic standards. He finds that if the inquiry approach for obtaining knowledge becomes part of a general approach for learning and schooling, we can witness a growing acknowledgement of democratic values and progress in all aspects of social life.

Dewey did not pay particular attention to mathematics. But Lakatos (1976) did. His interest was to provide a general characteristic of the logic of mathematical discovery. This logic we can also refer to as an inquiry process. Lakatos was eager to relate his conception of mathematics to the overall Popperian outlook. Thus Popper (1965, 1972) had characterized the logic of scientific discovery in terms of conjectures and refutation. In parallel to this, Lakatos characterized the logic of mathematical discovery in terms of *proofs* and *refutation*s. We find, however, that many more notions are necessary in order to provide such a characteristic.

We are in the situation that Dewey provides a general characteristic of inquiry that relates directly to the natural sciences and simultaneously opens into education. Lakatos provides a characteristic that relates directly to mathematics, but

---

[1] See also Alrø and Johnsen-Høines (2010).

only briefly addresses implications of his investigations for mathematics education.

Many educators have celebrated inquiry processes with reference to both Dewey and Lakatos. Sometime these processes have been celebrated as providing a unique quality to education. They have been treated as if they were accompanied by their own justification.

We find it important to try to characterize inquiry processes in mathematics education.[2] However, we are not going to assume that such processes as such have unquestionably attractive educational value. The relevance of inquiry in mathematics education is a much broader issue than the characteristic of such processes. However, now we concentrate on providing such a characteristic.

The Context of Our Data

The empirical context of our research was set by a teaching practice course where prospective teachers were engaged in dialogue activities with reference to inquiry, planning, and reflection. In this context Raquel, coauthor of this chapter, assumed the position of practitioner-researcher, as Jarvis (1999) considers the term.

The people involved in the data production were four prospective teachers whose pseudonyms were chosen by themselves: Amanda, Isabela, Lyriel, and Valentina; one supervisor of the teaching practice course called Isolda; and Raquel. In the following we will refer to the prospective teachers as the students, and sometimes to Isolda and Raquel as teachers.

Counting Sticks and Squares: An Example of Inquiry Process

It was a cold winter Thursday evening. Amanda, Isabela, Lyriel, Valentina, Isolda, and Raquel met at the Teacher Training College. Everybody was a bit tired. Amanda, Isabela, Lyriel, and Valentina had all been working (some of them teaching) during the day; Isolda had joined a long afternoon meeting with her colleagues, while Raquel had been preparing the last details of this class. It was early evening and the classroom was illuminated, making it appear even darker outside.

The meeting was planned to last for about 2 hours. Isolda and Raquel had prepared an activity having to do with counting sticks and squares. The planning of the activity had been conducted in collaboration with Denival Biotto Filho, who is a colleague of ours.

The question presented was, *What is the smallest number of sticks necessary to use in order to construct a square of squares?* Figure 1 indicates the landscapes of investigation as formed in Raquel and Isolda's minds when planning the task.

[2] For a broad discussion of inquiry processes in mathematics education see, for instance, Jaworski (2003, 2006) and Jaworski et al. (2007).

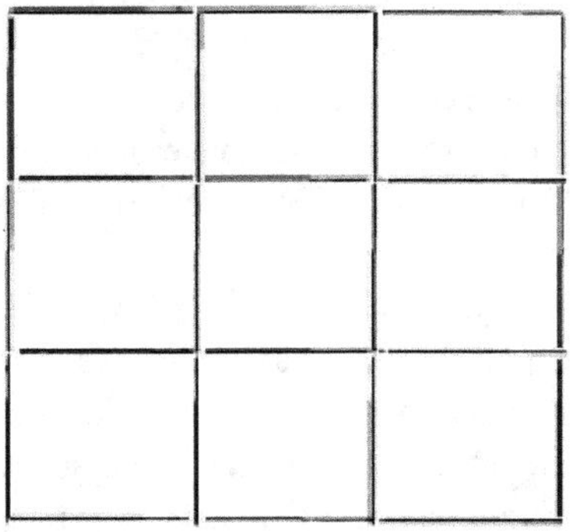

FIGURE 1. A square of squares of sticks.

The students, however, were not presented for the Figure 1. Their starting point was the verbal formulation, and their first step was to *visualize* the task by means of drawings. The four students were working as one group, although in this visualizing process they each had a paper in front of them. Among the drawings we found that the squares constructed by Amanda were constituted by different size squares, like shown in the Figure 2.

Making drawings was a way of *thinking aloud*. This expression can be taken literally, as one in fact can present thinking in words. It can also be taken metaphorically, meaning that one makes one's thinking public through drawing or gestures. At the same time, thinking aloud, literally as well as metaphorically, makes it possible to think further.

Making a drawing is a way of *experimenting*. One outlines something, one makes additions, one erases something. And one gets a new piece of paper. The

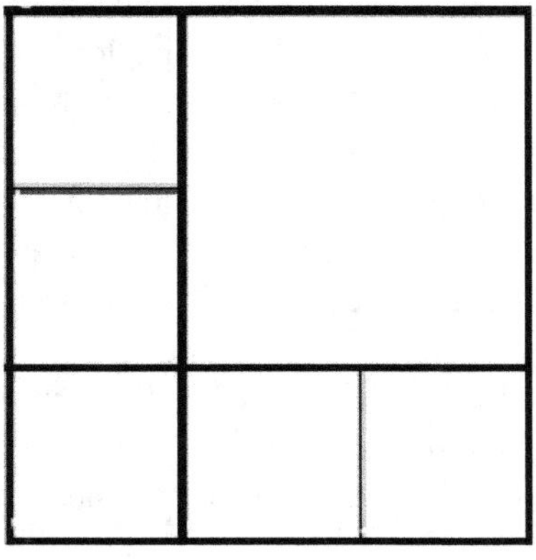

FIGURE 2. A square formed by different squares.

drawings made by Valentina illustrated the process of *experimenting*. The first lines were thin and feeble. Soon followed lines that demonstrated a kind of determination. Finally, on top of everything, she made some almost aggressive lines, which the pencil went over several times in order to make them thicker and darker and to eliminate the figures that had already been drawn.

The students' drawings provided *challenges*, also to Isolda and Raquel. The idea that the squares could be of different sizes had not occurred to them during their planning, which had taken departure in the pattern shown in Figure 1. Naturally, the inquiry could have continued without assuming that the small squares should be of the same size. Such an assumption might, however, be introduced later as part of the inquiry process. Thus a simplification is a relevant step to make in an inquiry process.[3] However, instead of letting the group proceed along its own path, Isolda and Raquel intervened, saying that the little squares that formed the square in question should be of the same size.

Even with different lengths, Amanda used pencils and pens to construct squares, now all of them with the same size, at her study table. But it did not work, as the table had some slope and pencils and pens rolled down at the floor. The pencils and pens remained there as the students started to construct the squares on this plane surface. Suddenly, Isabela realized that the room's floor was coated by a parquet of squares of the same size and pencils and pens were put away. This was a *discovery* that had immediate impact on the process. The floor provided an obvious *visualization* of the task. The whole group looked down at the floor. Looking from above, one saw four heads down there, and legs and arms moving back and forth in order to indicate the sizes of the bigger squares. The group was literally getting into a landscape of investigation. They were *visualizing* and *experimenting*.

The act of studying the floor also symbolized that *intentions* was brought into the process. An inquiry process cannot be simply imposed on students. One can give assignments to students; one can give them a task of completing some exercises; one can also assign them to do an inquiry. However, defining for an inquiry process is that the students get involved; that it becomes they who drive the process; and that they assume some ownership of the process. We see them studying the floor as symbolizing that now they were into it. They were acting with *intentions*.

When Lyriel had *visualized* bigger squares with sides of 1, 2, or 3 units, she promptly concluded that "4 is the smallest number of sticks necessary for constructing a square formed by squares," referring to a square formed by sides the length of a stick, the unit. This was an insight that became immediately *perceived*. Isolda then followed up with the question, "What if we think of any square formed by squares?"

[3] Simplification is, for instance, suggested by Polya (1957).

Length of the side of the square	Number of sticks
1	4
2	12
3	24
4	40

FIGURE 3. Table suggested by Amanda.

This question presented a *challenge* and moved the inquiry forward. Thus, students returned to the squares drawn on the paper and counted the number of sticks. Amanda suggested presenting information in a table, as shown in the Figure 3. Providing this figure is a specific way of *thinking aloud.* This also made the *challenge* more specific. The question became "What is the relationship between the numbers in the two columns?"

Isabela made a suggestion different from what Raquel had been imaging during the preparation of the task. Isabela proposed identifying the number of sticks by considering how the squares were constructed. She explained that she was trying to write the number of sticks by observing that each constructed square was formed by four sticks. In her own words, "The number of sticks is a multiple of 4, but I must subtract sticks that are counted twice." Thus, for instance, the square of side 2 was formed by 4 unit squares of 4 sticks each—16 sticks in total—and, discounting the repeated 4 sticks, one got to 12 in the table.

Thus, Isabela described the relationship with direct references to the visual representation of the situation: she drew squares, indicated the repeated sticks, and tried to rewrite the numbers in the second column as a subtraction between a multiple of four and certain repeated number of sticks. Thus, the relationship became expressed by Isabela by showing on the figure which repetitions occurred. She was smiling and nodding and pointing again.

When Raquel was planning the activity, she had identified this relationship in terms of an algebraic expression. In her preparation, she did not consider what could be read from the visual representation of the situation. She became *surprised* by Isabela's approach. In a very short period of time, while Isabela was presenting her initial thinking, many ideas were passing through Raquel's mind. She wondered what would happen if she did not follow Isabela's perspective. In this case Raquel would be moving away from the inquiry dialogue. This could not happen, as we believe that in an inquiry process it is important to explore the perspectives of participants, and also to be willing to suspend one's own perspective. When Raquel decided to follow Isabela's perspective, she declared that "it can work" and felt *challenged.*

From then on, Raquel entered in a risk zone (Penteado, 2001) and the path became unknown also to her. Raquel had left her role of the well-prepared teacher who had foreseen what was going to appear in terms of mathematical solutions. But inquiry processes are not predictable. Raquel realised that and became totally involved, not any longer as a teacher but as one of the students, curious and wanting to *discover* something new.

Isabela *advocated* for her procedure, and the group *listened* carefully. The group returned to the squares they had been drawing and *experimented* how to put Isabela's procedure into practice. As all participants became convinced that the perspective provided a real possibility of action, Isabela went to the blackboard and assumed the ownership of the inquiry. The involvement of the other students, in different ways, demonstrated their active participation in the activity. This was an experience of shared *intentionality* among everybody: Amanda, Isabela, Lyriel, Valentina, Isolda, and Raquel.

Let us just summarize a bit of the thinking that took place. Let us call the number of sticks necessary for making a big square for S, and think of S as a function of the length l of the side of the square. This was, however, not the formulation used immediately in the process. Here expressions like "depends on" were used. As already observed by Lyriel we have $S(1) = 4$. The next step was to acknowledge that $S(1) = 4 \times 4$ minus the overlap.

To identify the overlap from the drawing was somehow easy enough, and it was done for the remaining squares indicated in the table. Thus, the group decided to rewrite numbers in table's second column. As Isabela was at the blackboard, she conducted this rewriting with the help of teachers and colleagues until the group got to Figure 4.

But how could this rewriting become turned into a formula? The students looked at the number sequences: 4, 9, 16 and 4, 12, 24. Isolda suggested rewriting the first line in order to get the beginning of these sequences explicit. Thus, $S(1) = 1 \times 4 - 0$.

Isabela began to express what she was *visualizing*: "Ahhh we need to find where side l is in here." She was pointing at the expressions in the table. The group agreed, and Raquel said, "1, 4, 9, 16 . . . what is the next?" Valentina

Length of the side of the square (l)	Number of sticks (S)	
1	4	$= 4$
2	12	$= 4 \times 4 - 4$
3	24	$= 9 \times 4 - 12$
4	40	$= 16 \times 4 - 24$

FIGURE 4. Rewriting the table's second column.

Length of the side of the square (l)	Number of sticks (S)	
1	4	$= 1 \times 4 - 0 = 1^2 \times 4 - (1^2 + (-1) \times 1)$
2	12	$= 4 \times 4 - 4 = 2^2 \times 4 - (2^2 + 0 \times 2)$
3	24	$= 9 \times 4 - 12 = 3^2 \times 4 - (3^2 + 1 \times 3)$
4	40	$= 16 \times 4 - 24 = 4^2 \times 4 - (4^2 + 2 \times 4)$

FIGURE 5. The last table's format.

promptly answered, "25." Easily Isabela identified l^2 multiplying 4 and stated, "We have here the beginning of our function!"

Visualizing how the sequence 0, 4, 12, 24 was related to side l seemed to be difficult. Lyriel *perceived* that the number 24 in the fourth line already had appeared in the third line as the number of sticks of side 3 square. The same thing happened with the number 12 in line 3 and 2, and the number 4 in line 2 and 1. Isolda *became surprised* and with a smile in her face *challenged* the group: "Thus we can try to write the second part (second sequence) related to what appeared previously in a recursive way." This was not so easy, and Lyriel suggested another idea: "24 is 4^2 plus 8, and 8 could be 2 times 4." She wanted to write l^2 as it was used in first part of the formula. As the group did not get her idea, Isolda went to blackboard and gave a break to Isabela. Lyriel then tried to *reformulate* her own ideas while the group *listened* carefully. One more equal sign was added to the table as shown in Figure 5.

This adding has started from the fourth line in Figure 5 up to the first one. As Lyriel suggested using l^2 for the second part of the expressions, a number should be added to l^2 to result in 0, 4, 12, and 24. This number appeared in all lines in the table as a tentative step of writing side l again. Isolda used a different color to facilitate the *visualization* of the length of the side in each expression. The group thought factor -1 a bit strange, but went on trying to find the formula. Isolda noticed that the number multiplied by length of the side between parenthesis changed in each line and *challenged* the students to find how this changing occurred: "What is 2 from 4? What is 1 from 3?" Valentina quickly answered, "l minus 2." Students agreed and in chorus they were "singing" the formula to Isolda, who wrote it on blackboard $4\,l^2 - (l^2 + (l - 2)l)$. Simplifying the expression, the group identified the function S given by $S(l) = 2(l(l + 1)$, which expresses the number of sticks according to the length of the side of the bigger square to be constructed. After *discovering* the formula, the group tested it to specific cases in the table and checked that it really worked.

When the formula $S(l) = 2(l(l + 1)$ was written up there on the blackboard, it almost felt as if there were more light in the classroom. Well, it could be that it had just grown more dark outside. It was evening, and the nice feeling of having *discovered* something was now mixing with the feeling of being hungry. However, the students had more classes, Isolda more teaching, and Raquel a long way to drive back home.

Later the group became involved in another inquiry activity, and evaluating both processes, Isabela stated, "I like these kinds of activities where we construct something. It seems that I *discovered* that! It does not seem that a mathematician formulated it 300 years ago. It seems I *discovered* that magic formula. That is the sensation I feel when I do this thing. For me, this is good."

Naturally the inquiry process could have been continued. What about thinking of triangles and not of squares? Or about hexagons? One could also move from 2 dimensions to 3 dimensions, and consider the functions $P(l)$, which refers to the number of sticks necessary for making a bigger cube out of smaller cubes with the sides the length of l. One can also consider the relationship between $S(l)$ and $P(l)$ We already have that $S(1) = 4$, and looking at a dice we easily count that $P(1) = 12$. But could there, for instance, be some nice expression of $P(l)/S(l)$? The same could be investigated with reference to triangles in 2 dimensions and pyramids in 3 dimensions. And why not move into 4 dimensions? Yes, how many sticks are necessary in order to construct a 4-dimensional dice?[4] And by the way, what kind of sticks are we talking about here? How do we visualize this construction? If $R(l)$ refers to the number of sticks necessary for construction a 4-dimensional cube with the sides the length of l, we have already a hard task in calculating $R(1)$. Sure it would be interesting to consider the relationships between $S(l)$, $P(l)$, and $R(l)$. There are really many new *challenges* to be *discovered.*

Ahh, sorry, now we, the two authors, are becoming all too engaged in the inquiry process. We are almost forgetting that we in fact are writing a chapter. But let us now concentrate on bringing this chapter to a proper conclusion.

Inquiry Gestures

We have now been through an example of an inquiry process. Throughout this presentation we have used a range of notions, some of which we have put in italics. They are *surprising, intending, thinking aloud, listening, advocating, visualizing, perceiving, experimenting, challenging,* and *discovering.*

These notions represent *inquiry gestures* and we are now going to comment on each of them. In each inquiry process the gestures might occur in different orders and each gesture might be repeated many times. The following order is just made in order to provide a certain flow in the presentation:

Surprising is an emotional gesture. It can take place in different moments in inquiry. Thus Raquel became surprised when Isabela presented her perspective. Another surprising moment in inquiry was when the participants discovered that it was in fact possible to put the l into a formula. This discovery feeling can be

[4] See Rønning (2010).

accompanied with a smile or utterances as "ahhh, that is it!" Moments of surprises are crucial for bringing the inquiry process further.

Intending. We see intending as a defining element in an inquiry process. A vibrant involvement by the students is a sign that they have intention in doing the inquiry. They have accepted the teachers' invitation and get involved. During the process, we could see different clusters of intentions. On the one hand, the teachers were intending to promote a dialogical experience for the students. On the other hand, the students were intending to explore the presented landscape of investigation, with different degrees of involvement. An expressive moment of establishing intentions in the process took place when students bowed towards the floor helping each other visualizing the challenge.

Thinking aloud refers to the verbalization of reasoning, ideas, and perspectives to make them public, and thus allow them to be investigated. During the process, thinking aloud was taking place in multiple ways: by talking, making drawings, placing pens and pencils, and by gesturing.

Listening shows that one considers and wonders about what others are suggesting. As thinking aloud, so also listening is crucial for establishing an inquiry process as a collective undertaking. Helle Alrø and Ole Skovsmose (2002, p. 62) define active listening as "asking questions and giving non-verbal support while finding out what the other is getting at." When Isabela was presenting her ideas, the group was listening. Teachers and students asked questions and tried to reformulate the ideas. Reformulating represents an active part of listening and is an attempt to grasp what the other is saying. When Raquel perceived that the inquiry process could take a surprising turn, she listened and reflected about the potentials of Isabela's approach.

Advocating is used in order to defend or reject ideas. In our empirical observation, when Isabela was presenting her perspective and answering questions, she was advocating her suggestion for how to discover a relation between quantities in the square situation. This gesture reminds us of the category of "proving" in Lakatos proofs-and-refutations characteristic of mathematical discovery. Advocating can be seen as a more general form of proving. It is tentative, an observation which Lakatos also made with respect to proving. Advocating can also result in a refutation.

Visualizing is crucial in an inquiry process, and immediately after the task had been presented to the students, they started making sketches. However, visualization does not only mean to see objects; visualizing also takes place when students and teachers used their bodies to draw imaginary squares. Legs and arms were making squares on the parquet floor, and all these body movements helped the group in visualizing the situation.

Perceiving different perspectives is essential in an inquiry process. In order to turn a preliminary idea into a real perspective of action, the idea needs to be grasped and explored. The potentials of the idea need to be perceived. Thus Lyriel perceived that there was a recurrence in the table's numbers. Raquel perceived that inquiry process could take a different path from her own original idea.

Experimenting means trying to turn one's imaginings into actions. The students experimented when they tried to draw squares in different ways. The act of counting sticks, trying to find some relation between the constructed squares and number of sticks, and raising some hypothesis are also forms of experimenting. Somehow some lines of experimenting were blocked by the teachers when they directed students to draw specific squares. Even if students had not reached any conclusions, they could have come to experience the nature of the difficulties. Experimenting can be seen as the basic category through which Dewey wanted to describe inquiry processes. Experimenting applies broadly to the domain of natural sciences; it applies as well to very many daily-life examples of inquiry processes. We find that it also applies to mathematics.

Challenging. This gesture means trying to push ideas in a new direction or questioning what is assumed. The challenge, however, cannot be done in just any way. It should suit the students' current conceptions, and it cannot be too much or too little. Challenging makes inquiries move forward. It clearly happened when Isolda asked the students, "What if we think in any square formed by squares?" Challenging is a more general formulation of Lakatos' category of refutation. A refutation constitutes a new challenge.

Discovering is almost a defining feature of an inquiry. It does not matter in what sense the discovery might be new in any objective sense, but it was new for those who discovered it. Thus, Isabela stated, "It seems that I discovered that!" It should also be noted, that it is not only results that can be discovered. One can also discover new possibilities, new challenges, new questions. One can discover mountains that one is not able to climb.

Instead of Conclusions

We have now presented a range of inquiry gestures. We find that by being aware of such gestures one might become able to identify inquiry features of educational processes.

However, let us return to the claim made by many mathematics educators: that an inquiry-based mathematics education can be considered a goal in itself, as "inquiry" is a process with many intrinsic good qualities. Such a claim can certainly find support in both Dewey and Lakatos' conceptions of inquiry. As already mentioned, we do not want to assume this position. We find that we have to address the relevance and the benefit of inquiry processes in mathematics education with reference to each specific context.

Nevertheless, it is important to be able to identify such processes, and we hope that the inquiry gestures that we have presented here might serve this purpose.

REFERENCES

Alrø, H., & Johnsen-Høines, M. (2010). Critical dialogue in mathematics education. In H. Alrø, O. Ravn, & P. Valero (Eds.), *Critical mathematics education: Past, present and future* (pp. 11–21). Rotterdam, The Netherlands: Sense.

Alrø, H, & Skovsmose, O. (2002). *Dialogue and learning in mathematics education: Intention, reflection, critique*. Dordrecht, The Netherlands: Kluwer Academic.

Dewey, J. (1963). *Experience and education*. New York, NY: Macmillan. (Original work published 1938)

Dewey, J. (1964). The relation of science and philosophy as a basis of education. In R. D. Archambault (Ed.), *John Dewey on education: Selected writings* (pp. 15–19). Chicago, IL: University of Chicago Press (Reprinted from *School and Society*, *48*, 470–473).

Jarvis, P. (1999). *The practioner-researcher: Developing theory from practice*. San Francisco, CA: Jossey-Bass.

Jaworski, B. (2003). Research practice into/influencing mathematics teaching and learning development: Towards a theoretical framework based on co-learning partnerships. *Educational Studies in Mathematics*, *54*(2/3), 249–282.

Jaworski, B. (2006). Theory and practice in mathematics teaching development: Critical inquiry as a mode of learning in teaching. *Journal of Mathematics Teacher Education*, *9*(2), 187–211.

Jaworski, B., Fuglestad, B., Bjuland, R., Breiteig, T., Goodchild, S., & Grevholm, B. (Eds.). (2007). *Læringsfellesskap i matematikk* [Learning communities in mathematics]. Bergen, Norway: Caspar Forlag.

Lakatos, I. (1976). *Proofs and refutations: The logic of mathematical discovery* (J. Worrall & E. Zahar, Eds.). Cambridge, UK: Cambridge University Press.

Penteado, M. G. (2001). Computer-based learning environments: Risks and uncertainties for teachers. *Ways of Knowing*, *1*(2), 23–35.

Polya, G. (1957). *How to solve it* (2nd ed.). Princeton, NJ: Princeton University Press.

Popper, K. R. (1965). *The logic of scientific discovery*. New York, NY: Harper and Row (Original work published 1934)

Popper, K. R. (1972). *Conjectures and refutations: The growth of scientific knowledge* (4h Rev. ed.). London, UK: Routledge and Kegan Paul. (Original work published 1963)

Rønning, F. (2010). Hvor mange kanter har en firedimensjonal terning? [How many edges does a four-dimentional dice have?]. *Tangenten*, *2*, 5–10, 17.

PART 2

FOREGROUNDS AND POSSIBILITIES

CHAPTER 5

FOREGROUNDS AND THE POLITICS OF LEARNING OBSTACLES

In 1954, Hendrik Verwoerd made the following statement in his address to the South African Senate:

> When I have control over Native education I will reform it so that the Natives will be taught from childhood to realise that the equality with Europeans is not for them . . . People who believe in equality are not desirable teachers for Natives . . . What is the use of teaching the Bantu mathematics when he cannot use it in practice? This idea is quite absurd. (as cited in Khuzwayo, 1997, p. 9)

According to Verwoerd and the apartheid regime, the paramount task was to make sure that Blacks were prevented from climbing the social ladder. Being excluded from mathematics also meant being excluded from the possibility of advancement in society.

In other words, learning obstacles for a certain group of students can be established in explicit ways such as being subject to an absurd policy. Here we are a long way away from the epistemic notion of learning obstacles, analyzed in terms

Critique as Uncertainty, pages 59–72.

of students' preconceptions, if not misconceptions, of some mathematical notions and ideas. The epistemic interpretation of "learning obstacle" is *not* the only one possible. However, processes of exclusion in education can be dressed up in such a way that their political dimension becomes hidden and ignored. It could appear that exclusion is not imposed on students. Instead, exclusion may appear as a consequence of some students' so-called low achievement.

When mathematics education operates as part of social mechanisms, providing or justifying certain forms of inclusion or exclusion, it comes to serve as a gatekeeper. According to Volmink (1994), "to deny some access to participate in mathematics is . . . to determine, a priori, who will move ahead and who will stay behind" (pp. 51–52).

This statement can be read as a dramatic description of the role of mathematics education as institutionalizing a distinction between those who are included and those who are excluded. Bourdieu (1996) refers to a "state magic," by means of which the state assigns some authority to a group of people, by referring to that group's good performance in school, and in particular to the performance in mathematics.

To me, the notions of inclusion and exclusion are as relevant for the discussion of mathematics education as the notion of achievement. Is it initially high or low achievement that produces social inclusion or exclusion? Or do socio-political processes of inclusion and exclusion manifest themselves in schools and in the mathematics classroom as high or low achievement? Needless to say, the relationship between high/low achievement and inclusion/exclusion is more complicated than that, but I find it essential to consider carefully how social processes of inclusion and exclusion might appear in the mathematics classroom. In this way, it might be possible to discuss the politics of learning obstacles.

I am going to consider such issues from a conceptual perspective. In particular, I am going to draw attention to the notion of *each student's foreground* as a conceptual construct that might facilitate a discussion of the "politics of learning obstacles."

The introduction of the notion was inspired by some observations I did years ago of 7-year-old students doing mathematics. At the beginning of each lesson, the teacher carefully explained the new task. Following this, the students had to do some exercises from the textbook. Suddenly several students seemed completely occupied in their own work, and when I tried to engage the boy sitting next to me in a conversation he simply ignored me. I soon realized the reason. After each student had completed the exercises, they went to the teacher's desk, and the teacher commented in an encouraging way about what each student had done.

However, there was another agenda in operation. As the students lined up, a public stratification took place: Who was first, second, third? A competition took place among the students, at least among some of them.

I also observed a group of girls who seemed to be ignoring this rush. They did address the exercises, but they worked at their own pace and with a lot of small interruptions. They seemed, somehow, to have renounced speed, realizing, perhaps, that whatever they did they would never become among the first in the line. It was common knowledge that sometimes Peter (who I was sitting next to) was the first, sometimes Anna, sometimes Maria, and a few times a fourth one, but the candidates for winning the game were a minority.

What could one do better than establish one's own priorities? The group of girls had started the first problem, but soon they needed to erase a wrong number and spent time choosing the proper eraser. Should it be the one with a smell of strawberry? Pencils had to be sharpened; comments and smiles had to be exchanged; and in these ways time would pass. Somehow, they protected their integrity very well. Had they, instead, struggled in vain to get first in line, lesson after lesson, their self-esteem might have been affected.

To understand their approach to mathematics I explored how they interpreted their opportunities. I found that they constructed their foregrounds as expressions of what they experienced as their opportunities, and that these foregrounds set the conditions for their engagement in, as well as their resistance towards, mathematics.

Originally, I developed the construct of foreground from a philosophic perspective and used it in the interpretation of learning processes taking place in Danish classrooms (Skovsmose, 1994). However, it has turned out to be useful in other, different, educational contexts:

- for a cooperative project conceptualizing and establishing a mathematics education with an explicit concern for democracy (running since 1994) between universities in Denmark and South Africa—discussing educational consequences of the apartheid regime's policy and of the damage to opportunities in life that this regime caused for Black students
- for exploring further, also since 1994, the ethomathematical research program in extensive cooperation with researchers from Brazil.

Thus, I do not consider the construct "foreground" to be relevant only when we consider the situations with reference to which the construct emerged. Instead, I consider the construct relevant for any discussion of learning that includes a discussion of learning obstacles.

Privatising or Politicising Learning Obstacles?

Let me illustrate what I mean by *politics of learning obstacles* by summarizing one aspect of White research on Black education carried out during the apartheid past of South Africa. Interpretations of achievements in school were a big issue, as particular interpretations could help to "explain away" the brutality of the apartheid regime. Racism was an all-embracing category. Both "classic" and what ironically can be referred to as "progressive" racism argue that it is not the school that has to be blamed for the weak performance of the Black child, but that the child brings the cause for his or her weak performance into school.

In the basic assumption of *classic racism*, performance in school was accounted for by referring to certain "facts." That Black children did not perform as well as White children had to be understood in terms of biological structures, established thousands and thousands of years back in time. Certainly, such an explanation provided a solid distance between the apartheid regime and the causes for what was observed in classrooms.

In particular, Black children's learning obstacles had nothing to do with the school structure, and certainly nothing to do with apartheid politics. These obstacles had to be found in the Black children themselves. They bring their own deficiency along with them, right into the school. The children are inevitably linked to their own poor performance, which is just a different expression of their color of skin. Thus, the political dimension of school performance is efficiently explained away.

In *progressive racism*, the idea that social aspects play a fundamental part in a person's intellectual and emotional development, and not the biological framing, did lead to new priorities within White research on Black education. Instead of searching for a biological explanation of the weak performance of Black children, social factors, intrinsic to "Black culture," could be identified. In his study of White research carried out during the apartheid era of the Orange Free State University, Khuzwayo (2000) has uncovered what such research included: How do you explain observations of Black children's weak performance in school? One suggestion was that reasons had to be sought for in their social background.

In one study from the Orange Free State University, an explanation was suggested in terms of family traditions and, in particular, in terms of the dominant role of the father in the Black family. According to this study, this aspect of the Black family helps to explain that creativity, and also the mathematical creativity of the Black child, was eliminated. Thus, family structures became a main factor in the explanation of the Black child's performance in school. The problem is thus to be found in the cultural background of the child. In other words, Black culture (and therefore not any suppressing White culture) produces the learning obstacles for Black children.

Black children's problems in school are established beforehand and are not to be located in the school structure. The Black children themselves bring their learning obstacles to school. The best the school can do is to compensate for such cultural deficiencies. Applying Bourdieu's (1996, pp. 10–11) formulation, we could claim that racism establishes categories of perception and forms of expression which "suppress or repress the social dimension of both recorded and expected performances and . . . dismiss any questioning of the causes."

Many deficiency theories (theories of the deprived child) follow the approach of racism in explaining away the sociopolitical dimension of school performances, by *privatizing and personalizing* the causes of such performance.

Bopape (2002) has made a study of mathematics education in the most desolate parts of South Africa, and he has showed me what a school might look like; broken windows; doors and all electric installations missing. There was a hole in the roof—maybe the tiles had been removed by somebody who found that his or her house needed the tiles a bit more than the school building did. When it was raining, the children had to get away from this part of the classroom. The classroom would be either too hot or too cold or too wet. It looked like a place where teacher and students would meet having a shared intention to leave this ugly place as soon as possible. What seems to be the most obvious learning obstacle to the children in this school: their colour of skin, their dominant father, or the hole in the roof?

What does this mean for mathematics education? It is all too obvious. When we enter the classroom, the first thing that hits us is the evident physical learning obstacles. The learning obstacles are right there in front of our eyes, and on top of our heads. To me, this hole in the ceiling, not referred to in much research in mathematics education, calls any deficiency theory of the child into question (see also Ginsburg, 1997; Gorgorió & Planas, 2000, 2001).

How could it be that this hole in the roof has not been seriously addressed by mainstream research in mathematics education? Learning obstacles can be looked for in the actual situation of the children and with respect to the opportunities which society makes available for the children. The actual distribution of wealth and poverty includes a distribution of learning possibilities and learning obstacles. This distribution is a basic political act. Paying attention to this means reestablishing the politics of learning obstacles. Civil and Planas (2004) address the notion of learning obstacle in a similar way: "Learning obstacles are not related to students' preconceptions and misconceptions of some mathematical ideas, but to forms of excluding some groups from mathematics education" (p. 12). And they emphasize that learning obstacles go "much deeper than classroom dynamics" (p. 12).

During the rest of this article, I shall try to explore in greater detail the nature of learning obstacles by considering the notion of *each student's foreground.* By means of this notion I shall try to illustrate that learning obstacles can take the form of a ruined foreground, and that ruining the foreground of a certain group of children is a sociopolitical act. That a foreground is spoiled does not mean that there is no foreground, but that the foreground appears to be without attractive and realistic opportunities. Learning can, to a student with a shattered foreground, appear pointless: Why should I try to learn this? Why should I pay attention to mathematics? I cannot see any meaning in this. A ruined foreground does not easily support hopes. However, before getting to these ideas, I want to take a look at the notion of background.

Meaning and Background

Meaning has been discussed with reference to the notion of culture. Meaningful mathematics education has been sought for by relating classroom activities and possible contextualizations of exercises to the students' culture and background. Bringing the students' *cultural background* into the classroom as a resource for contextualization seems relevant for bringing meaning to the mathematics classroom. To draw on students' background appears to be a simple and obvious idea. Nevertheless, I am not sure that we are dealing with a "simple truth."

Culture can refer to many things, and cultural background can operate in different ways in the classroom. Culture can refer to tradition and folklore. But when expressed by the apartheid system, a notion like Zulu culture comes to operate in an oppressive way. It could constitute a trap. An "appreciation" of Zulu culture might be associated with the assumption that people belonging to this culture stand outside the Western development. The Zulu culture could be picturesque and include, for instance, dances with traditional weapons. The notion of culture could get a negative connotation, referring to people "out there" and "down there," supporting the view that people with such a "different" culture had better stay in their homelands.

It appears problematic if an appreciation of culture results in a constant appreciation of traditions. Culture is changing and developing; it includes new elements, good and problematic ones, in a complex mix. "Culture" is a contested concept. As a consequence, it is not a straightforward act to search for meaningful mathematics education in the cultural background of the students.

In much ethomathematical literature the intimate connections between culture and mathematics has been emphasized. D'Ambrosio (2001) explains that by *ethno* he refers to the fact that mathematics is acted out in many different ways in different cultures and by different groups. Mathematics is always socially embedded. The ethnomathematics perspective has had implications for the practice of mathematics education: it has been emphasized that it is important to consider the background of the students when we try to constitute meaningful mathematics education.

I certainly agree with the point that making mathematics education meaningful is essential. I also agree that meaningfulness has to do with the cultural background of the students, but I would argue that meaningfulness has much to do with another dimension as well, namely the foreground of each student. To me, "cultural background" should not remain the only key notion when meaningfulness in mathematics education is discussed.

Learning as Action

I find it problematic to try to explain the performance of somebody by, first of all, referring to the background of the person. This could be any kind of performance, and also (high/low) achievements of students in a mathematics classroom. "Referring only to the background" is a strategy by means of which the political nature of learning obstacles can be eliminated. If, instead, we try to explain the performances in terms of the background, the here-and-now situation, and the foreground of each student, then the political nature of the learning obstacles becomes more obvious.

By the *foreground* of a person, I understand the opportunities, which the social, political, and cultural situation provides for this person. However, not the opportunities as they might exist in any socially well-defined or "objective" form but the opportunities as perceived by a person. Nor does the background of a person exist in any "objective" way. Although the background refers to what a person has done and experienced (such as the situations the person has been involved in, the cultural context, the sociopolitical context, and the family traditions), then background is still interpreted by the person. Taken together, I refer to the foreground and the background of a person as the person's *dispositions*[1] (and for simplicity I include the meeting point between foreground and background, the present situation, in the foreground).

These dispositions need in no way be homogeneous entities. Dispositions can incorporate conflicts and contradictions. A person can conceptualize different sets of aspirations and foregrounds; and his or her background can be structured by conflicts. And, certainly, a person's aspirations need not square with his or her background.

Dispositions should not be taken to be characteristic of only an individual. It makes sense to talk about the dispositions of a group with shared background and foreground. Dispositions provide resources for action—not in the sense that dispositions "cause" actions, but dispositions embody propensities that become manifest in actions, choices, priorities, perspectives, and practices. Dispositions are continuously reworked and remolded. They are dynamic. They are relational or interpersonal characteristics. They are sociologically structured. This being said, I continue to talk as well about a person's dispositions as those which are reworked and remolded in the person's interactions with other people, but which the person can draw upon as his or her resource for action.

According to many philosophic interpretations, and mine also, intentionality or intentions-in-action is a defining element of an action. In order to understand a person's action, we have to consider his or her intentions. Behaviorism proposed that actions could be identified with their physical appearance. In this way a certain conceptual framework was suggested for how to explain actions—in terms of events taking place before the actions are carried out.

---

[1] My use of disposition can be compared to Bourdieu's concept of habitus: "The habitus is the set of dispositions which incline agents to act and react in certain ways. The dispositions generate practices, perceptions and attitudes which are 'regular' without being consciously co-ordinated or governed by any 'rule.'" (Thompson, in the *Introduction* to Bourdieu, 1991, p. 12). However, the notion of foreground does not play a vital role in Bourdieu's notion of habitus, which appears to be more related to background in my interpretation. While background (and habitus) concerns experiences, foreground has to do with expectations, although certainly mediated by experiences. Nevertheless, I find Bourdieu's concept of habitus extremely interesting as part of the clarification of both the notions of foreground and background.

Several alternative interpretations have been suggested indicating that an action is more than its behavioral appearance. Following the idea of Brentano (1995), it has been suggested that human consciousness can be characterized in terms of its directedness, its intentionality. I propose an interpretation of action that tries to grasp the specificity of action through its intentionality. This makes intention-in-action a crucial construct.

The intentions of a person are not simply grounded in his or her background, but emerge also from the way the person sees his or her possibilities. In other words, I see intentions as rooted in a person's dispositions. Intentions express expectations, aspirations, and hopes. Intentions make up a constitutive part of any action, just as actions without intentions degenerate into simple physical movements. Actions become not simply caused by the past but represent forms of grasping the future. When we want to try to understand how and why a person is acting, then it is important to get an understanding of the person's disposition.

I see learning as action. This, however, does not indicate a simple description of learning in general, as there are many forms of learning that I cannot describe as action (but instead, in my terminology, a forced activity). Situations where somebody has to learn some routines, like soldiers learning how to go marching in rows, I do not refer to as learning as action. Nor do I consider learning as action, when learning, so to say, takes place beneath the conscious level of the person, like, without noticing it, adjusting oneself to certain habits when put in a new cultural context. Learning as action cannot be forced upon somebody, but students can be invited into situations where they can be involved in processes of learning as action.

Such learning presupposes that students establish intentions-in-learning, as intention-in-action is a defining element in an action. This means that the students see meaning in what they are doing. Maybe it is even better to think of learning as interaction: learning means doing things together.

However, a particular situation or a particular way of organizing teaching-learning processes can prevent students from acting as learners. This was what I observed in the classroom with students reacting to a public stratification. Some engaged in the activities and put extra effort into what they were doing, while others, including the group of silent girls, withdrew their intentions from the "official" activities of the classroom and redirected them into "alternative" activities. In order to interpret the activities of the group of "silent girls," I had to consider not only their backgrounds but also what they saw as their opportunities. Their foregrounds might prevent them from putting effort into the designated activities.

I see achievement (low or high) as being related to the opportunities that the school structure and the sociopolitical context in general make open for the students to perceive as their opportunities. This provides a different interpretation of learning obstacles. Such obstacles have not only to be sought in the historical past of the person but also with reference to the opportunities that the social and political system makes available for the person. In particular, the apartheid system destroyed the future of Black children, and this explains some of their behavior in school. *When a society has ruined the future of some group of children, then it has also obstructed the incitements of learning.* A ruined future can be the most brutal form of learning obstacle. This interpretation of learning obstacle accuses the society in question. A spoiled foreground is a dramatic learning obstacle, and spoiling a foreground is a sociopolitical act.

Meaning and Foreground

In order to establish meaning in education, students should be involved in meaning production, and each student's foreground is an essential resource for this production. I decided to talk about *meaning production*, following a terminology presented by Lins (2001). However, I do not follow the theoretical framework also presented by Lins, as I restrict my use of meaning production to a metaphoric sense, emphasizing that meaning is produced and constructed.

Let us imagine that we, as mathematics educators, arrive in a village far away "beyond the mountains," and we find that in this village there are chickens everywhere. We realize immediately that all kinds of mathematical activities can refer to these chickens, such as counting, selling, buying, and cooking. What a perfect situation for making sense of mathematical notions! We construct tasks that relate to counting and selling and buying and cooking chickens. The students will be familiar with it all. This is part of their background. Nevertheless, this need not make sense to students. They may be more interested in, say, pilot's mathematics, although they have only seen an airplane passing by up there high in the sky making a fine white line, signifying that there are many different places to go. When we, as mathematics educators, come to this village, we must not only consider the students' background, but we must also consider their hopes and aspirations. We must consider where they want to go. *Meaning not only represents the past. It also represents the present and the future.* Each student's foreground is a principal resource for meaning production.

To me, the notion of meaning production emphasizes that meaning is not a referential property.[2] Meaning is produced; it becomes an aspect of acts—and not just an aspect of concepts. This is in line with the ideas indicated by Wittgenstein (1958) suggesting an interpretation of meaning in terms of use. Following this direction, I see meaning as an aspect of acts, and meaningful education means that students are invited to engage in meaningful learning acts. Meaning is produced by students, and by cooperation among students and between students and teachers. This production is resourced by the dispositions of the students. Therefore, "meaning of learning," "meanings for students," and "each student's meaning

[2] With reference to classic philosophic interpretations, meaning is assumed to be a referential property. The meaning of a concept is the set of objects or entities to which the concept refers. According to this interpretation, when we in mathematics education are able to expand the references of notions such as function, exponential function, and metric space, we provide extra meaning of such concepts.

production" must be investigated and interpreted with reference to the dispositions of the students (including their background and foreground).

Meaning production takes place in terms of what the students see as their opportunities, including motives, perspectives, hopes, and aspirations. It gets its extra fuel from the foreground of each of the students. Meaning production, however, can also be obstructed. If meaning production takes place with reference to the foreground of the each student, then ruining the foreground for certain groups of students creates a real learning obstacle. To a Black child in apartheid South Africa, what could be experienced as the point of struggling with mathematics, when jobs which presupposed mathematical skills, like engineering, for instance, were not for Blacks? What was the point for many girls in Denmark (or other Western countries) in a not-so-distant past of concentrating on studying physics and mathematics when jobs, demanding such skills, were apparently for men? Based on her studies, Wedege (1999) claims that the habitus (as interpreted by Bourdieu) of a young woman in Denmark in the 1930–1940s does not automatically encompass a disposition for learning mathematics, nor does it generate a conception of mathematics as a relevant subject. How do children of immigrant parents conceptualize their opportunities in today's Danish (or European) society, and how does that affect their attitudes towards learning? How does an apparent marginalization of some groups of people affect some groups of students' foreground and therefore their possibilities for being involved in certain learning processes? How does the emerging racism in Western Europe provoke learning obstacles to certain groups of students?

As each student's foreground is one principal resource for meaning production, then ruining a foreground becomes a learning obstacle. This has to be discussed in terms of the sociopolitical situation of the students. As I see it, by considering the processes of globalization and ghettoizing, we get an idea of the basic condition of learning for many children around the world. Here we find fundamental forms of establishing or ruining foregrounds.

Globalization Includes Exclusion

Globalization refers to inclusion, but also to processes of exclusion. Castells (1998) discusses the rise of what he calls the Fourth World as an aspect of globalization:

> This widespread, multiform process of social exclusion leads to the constitution of what I call, taking the liberty of a cosmic metaphor, the black holes in information capitalism. These are the regions of society from which, statistically speaking, there is no escape. (p. 162)

Furthermore,

> The Fourth World comprises large areas of the globe, such as much of Sub-Saharan Africa and impoverished rural areas of Latin America and Asia. But it is also present

> in literally every country, and every city, in this new geography of social exclusion. (p. 164)

One essential question concerns the possible foreground of each student living within or close by the Fourth World. What aspirations could they have? How do they see their opportunities? What opportunities could be created for them in their process of schooling (if they have the opportunity to go to school)? What can be created within mathematics education? Schools have got a tricky position in the "new geography of social exclusion." This geography determines the foregrounds of the majority of students around the globe. It determines the disposition of groups of people, and as a consequence it structures motives for learning and for meaning production.

"Schooling" can be seen not only as a support for entering the network society, but it can also become a gatekeeper and an "excluder" from the network society. If students perceive their foreground as ruined, this could easily turn into low achievement, which, in turn, could confirm their exclusion. In this way, schooling can mean a preparation for the dumping of people into the Fourth World. The remarkable statement of Verwoerd thus represents a clear indication of what it means to put people into a ghetto, in this case represented by homelands. The grand idea of apartheid was built upon the notion that people should be separated, and that Black people had no role to play in White society beyond the level of unskilled labor. The existence of ghettos of the network society seems to indicate that the informational economy does not need everybody. Only a part of the global population fits into the networking—the rest had better be left in their "homelands."

In the days of apartheid, learning obstacles had their obscure significance, but recent processes of globalization also cause dramatic learning obstacles to many students. What does it mean for each student's aspirations and hopes to be located close to the black holes of the informational society? Castells (1998) refers to the millions of homeless, incarcerated, prostituted, criminalized, brutalized, stigmatized, sick, and illiterate persons, who are literally expelled from the informational economy with no functional role to play. From a certain economic perspective, they appear disposable. They are worth nothing as consumers; they have no value as possible human resources for production. Other and sufficient human resources are available. This is one brutal aspect of the informational economy and of the "new geography of the social order." Children experience this new order, and many as something highly problematic. This influences their foreground and therefore their dispositions and motives for learning. To me, a ruined foreground is a learning obstacle. As a consequence, I find it important to discuss learning obstacles with references to globalization and ghettoizing as represented by the Fourth World.

Searching for the Foreground

If we want to understand the learning activities of students and the way they engage in learning processes, it is important to understand what they consider to be their foreground and their possibilities in life.

I do not see any contradictions in assuming, for example, that working with computers and playing with dynamic geometry can be meaningful when we deal with marginalized students living close to the Fourth World. Working with computers may be miles away from their cultural background. But this background does not represent the only essential parameter in their meaning production. Their foreground is important as well. So, an essential question is how activities, say, including playing with dynamic geometry, may capture their hopes and aspirations. It is important that mathematics education provides opportunities, and when these appear to be real opportunities, seen from the students' perspectives, they can become active in their processes of learning.

There is no point in mathematics education celebrating a certain kind of technological culture or celebrating the network society. There is no point in mathematics education presupposing an appreciation of any "high-tech" culture. The point is simply that mathematics education should not organize resources for meaning production in such a way that students are parked in a Fourth World.

Valero (2002) suggests a change in interpretations of learners' activities in the classroom. Instead of emphasizing the cognitive interpretation, Valero suggests a sociopolitical interpretation. She refers to a turning point for her reflections taking place during her observations in a school in Bogotá. On one occasion the teachers were missing, and Valero took over the mathematics lesson:

> I was supposed to give them the exercise worksheet that they had started to solve the previous session. While many in the class worked, two male students engaged me in a chat. These two boys were supposed to be doing the mathematics. Instead, they looked at the worksheet and laughed. They called me for help, but in reality they were curious to know about my intentions and motivations to be in their school. (p. 490)

The students could not understand why Valero had gone to their poor school and chosen to talk with poor students. The conversation made Valero emphasize that there were reasons to study, and also to study mathematics. Then one of the students answered, "The only class I would like to pay attention to is English, because I want to get out of this fucking country and to go to the U.S."

It is difficult to find a clearer expression of the idea that reasons for learning are to be found in the foreground of the students and that a ruined foreground is a learning obstacle of enormous dimensions. Students may be well aware of this. Learning obstacles are interpreted by students, and the sociopolitical and economic dimension of the learning obstacles might turn into a student's or a group of students' resistance towards learning. Thus, Planas and Civil (2002) observe,

"If students cannot see any perspective in what they are doing, then we cannot expect any meaningful participation of students" (p. 15).

It is not possible to claim that mathematics education can fundamentally affect the sociopolitical processes of inclusion and exclusion. Processes of globalization are stronger than educational processes. They influence the dispositions of many students, their motives for learning, as well as the learning obstacles they come to face. Nevertheless, I see the task of mathematics education as being to provide opportunities for students. Mathematics education has to realize the foreground of the students as a resource for any meaning production. To look for each student's foreground, and to try to relate learning activities to this, becomes an ongoing challenge to any mathematics education.

NOTE

This paper was first published in 2005 in *For the Learning of Mathematics*, *25*(1), 4–10.

REFERENCES

Bopape, M. (2002). Mathematics school based in-service training (SBIN- SET): A study of factors contributing towards success or failure of SBINSET in the South African school context (Unpublished doctoral dissertation). Department of Education and Learning, Aalborg University, Aalborg, Denmark.

Bourdieu, P. (1991). *Language and symbolic power* (Edited with an introduction by John B. Thompson). Cambridge, UK: Polity.

Bourdieu, P. (with de Saint Martin, M.) (1996). *The state nobility: Elite schools in the field of power*. Cambridge, UK: Polity.

Brentano, F. (1995). *Psychology from an empirical standpoint* (A.C. Rancurello, D.B. Terrell, & L. L.McAlister, Trans.). London, UK: Routledge. (Original work published 1874)

Castells, M. (1998). *The information age: Economy, society and culture. Volume III, End of millennium.* Oxford, UK: Blackwell.

Civil, M., & Planas, N. (2004). Participation in the mathematics classroom: Does every student have a voice? *For the Learning of Mathematics*, *24*(1), 7–12.

D'Ambrosio, U. (2001). *Etnomatemática: Elo entro as tradicões e a modernidade*. Belo Horizonte, Brazil: Autêntica.

Ginsburg, H. (1997). The myth of the deprived child: New thoughts on poor children. In A. Powell & M. Frankenstein (Eds.), *Ethnomathematics: Challenging Eurocentrism in mathematics education* (pp. 129–154). Albany, NY: State University of New York Press.

Gorgorió, N., & Planas, N. (2000). Researching multicultural classes: A collaborative approach. In J. Matos & M. Santos (Eds.), *Proceedings of the 2nd International Mathematics Education and Society Conference* (pp. 265–274). Lisbon, Portugal: Centro de Investigaçao em Educacao da Faculdade de Ciências, Universidade de Lisboa.

Gorgorió, N., & Planas, N. (2001). Teaching mathematics in multilingual classrooms. *Educational Studies in Mathematics*, *47*(1), 7–33.

Khuzwayo, H. (1997). *Mathematics education in South Africa: A historical perspective from 1948–1994* (Unpublished doctoral dissertation). Department of Mathematics, Physics, Chemistry and Informatics, Royal Danish School of Educational Studies, Copenhagen, Denmark.

Khuzwayo, H. (2000). *Selected views and critical perspectives: An account of mathematics education in South Africa from 1948 to 1994* (Unpublished doctoral dissertation, contact e-address: hbkhuzw@pan. uzulu.ac.za). Aalborg University, Aalborg, Denmark.

Lins, R. (2001). The production of meaning for algebra: A perspective based on a theoretical model of semantic fields. In R. Sutherland, T. Rojano, A. Bell, & R. Lins (Eds.), *Perspectives on school algebra* (pp. 37–60). Dordrecht, The Netherlands: Kluwer Academic.

Planas, N., & Civil, M. (2002). Understanding interruptions in the mathematics classroom: Implications for equity. *Mathematics Education Research Journal, 14*(3), 169–189.

Skovsmose, O. (1994). *Towards a philosophy of critical mathematics education.* Dordrecht, The Netherlands: Kluwer Academic.

Valero, P. (2002). Reform, democracy, and mathematics education: Towards a socio-political frame for understanding change in the organization of secondary school mathematics (Unpublished doctoral dissertation). Department of Curriculum Research, The Danish University of Education, Copenhagen, Denmark.

Volmink, J. (1994). Mathematics by all. In S. Lerman (Ed.), *Cultural perspectives on the mathematics classroom* (pp. 51–68). Dordrecht, The Netherlands: Kluwer Academic.

Wedege, T. (1999). To know—or not to know—mathematics, that is a question of context. *Educational Studies in Mathematics, 39*(1/3), 205–227.

Wittgenstein, L. (1958). *Philosophical investigations* (2nd ed.; E. Anscombe, Trans.). Oxford, UK: Basil Blackwell. (Original work published 1953)

CHAPTER 6

JUSTICE, FOREGROUNDS, AND POSSIBILITIES

What does social justice mean? What does social justice mean in an unjust society? What does it mean to try to establish an education for social justice if educational institutions serve important functions in regimenting and controlling people? Such questions points to the *paradox of social justice*: Education for social justice is necessary considering the profound inequality in society; however education for social justice is an illusion as well as it is impossible to achieve socioeconomic changes through education.[1]

There are different kinds of paradoxes and different ways of facing them. One can see a paradox as an ultimate fact, which blocks any way forward. A paradox can signify a dead end. However, one can also see a paradox as a challenge. Maybe not as something that can be solved but nevertheless as something that can be acted upon. In the following, I want to address the paradox of social justice and indicate how one can see it as a challenge.

[1] An argumentation that sharpens the paradox of social justice has been presented by Pais (2012). The notion of social justice with respect to mathematics education has been addressed by many; see, for instance, Ernest, Greer, and Sriraman (2009); Gates (2006); Greer, Mukhopadhyay, Powel, and Nelson-Barber (2009); Gutstein (2009); Knijnik (2009); Sriraman (2008).

Critique as Uncertainty, pages 73–86.

In order to do so, I want to relate three notions: justice, foreground, and possibility. Justice has a long history in philosophy and has been pondered upon by brilliant minds. The notion of foreground is feeble and without any noticeable history. The notion of possibility draws on a range of everyday formulations, and I would argue operates across a range of discourses. So, the paper deals with three very different notions: a well-established one, a feeble one, and an everyday one.

Justice

There are many philosophical discussions trying to clarifying what justice could mean. I will make a few simple observations, being completely aware that any simplification is an illusion.

According to Plato, mathematical notions like triangle, line, and point represent mathematical ideas, located in the world of ideas. Access to this world is restricted, as only rationality provides this. All mathematical notions and properties have to be discovered as features of the world of ideas. And, according to Plato, ethical notions and principles have to be discovered as well. They are just to be found in other provinces of the world of ideas. There is a long tradition in philosophy, including Socrates' conversations at the Agora in Athens, which embark on glorious journeys into the world of ideas. Thus, in *The Republic* (2009), we follow a journey searching for the idea of justice.

Many religious considerations about what is right and wrong also exemplify the idea that ethical principles are to be discovered—maybe not by exploring the world of ideas, but by studying holy texts. Here, ethical dogmas have been laid down, often in a cryptic format; but they can still be seen as discoverable. A basic principle of Catholicism has been that orthodoxy has to be made explicit through a particular authority that is able to conduct the decryption and therefore represents the authority of God.

There is a secular variant of the idea that principles of justice and ethical principles in general have to be discovered. Utilitarianism represents such an approach: What is a good and what is bad can be identified through calculations. In order to calculate whether an action is right or wrong, one has to add up the total amount of "pleasure" that it implies as well as the total amount of "pain." The comparison of these two sums gives the answer. The importance of utilitarianism is to establish ethical questions in a solid distance from any orthodoxy by turning them into empirical questions. Ethics becomes a secular affair. It is in the hand of human beings to discover ethical standards, including what social justice could mean.

Marx did not show much interest in ethics, but he was eager to point out that ideologies represent a way for the bourgeoisie to keep control, and religious labeling of what was "good behavior" is nothing but a capitalist mechanism for control and suppression. Marx wanted to brush aside such ideological superstructures, serving as a cover-up for a most unjust society. He wanted to identify the real economic moving force of progress. If we see Marx' identification of class struggle as such a force and assume that siding with the working class is an ethical principle, then Marxism also represents the idea that ethical principles have to be discovered. According to Marxism, this discovery does not make part of any orthodoxy, nor of any straightforward empirical investigation. Instead the discovery has to be based on in-depth theoretical excavations of the laws of economic development.

There is a different way of seeing ethical principles: They might not be discovered but they might be constructed. Ethics might be a genuine human construction. This is the view that I want to adopt. However, it could mean very many different things. One could imagine that ethics becomes engulfed by anarchism, leaving behind only a Darwinist struggle, meaning what to call "good" and what to call "evil" becomes a question of the survival of the fittest. To try to grasp what this could mean, one could glimpse into Nietzsche's book *Beyond Good and Evil* (1998). Here, nothing is left of ethical principles as representing some extra-human standards. Ethics turns out to be nothing but a secretion of history, and what is called justice at a particular historical moment is only an expression of the power dominating at that particular moment. Justice is strictly momentary. It is a human construction, and a most suspicious one.

There is, however, other ways of considered justice as a human construction. In the *Theory of Social Justice*, John Rawls (1999) presents an idealized version of such a constructive process. We can imagine a union for the construction of principles for a social just society. The point is that the construction of these principles cannot be guided by reference to some standards extrinsic or *a priori* to the union taking place. The principles for the constrictions have to be specified through the format of the meeting. Rawls suggests that the meeting takes place under two conditions. The first being that the people who join the meeting are going to become members of the society in question; they have to live through all the principles which they are going to impose on society. The second condition is that they are not going to know what role in society they will be assigned: They will not know if they become academics or workers; rich or poor; men of women; adult or children; ill or healthy; they will simply not know anything about their possible roles, except that they will be assigned one. Giving these two conditions, one can stipulate the result of the negation as signifying social justice. These two conditions are naturally impossible, but they illustrate the idea that ethics has a discursive basis and that principles of social justice are constructed from a particular discursive basis. Jürgen Habermas (1995) follows a similar idea, as he also tries to locate the sources for ethical principles in the discursive format. Any discourse ethics represents the idea that conceptions of social justice emerge through human interaction and communication. This has implications for which notions to include in ethical frameworks.

So, there are many ways of seeing justice as constructed. This construction may take an anarchistic format and become an expression of power. One can instead see justice as a result of negotiations and as emerging through a deliberative consensus. And one can imagine a range of mixed formats of the construction of ethical principles, where power and particular interests mix with idealized discursive processes. Not any adequate clarification of justice as a construction; anyway, let us move on to the next notion.

Foregrounds

According to Wiktionary (http://en.wiktionary.org/wiki/foreground), there is not much to say about the word foreground: It refers to the "elements of an image which lie close to the picture plane" and to the "subject of an image, often depicted at the bottom in a two-dimension work." With reference to computing, foreground can also refer to the "application the user is currently interacting with."

I see a foreground as possibilities that the socioeconomic situation makes available for the person. At the same time, the foreground represents what the person might consider possible. Foreground might include realistic features, but also aspects of desperate hoping, if not illusions. Foregrounds might appear bleak, not revealing much for the person to hope for.

We can talk about the foreground of a person, but as well about the foreground of a group of people. We can talk about the possibilities that children from wealthy neighbourhoods might have for further educations. We can consider what possibilities children from poor neighborhoods have. We can try to compare conditions for further education for different groups of children in different countries, or in different historical periods.

Foregrounds are not static. One can come to see new possibilities; one may also come to realize that certain possibilities are out of reach. Different political initiatives might rework the landscapes of possibilities for both individuals and for different groups of people. I will try to present different features of foreground by exploring it as a psychological as well as a socioeconomic concept.

As a *psychological* concept, foreground can be related to *intentionality*, *action*, and *learning*. The concept of intentionality has played an important role in Franz Brentano's work in psychology.[2] Brentano's most important work is *Psychologie vom empirishe Standpunkt* (*Psychology from an Empirical Standpoint*; 1995b), which was published in 1874. He gave particular importance to the notion for intentionality, which represents a directedness of the mind. This directedness was pointed out as the defining characteristic of the human mind.

Brentano inspired many, including Edmund Husserl, who from 1984 to 1886 followed a series of Brentano's lectures. Husserl gave the notion of intentionality a prominent position in phenomenology, and he deals with the notions of intention and intentionality in many places in his writings.[3] Through the notion of intentionality, the analysis of human phenomena and human experiences moved away from any naturalistic format. Here, naturalism refers to the idea that the notions through which one tries to describe and analyze human phenomena are the same

[2] See, for instance, Brentano (1995a, 1995b).

[3] See for instance, Husserl 1998).

as needed for describing and analyzing, say, physical, chemical and biological phenomena. Behaviorism exemplifies this approach within psychology.

Contrary to behaviorism, phenomenology emphasizes that human actions are principally different from mechanical operations, and intentionality refers to a genuine human quality. Human beings can intend something; they can have aspirations, hopes, desires; they can make plans and follow up on such plans. According to behaviorism, however, such phenomena can be reduced to complex mechanical phenomena. In particular, a notion like intention is just a condensed expression of a complex range of phenomena which, when analyzed carefully, all can be revealed to be of a cause-effect format. According to phenomenology, such a reduction is impossible with respect to intentionality, which represents a nonmechanical feature of the human psyche.

According to behaviorism, the notion of motivation plays an important role in the interpretation of action. Motivation can be seen as a relationship between the person's background and his or her action. This conception facilitates a mechanical perspective on human actions, fitting actions into a cause-effect pattern of explanations. Thus, according to a mechanical framework, an act has to be explained by reference to something taking place before the act. Explanations have to take departure in the background of the act. However, the notion of intentionality brings a different perspective on human action. Actions included directedness. They can be planned, they can express aspirations; they can include decisions; they can represent resignation. Such observations make it important to understand the person's intentions in order to understand his or her actions. Within this framework, an act has to be interpreted with reference to the intentions that make part of the act. As a consequence, an act has to be interpreted with reference to the acting person's foreground, which provides a landscape for directedness of intentions. It represents a panorama of motives for actions—a motive can be seen as relationship between foreground and an act.

One can talk about the act of learning. In order to interpret processes of learning, or any phenomena related to learning as, for instance, the difference achievements of different groups of students, it becomes important to consider what could be the students' possible intentions of learning. What features of their foreground could their intention for learning be directed at? What could be their aspirations for learning? What kind of hope might make part of their learning acts? One could also ask questions like, What could obstruct students' aspirations for learning? What could wipe out their hopes? Why might they show resignation with respect to learning? What features of their foreground might tend to obliterate motives for learning.

In order to understand, for instance, students' conception of what could be meaningful, it is important to pay attention to their foreground. Often meaningfulness has been interpreted as a relationship between activities in the classroom and phenomena outside the classroom that the students are already familiar with. However, I see meaningfulness as *also* referring to the relationship between what is taking place in the classroom and the foreground of the students. Meaningfulness is important for the students' learning acts. To be meaningful, a learning act needs to be loaded with intentions. It needs to be directed towards something. If not, learning coagulates as drill activity—in fact much learning is school has coagulated as such.

While the notion of background refers to the history of the person, the notion of foreground refers to the landscape of possible directions for actions. My point is not to try to ignore the background but to consider the foreground, when one tries to understand the person's actions. Naturally, one can expect a complex dialectic to take place between foreground and background. This may become more obvious when we see foreground in socioeconomic perspective.

It is important not only to see foreground as a psychological but also as a *socioeconomic* concept, including a range of social, economic, political, and cultural features. A foreground is far from any private affair. A foreground is not just structured by person's experiences. It should also be seen as being structured by a person's socioeconomic background.

In order to clarify this point, we could speculate about the possibilities provided for young people living in an impoverished neighborhood within a huge metropolis, or for a person living in a rich neighborhood in the same metropolis. What is the foreground of young people living in a neighborhood dominated by competing gangs? What is the foreground of a person living in a small city partly isolated at the countryside? What is the foreground of an indigenous student living in a community close to a huge city in Brazil? What is the foreground of immigrant students, for instance, belonging to one of the many groups arriving to Europe? What is the foreground of Mexican immigrants in the United States? What is the foreground of Palestinian students living a neighborhood surrounded by an intimidating Israeli wall? What is the foreground of a girl brought up in a radicalized Muslim community?

A foreground is structured through a range of socioeconomic factors. As an illustration of what this could mean, I can refer to the World Bank Report (2006) of *Equity and Development*, which is introduced by a short description of two South African children born on the same day in 2000. Nthabiseng is Black and born in a poor family in a rural area, while Pieter is White and born in a wealthy family. These differences have a tremendous impact on the future of these two children. Thus, the World Bank Report points out that Nthabiseng has 7.2% risk of dying in the first year of her life, while Pieter has 3%. The expectation of life is different: according to statistics, Pieter can look forward to 68 years of life while Nthabiseng can look forward to only 50 years. Pieter can expect to complete 12 years of formal schooling while statistics tell us that Nthabiseng only can expect 1 year.[4]

[4] See, World Bank (2006, p. 1). Renuka Vithal referred to this formulation in her lecture at the *Symposium Mathematics Education, Democracy and Development: Challenges for the 21st Century*. Faculty of Education, University of Kwazulu-Natal, Durban, April 4, 2008.

Naturally, such statements have only statistical significance. Thus, Nthabiseng's 1 year of schooling can be seen as a statistical outcome of the fact that some children from her community can expect more years of schooling, while many never get to attend school at all. Naturally, Nthabiseng may happen to live longer than her 50 years, and Pieter may get killed in a traffic accident while he is still young. A socioeconomic structuring of a foreground is not made up by a set of necessities but by a grid of possibilities.

A foreground is a collective phenomenon. A larger group can share the same foreground as they can be submitted to the same grid of possibilities. They can share statistics with respect to parameters that run into the future. On top of such shared statistics, foreground as an individual phenomenon may include impossible dreams, realistic expectations, and frustrations in a contradictory mix.

In many cases, I want to talk about a ruined foreground. This is the case when the socioeconomic reality does not, statistically speaking, make space for many possibilities. During the apartheid era in South Africa, the foreground of Black peoples were ruined due to the explicit rules of apartheid: Black people could not be owners of any piece of land; they had no access to certain areas strictly reserved for White people; they had no access to adequate health services; they were forbidden to enter a range of studies, becoming an engineer for instance; they were forbidden to vote. So, although South Africa is today a democracy, the shadows of apartheid are still lingering over the country. The different opportunities in the lives that exist for Nthabiseng and Pieter are dramatically different. Although apartheid has not survived as a legal system, it has been projected into the future in terms of huge socioeconomic differences.

Socioeconomic differences construct very different foregrounds for different groups of students. Such differences provide different panoramas of possibilities, and, as a consequence, different possibilities for aspirations, hopes, and intentions. Different foregrounds make different acts possible, including different ways of acting as a learner.

The notion of foreground relates to the notion of background. One can see the background of a person as a determining factor for his or her foreground, and to some extent this makes sense. Certainly the background makes part of the structuring of the foreground. Thus, the backgrounds of Nthabiseng and Pieter frame their foregrounds. However, one needs to be careful when stipulating that the *personal* background determine the person's foreground. A foreground is not structured as an individual undertaking but also through socioeconomic processes that, as in the case of Nthabiseng, could include brutal forms of exploitation.

Possibility

The word possibility is popular. It can make part of all kinds of formulations: It is a good possibility to choose ice cream for dessert; it is possible that the train will be much delayed; the possibility that Barcelona will lose their next match is small; it is possible that I get hold of the newspaper later, and so on. The adjective "possible" or the noun "possibility" often enters everyday conversations and discussions. They are most welcome participants in everyday small talk. The words can also enter more grandiose political discourses: It is possible that the political development in Iraq will stabilize; it is possible that the peace process will be successful; it is a possibility that Brazil will become more economically powerful. "Possibility" has a prominent cousin, namely, "probability." This notion has entered sophisticated mathematical discourses, and it has become dressed up in the most advanced symbolism. Within the philosophy of mathematics there is established a huge domain called the philosophy of probability, while it is difficult to find any philosophy of possibility.

However, let us try to do something about "possibility" by jumping right into the main principle of existentialism. According to Jean-Paul Sartre (1989), human beings have no essence. Existence comes before essence, meaning that human beings are free to do what they want to do. We are doomed to be free; we are doomed to make choices; and we are doomed to be responsible for our choices. Through our choices we construct who we are: our character and essence. Whatever we do, we could have done differently. There is no essence that determines one's existence. Instead, one's essence is completely homemade. Essence is a self-construction as existence comes before essence. The metaphysics of existentialism assumes that we are facing an abyss of contingencies.

I do not share this grandiose wall-to-wall metaphysics. However, I find it inspiring. I see human beings as located within a network of not only propensities but also of contingencies, and I agree with existentialism that we, in many cases, are not forced to do what we are doing. We have some freedom. Our life conditions include both propensities and contingencies. Existentialism has identified a huge contingency as defining the human condition. I consider this a metaphysical Fata Morgana. But the alternative to an all-embracing contingency is not a complete determinism. Foregrounds might be crisscrossed by propensities, some of which might deepen into necessities. At the same time, foregrounds might include contingencies. As a consequence, I find that notion of possibility to be crucial for understanding the human condition. Some possibilities might be almost out of reach; some within; some might be established during time; some might be eliminated. Possibilities are not to be found in any prefabricated metaphysical package, as suggested by existentialism. They need to be meticulously constructed.

The construction of possibilities can take place in many ways. Let me just indicate some. Classic Marxist analysis brings about an idea about what it could mean to provide new possibilities for the suppressed working class trapped by the capitalist logic of economic exploitation. Here, the construction of possibilities for the suppressed becomes portrayed as a once-and-for-all revolution. As a moderate form of this grandiose revolutionary creation of possibilities, one can see the original socialist projects of providing new possibilities for the exploited through series of step-by-step evolutionary changes taking place within the limits of a not-yet-revolutionized society. Whatever being the case, possibilities can be seen as sociopolitical constructions.

The construction of possibilities can take a psychological format. One can come to recognize that one can do something, or not do something, and that certain choices are possible. In order to be genuine, a possibility needs to be experienced as such. Possibilities can be seen a psychological constructions.

However, many constructions of possibilities are located somewhere between the socioeconomic and the psychological format. The educational construction of possibilities is one of them. One can see schooling as a way of opening possibilities for the students. Although educations might not change the socioeconomic framing of society, it might still create some new possibilities for young people from an impoverished neighborhood within a huge metropolis. It might establish some alternatives for young people living in a neighborhood dominated by competing gangs. It might create possibilities for young people from isolated communities. It might open different possibilities for indigenous students or for immigrant students. Maybe also for Palestinian students, ghettoized in the most brutal way. Education might also bring some alternatives to girls from a radicalized Muslim community. I am not claiming that education will be successful in all such cases. I just want to point out that education might provide some new possibilities to some students in very different situations.

At the same time, schooling can include a most brutal form of selection, providing many possibilities for the few, and a few possibilities for the majority. Schooling may provide possibilities within a certain limited format: being meticulously adjusted to the demands of the labor market as set by the capitalist format of the informational economy. Within education, the construction of possibilities is contested. Education might provide possibilities for some students in terms of political empowerments; it might bring value to some students by opening new job possibilities. It might provide new possibilities for companies to hire better qualified and at the same time submissive workers. Education might provide all kind of possibilities. One cannot assume any intrinsic goodness associated to these possibilities. In this sense, the construction of possibilities in education is contested. This also applies to mathematics education.

Uncertain Observations

Time has come for trying to mix the classic, the feeble, and the popular notions: justice, foreground, and possibility.

First let us try to relate foreground and possibility. It is important within education to think of the construction of possibilities in terms of adding new features to the foregrounds of the students. This construction need not be of any extravagant format. It can have to do with small-scale constructions within day-to-day educational activities. The construction of possibilities can be seen as an activity for the classroom. Let me try to illustrate what I have in mind by a few examples.

In Chapter 2, Miriam Godoy Penteado and I discussed social justice with reference to students many of whom came from poor neighborhoods.[5] We considered, how might we work for social justice in such a situation? Whatever we might be doing as educators we cannot do much to the socioeconomic situation which holds the students in a firm grip. However, even if one cannot, through particular educational activities, change overall socioeconomic conditions, one can still change many things for many students. The possibility of providing such change brings a new dimension to the notion of meaningfulness.

If we consider many of the discourses that have been elaborated with respect to meaningful education and about bringing the students' reality into the classroom, one could easily be guided to search for educational activities with a close relationship to the students' experiences. Thus, inspiration from the ethnomathematical approach has brought about a range of activities which relate directly to the students' background. The assumption has been that meaningfulness emerges from relationships established between the suggested activities and the students' background. However, this does not necessarily make good sense. The students Miriam and I worked together with had little experience with computers, nevertheless we worked with computers, and the students were extremely interested. Working with computers made sense to them. Some of the students were simply happy to touch a keyboard. One student was, for the first time, experiencing that the movements she made with her hand on top of a mouse caused movements at the screen. At first she found it difficult to coordinate the two so different movements, but it became a victory when she got control. Such a victory we see as a small, almost infinitesimal, addition to her possibilities. She made a small step towards realizing that working with computers could also be for her. We worked with traditional geometric properties, like the sum of the angles of a triangle. This way the students experienced that they could face challenges including some more abstract reasoning. Through such infinitesimal additions, education could change students' foregrounds and make new forms of directedness possible. Many examples illustrate how educational practices might provide new possibilities for students, for instance, Gutstein (2003, 2006).

Getting to school, Nthabiseng might get new possibilities in life. We cannot expect education to change the statistics which in the most sinister way reaches into her foreground. Nevertheless, education might change something by modulating propensities and turning some contingencies into attractive possibilities. This applies to all Nthabisengs in the world.

Now let us move on to "justice." Classic philosophic discussions relate justice to the notion of equality. It has been pointed out that social justice presupposes equality for the law; it presupposes equality in voting; it presupposes equality with respect to both rights and responsibilities. This could be seen as a liberal discussion of social justice. We can continue the argument: Social justice presup-

[5] See also Skovsmose and Penteado (2011).

poses equal opportunities with respect to schooling; with respect to health; with respect to welfare. Social justice presupposes the elimination of poverty. It presupposes a departure from the logic of capitalism. This line of thought also brings about a change in the notion of equality. When one wants to emphasize that equality is supposed to include more than the liberal features, it has become common to talk about equity. For me, it is important to explore the meaning of social justice in terms of equity. Still, the world is filled with deep inequalities; one just needs to remember the life perspectives of Nthabiseng and Pieter. How does one establish any education for social justice in a world dominated by extreme forms of inequalities? Remembering the paradox of social justices, one could argue: As it is impossible to reach socioeconomic equity through education, an education for social justice is a liberal illusion. Such an illusion might diffuse into well-intended rhetoric, but will never materialize as equity.

One should not assume that there exists a ready-made theory of social justice to be applicable in education. I do not think it is possible to colonize education by any off-the-rack ethical principles. There are many forms of applied ethics: business ethics, computer ethics, environmental ethics, government ethics, marketing ethics, medical ethics, research ethics, sports ethics (see, for instance, Frey & Wellman, 2004). Educational ethics could be another form of applied ethics. However, I find it important to see also educational ethics as being constructed. Establishing an education for social justice is not a question of providing an importation of already elaborated ethical theories. Nor is it possible to claim that education for social justice is impossible by referring to some interpretations of what social justice might mean. Instead, I see the conception of justice as a construction, and education for social justice as an example of such a construction.

A crucial issue in such a construction is to try to formulate what is meant within education to work for social justice in an unjust society. I find that discussions about students' foreground and the construction of possibilities make part of such a clarification. An education for social justice can be seen as education where the construction of new possibilities for the students is considered important. It is an education which searches for alternatives to be included in the students' foreground. An education for social justice addresses both the propensities and the contingencies, making part of the students' foreground. In many cases, propensities can be hard to bend; thus, one cannot assume that education can change the statistics that states that Nthabiseng can expect 50 years of living. There are many features of an unjust society that no education, nor an education for social justice, can change. Nevertheless, some features might be changed through education. An education for social justice tries to make it possible for students to reach out for new possibilities in life. To me, this signifies an important element of an education for social justice in an unjust society.

All such observations, however, are uncertain. This is not only due to the preliminary nature of my comments. I find that uncertainty and doubt represent a profound human condition. Within the covering framework of modernity, many notions and ideas have developed as if they hold a convenient share of certainty. This applies, for instance, to the notion of critique that has been developed by Kant as well as by Marx, although in quite different directions: by Kant, as a philosophical category; by Marx, as a political category. Different as they are, however, both notions of critique have been promoted as if they were putting aside any form of doubt. However, I see any glimmer of certainty to be an illusion, both philosophically and politically. This also applies to any form of critical education, trying to explore what an education for social justice could mean.

I suggest that we leave any assumptions about obtaining any form of analytical certainty. And the form of critique is not only tentative; it is an expression of uncertainty. In my book *In Doubt: About Language, Mathematics, Knowledge and Life-Worlds* (Skovmose, 2009), I have discussed uncertainty and doubt as human conditions. This also applies to any more particular observations about justice, foreground and possibilities, and therefore I have called this section "uncertain observation."

Did the Paradox Disappear?

The paradox of social justice suggests that whilst education for social justice is necessary, it is simultaneously impossible to attain. I do not try to dissolve this paradox by questioning the *necessity* of working for social justice. For me, this is an educational obligation. I want instead to explore the other side of the paradox; that is, the *impossibility* of establishing an education for social justice.

Social justice has been discussed with reference to the notion of equity, and I agree completely with the relevance of this. My point, however, is that there are other notions to which social justice *also* can be related. I see the notion of an education for social justice as also evolving around the notion of possibilities and in particular, as possibilities might come to feature in the students' foreground. An education for social justice can be interpreted as an education which tries to create new possibilities for the students, keeping in mind that their life-worlds are formed within a range of socioeconomic parameters that remains untouched by educational effort. Creating new possibilities for students is not an impossible task, as foregrounds are crisscrossed not only by necessities but also by contingencies. Furthemore, the construction of possibilities can take many different forms. Thus, one can construct possibilities that appear like infinitesimal changes, but taken together, they can change the futures of some students.

I find it necessary to work for an education for social justice, and in particular, it is *necessary* to do so in an unjust society. I also find that it is *possible* to do so by adding new possibilities to the students' foreground. This is how I construct the paradox of social justice as a constructive challenge.

NOTE

This paper was first published in 2012 in Cotton, T. (Ed.), *Towards an education for social justice: Ethics applied to education* (pp. 41–62). Oxford, UK: Peter Lang.

REFERENCES

Brentano, F. (1995a). *Descriptive psychology* (B. Müller, Trans. & Ed.). London, UK: Routledge. (Original work published 1982)

Brentano, F. (1995b). *Psychology from an empirical standpoint* (2nd ed.; A. C. Rancurello, D. B. Terrell, & L. L. McAlister, Trans.; L. L. McAlister, Ed.). London, UK: Routledge (Original work published 1874)

Ernest, P., Greer, B., & Sriraman, B. (Eds.). (2009). *Critical issues in mathematics education.* Charlotte, NC: Information Age.

Frey, R. G., & Wellman, C. H. (Ed.). (2004). *A companion to applied ethics*. Oxford, UK: Blackwell.

Gates, P. (2006). The place of equity and social justice in the history of PME. In A. Gutiérrez & P. Boero (Eds.), *Handbook of research on the psychology of mathematics education: Past, present and future* (pp. 367–402). Rotterdam, The Netherlands: Sense.

Greer, B., Mukhopadhyay, S., Powell. A. B., & Nelson-Barber, S. (Eds.). (2009). *Culturally responsive mathematics education.* New York, NY; London, UK: Routledge.

Gutstein, E. (2003). Teaching and learning mathematics for social justice in an urban, Latino school. *Journal for Research in Mathematics Education, 34*(1), 37–73.

Gutstein, E. (2006). *Reading and writing the world with wathematics: Toward a pedagogy for social justice*. New York, NY; London, UK: Routledge.

Gutstein, E. (2009). Possibilities and challenges in teaching mathematics for social justice. In P. Ernest, B. Greer, B., & B. Sriraman (Eds.), *Critical issues in mathematics education* (pp. 351–373). Charlotte, NC: Information Age.

Habermas, J. (1995). *Moral consciousness and communicative action.* Cambridge,MA: MIT Press.

Husserl, E. (1998). *Ideas pertaining to a pure phenomenology and to a phenomenological philosophy: First book.* Dordrecht, The Netherlands: Kluwer Academic (Original work published 1913).

Knijnik, G. (2009). Mathematics education and the Brazilian Landless movement: Three different mathematics in the context of struggle for social justice. In P. Ernest, B. Greer, & B. Sriraman (Eds.), *Critical issues in mathematics education* (pp. 153–169). Charlotte, NC: Information Age.

Nietzsche, F. (1998). *Beyond good and evil: Prelude to a philosophy of the future* (M. Faber, Trans. & Ed.). Oxford, UK: Oxford Universioty Press (Original work published 1886).

Pais, A. (2012). A critical approach to equity. In O. Skovsmose & B. Greer (Eds.), *Opening the cage: Critique and politics of mathematics education* (pp. 49–92). Rotterdam, The Netherlands: Sense.

Plato (2009). The republic (B. Jowett, Trans.). *The Internet classisc archive*. Retrieved from http://classics.mit.edu/Plato/republic.html

Rawls, J. (1999). *A theory of justice*. Oxford, UK: Oxford University Press.

Sartre, J.-P. (1989). *Being and nothingness* (H. E. Barnes, Trans.). London, UK: Routledge (Original work published 1943.)

Skovsmose, O. (2009). *In doubt: About language, mathematics, knowledge and life-worlds.* Rotterdam: Sense Publishers.

Skovsmose, O., & Penteado, M. G. (2011). Ghettoes in the classroom and the construction of possibilities. In B. Atweh, G. Graven, W. Secada, & P. Valero (Eds.), *Mapping equity and quality in mathematics education* (pp. 77290). New York, NY: Springer.

Sriraman, B. (Ed.). (2008). *International perspectives on social justice in mathematics education* (Monograph 1). Charlotte, NC: Information Age.

World Bank. (2006). *Equity and development: World development report 2006.* Washington, DC; New York, NY: World Bank/Oxford University Press.

CHAPTER 7

RESEARCHING FOREGROUNDS

About Motives and Conditions for Learning

Denival Biotto Filho and Ole Skovsmose

By a foreground of a person, we understand the opportunities that the socioeconomic condition makes available for the person. However, a foreground is not merely defined through some social facts, but simultaneously through the person's interpretation of these opportunities. Thus, a foreground also refers to a person's experiences of life conditions.

A foreground provides motives for learning—or the opposite. Thus, a ruined foreground can cause the most severe learning obstacles. As a consequence, we see students' foregrounds as closely related to motives and conditions for learning. In particular, motives for learning mathematics have to do with what position mathematics might have within the students' foreground.[1]

[1] Several foreground investigations have been conducted. Thus, we can refer to Skovsmose, Alrø, and Valero (2008) investigating Indian students' foregrounds; Skovsmose, Scandiuzzi, Valero and Alrø (2008) addressing the foreground of students from a Brazilian slum; and to Alrø, Skovsmose, and Valero (2009), who investigated the foreground of students' situated in a suburb of a big city in Denmark with a high concentration of immigrant and low-income population. See also the discussions of foregrounds in Skovsmose and Penteado (2011) as well as in Chapter 5 and Chapter 6 in this book

Critique as Uncertainty, pages 87–94.

In the following, we will present an interview with two Brazilian students. On this basis we make some observations about the notion of foreground, and finally we relate foreground and learning.

An Interview

On a brilliant sunny morning, Denival met with two students, Julie and Mary (pseudonyms), 15 years old. The meeting took place in their school. Denival knew very well both students, thus in the past he has been their teacher during 2 years.

Denival brought his computer, and the students were curious to know what was going to happen. They had been told that Denival was going to talk with them about their future.

During the conversation, Denival showed them the questions on the computer, and talked about what they thought. After hearing them, Denival wrote the answers at the computer. Then he showed them the written responses and asked if they agreed with what was written. Sometimes they suggested changes, and in these cases, Denival rewrote the way they wanted. At some point, Denival asked if they would like to write, but they preferred him to continue writing. Denival made the choice of questions based on Skovsmose, Alrø, and Valero (2008). Below, we present a shortened version of the text built at this meeting.

Question 1: Where do you live? Talk about your city, neighbors. Imagine that you have to tell this to a person who lives far away from here—in Denmark, let's say.

Julie and Mary: We think Rio Claro is a small city. Here there is not much leisure. Also, there are only a few cultural activities. People here are not much interested in cultural activities. We also believe that Rio Claro is a very violent city. There are many examples which our friends and relatives gets robbed. Although we do not like Rio Claro, we believe that Brazil has good places to live. Because here there is a wide variety of cultures. Also, there is a variety and an abundance of food. It is a very beautiful country with magnificent scenery and untouched nature.

Question 2: What do you think about going to school? What do you like and what do you not like about school?

Julie and Mary: We like go to school. Sometimes we get lazy, because we get up early. Also, walking back in the midday sun is also very bad. But we like to go to school, because we have found good friends. We like the majority of teachers. The school where we are now is better than others we've ever been, because it is more organized, and the principal is more flexible. We can even bring the guitar!

Question 3: Who are your friends? What do you like to do with your friends? Do you talk about the future sometimes? What do you talk about?

Julie and Mary: We are friends. But we do not have many friends. We have some colleagues.

Julie: I have only three friends, including Mary.

Mary: I have Julie, which I consider a friend. I know I can count on her.

Julie and Mary: We are very sad. Our preference is different from most people with whom we have contact. We like theater, classical music, and it is not like common people around us. We talked about college, to marry, have children, the age at which we will do these things. Also, we like to talk about trivial things.

Question 4: What are you doing in mathematics class? What have you already learned in mathematics?

Julie and Mary: In the middle of the year, we had a new teacher, and because of this change, there was a delay in the schedule of the mathematics content. We think that our old teacher was a bad teacher. She believed in student pranks. She gave good scores to sycophant students. Now she is director at another school! Do you believe?! The current teacher explains well, but she is very nervous, especially when we do not understand. This teacher is striving to meet the content, and she gives a very traditional school.

Question 5: Outside of school, do you ever use numbers, do calculations, count, make estimations? If so, what kinds of calculations do you do? In what situations is it necessary?

Julie and Mary: Out of school, we used in the market calculations, in the bakery. Also, we do many calculations related to hours, to organize our busy days. Here where we live, our schedule is very full and our day is very busy. Are we busier here than in other places? Maybe, we think so. But we still do not work. Mathematics also helps us to establish rules, have logical reasoning in situations, and to understand how some things work. We do not have many questions whether the mathematics that we learn can be used. Sometimes we see that is useful later, and we are afraid of being ignorant.

Question 6: What do you want to be/do in the future? Where would you like to be living? You can say whatever comes to your mind.

Julie: I want to study music. I want to play in orchestras and have my own music school. My dream is to live in Germany and playing in the orchestra there. To study music, I wish go to a conservatory and maybe stay for a while and back later. In fact, I'm striving for that. I do music lessons three times a week, and I play double bass. Also, I want to have driver's license, buy my Chevrolet Impala, and having my family.

Mary: I want to study scenic arts in a university with practical and theoretical classes. I want to teach and build a theater group. But not like to plan too much. Because sometimes when I plan a lot, I get anxiety and fear does not work. I want to find a boyfriend in college. I also want to build an orphanage. To help children in need.

Julie: Mary, you could also be part of the "Doctors of Joy," who are clowns who visit hospitals to entertain hospitalized.

Mary: I do not think so, because I am very touched by the sick.

Question 7: Talking realistically, what do you think your future will be like?

Julie: I believe I can work with music. But I think it will not turn out as I wish. I think I'll have to work to support myself while I study. And I also know that the salary of an orchestra musician is low. So I guess I'll have two jobs. Or perhaps a relative sustains me. Surely, I cannot buy my Impala. How sad!

Mary: Here in Rio Claro, there isn't a course of scenic arts. And I do not want to go live in another city because I do not want to live away from my family. So I guess I'll have to choose another course that I become a teacher. And later, when I can support myself, I study scenic arts. On the other hand, I'd be happy just to be a teacher. In fact, there are many other courses I'd like to do. Sometimes I think about studying physical education. My cousin studied in this course and he liked it a lot, so I also want to study it.

Question 8: Do you see any relevance for mathematics (or knowing how to count, make estimates, relate quantities, etc.) for your future?

Julie and Mary: For us to attain our objects, we have to use mathematics in everyday life. We use mathematics to calculate, count, to

	employment tasks, to manage an orphanage, to manage a music school. Also, for the selection tests of the university.
Question 9:	Do you see any connection between the mathematics you are studying in school and what you would like to do in the future?
Julie and Mary:	No, just the entrance exam. At least, not what we are learning now. Perhaps a little math to manage the financial business, but that's all. Perhaps we only use basic math operations, but nothing that we are studying now.

Features of a Foreground

A person's foreground and background are closely related. Thus the cultural background certainly has an influence on the foreground of the person. Background has to do with the origin of people; it represents customs and "cultural baggage." In this section, however, we concentrate on features of a foreground.

Foregrounds are dynamic. They are always changing, although far from due to sociopolitical changes. Thus, a change in a foreground may occur when the person establishes new perspectives and expectations for the future. For children and adolescents, this dynamics becomes even more evident as, for them, the difference between fantasy and reality is less clear. Thus, we do not try to identify what is the foreground of a person, considered to be a given or a "natural" phenomenon; instead, we investigate and interpret the foreground as a flexible entity.

In the interview with Julie and Mary, the expectation of the future was expressed with much determination. Julie wanted to study music, and she expresses hopes of coming to study this in Germany. Mary wants to come to work as actor at a theater. Such aims are far from common ambitions. Our point with respect to foregrounds in general is that they can include many such visions, and that such visions can direct what one is doing now. Thus, Julie is now taking music lessons. Such visions, however, are not fixed. They make part of a "stream of consciousness" (to use an expression developed by Husserl, although for quite different purposes). The stream of visions is changing according to one's formulations of hopes and one's experiences of obstacles and frustrations.

Foregrounds are multiple. By this we mean that one could not expect that, at a given moments, one and only one foreground can be associated to a person. Foregrounds are, strictly speaking, plural. By the dynamics of foregrounds we emphasize that foregrounds are changing over time, while we by the multiplicity of foregrounds emphasize that a person could, simultaneously, envision different sets of possibilities, depending on the perspective he or she assumes.

Thus, when Denival asked Julie and Mary, "What do you want to be/do in the future? Where would you like to be living? You can say whatever comes to your mind" and then in making the following question: "Talking realistically, what do you think your future will be like?" he tried to explore some aspect of this multiplicity. Indeed, Julie considers that she would need to work while she are studying, and Mary is happy to consider other possibilities besides studying scenic arts, such as being a teacher or taking another course.

Foregrounds can be collective. One can think of foregrounds as both a collective and an individual phenomenon. Thus, Julie and Mary have different ambitions for the future, however there is also much collectivity in these expectations.

From a general perspective, one can claim that foregrounds represent the possibilities that a certain group of people are facing. Thus, one talk about the foregrounds of children from a slum in Brazil, children from the upper middle class, as somehow represented by Julie and Mary, young people from Indian communities, girls from Muslim families, and so on. This makes sense as one can see foregrounds as configured by socioeconomic and cultural parameters, although not parameters of a strictly deterministic nature, but they define tendencies. They represent obstacles and possibilities for the groups in question.

One can also think of foregrounds as being collective in a different sense, when one considers that foregrounds are based on interpretations of possibilities. Most often such interpretations are collective. Thus, one can think of the many conversations between Julie and Mary as forming not only their friendship but also their visions about the future. They talk about "college, to marry, have children, the age at which we will do these things." They are in an explicit way exploring their foregrounds, and by doing it together they construct possibilities together.

Foregrounds include hopes and stereotypes. Foregrounds include many things, but certainly also hopes. This is clearly illustrated by the interview. Hopes can be seen as one defining element of the stream of images, dynamic and multiple, that contribute to the composition of foregrounds.

Such hopes are, however, not genuinely individual. Hopes are formed through many collective parameters. Hopes have to do with what in general is considered a good life. The conception of a "good life" can, however, included very many stereotypes. All kinds of commercials contribute to the common discourse of happiness. Thus, Julie and Mary's comments about getting a driver's license, buying an Impala, finding a boyfriend at a college, can be interpreted as also expressing some stereotypes. Naturally, the comments could as well include a dose of self-irony, as when Julie talked about buying "my Chevrolet Impala."

Ruined foregrounds. Foregrounds are not only formed through attractive images and free-floating hopes. Foregrounds also represent hard realities. Thus, a range of foreground parameters are configured through the socioeconomic life conditions. Such parameters are operating at a macrolevel by defining the distribution of wealth and poverty. However, such parameters are also operating at a microlevel, defining life conditions within a local community and within a family. For students from disadvantaged social conditions, expectations help to make changes in their lives. That is, the foreground can motivate one to look critically at one's background and drive to make changes.

However, foregrounds can be ruined, due to both macro- and microlevel parameters. By a ruined foreground, we understand a foreground that does appear to contain only a few attractive possibilities. A ruined foreground can be a product of a background. A reality full of difficulties and problems can lead to social exclusion and ruin the hopes and expectations. We see a ruined foreground as a principle learning obstacle.

This is not the case for Julie and Mary. They seem to enjoy prosperous foregrounds, including visions and aspirations.

Foregrounds, ruined or not, configure motives. Actions are guided by motives, and motives are formed through features of a person's foreground. Thus, we find that there is a close relationship between foreground, motives and acting—and also for not acting, in case the foreground is ruined and does not include elements that can establish motives.

We see learning as one form of action. Thus, learning is performed by the student. Learning is driven by motives. However, as foregrounds can be ruined, motives for learning can also be ruined. In this sense, a ruined foreground can represent a tremendous learning obstacle, and this is what many children are facing.

With respect to Julie and Mary, one cannot see many such obstacles. One can, however, make an observation related to mathematics. They are not experiencing any relevance of the mathematics they are learning in school for the moment, meaning that their motives for learning mathematics becomes first of all instrumental: mathematics is relevant for the entrance exam. And in the case of Julie and Mary, this also provides strong motives for studying mathematics. This is a most common phenomenon: students only experience an instrumental value of mathematics. However, in case one's foreground does not make space for completing any further education, it becomes difficult for a student to find motives for learning mathematics.

Foregrounds and Learning

We find that foreground is a crucial notion for addressing students' motives, beliefs and attitudes.

Currently, several proposals highlight the importance of educating students to actively participate in their own learning, through the activities of investigative character. Some proposals even defend the student's participation in planning activities and assessments.

The student should be able to express their fantasies, dreams, utopias, which, although they may be associated with things impossible to reach, still may turn real. In all circumstances, it drives the desire for change and improvement in a given situation.

Thus, we believe that the students' foregrounds can provide motives for learning when the students accepted the invitation to an investigative activity, taking an active part in the selection of content and themes, planning, and choice of targets.

In classroom, the teacher does not deal with one foreground but with a different set of foregrounds. Thus, one cannot assume that all students in a room feel the same way, have the same cultural baggage, pass through the same problems, and have the same expectations. However, this diversity is not a limitation for the reconstitution of foregrounds. In fact, it is an advantage.

Just as a smile or a yawn can be "contagious," so expectations, hopes, desires, and intentions can influence others. The diversity of foregrounds presented in a classroom can be an advantage if they are revealed and discussed. This way foregrounds are reworked. A well-known proverb says, "By iron, iron itself is sharpened. So one man sharpens the face of another."

Pedagogical proposals that involve group work with students discussing concepts, task sharing, exchanging ideas and experiences can be excellent ways for reworking foregrounds. The teacher should assume the role of a guide. If the teacher is too directive, he will suppress the process of collaboration between students. On the other hand, if the teacher is absent, this may seem disinterested, which can also cause the indifference of the student. The teacher as a guide may participate in the redesign of the foregrounds, indicating possibilities and combining goals with the interests of students.

REFERENCES

Alrø, H., Skovsmose, O., & Valero, P. (2009). Inter-viewing foregrounds: Students' motives for learning in a multicultural setting. In M. César & K. Kumpulainen (Eds.), *Social interactions in multicultural settings* (pp. 13–37). Rotterdam, The Netherlands: Sense.

Skovsmose, O., Alrø, H., & Valero, P. (with Silvério, A. P., & Scandiuzzi, P. P.) (2008). "Before you divide you have to add": Inter-viewing Indian students' foregrounds. In B. Sriraman (Ed.), *International perspectives on social justice in mathematics education* (Monograph 1, pp. 209–230). Charlotte, NC: Information Age.

Skovsmose, O., & Penteado, M. G. (2011). Ghettoes in the classroom and the construction of possibilities. In B. Atweh, G. Graven, W. Secada, & P. Valero (Eds.), *Mapping equity and quality in mathematics education* (pp. 77–90). New York, NY: Springer.

Skovsmose, O., Scandiuzzi, P. P., Valero, P., & Alrø, H. (2008). Learning mathematics in a borderland position: Students' foregrounds and intentionality in a Brazilian favela. *Journal of Urban Mathematics Education, 1*(1), 35–59.

CHAPTER 8

INCLUSION-EXCLUSION: AN EXPLOSIVE PROBLEM

Renato Marcone and Ole Skovsmose

Inclusive education is widely celebrated today. With respect to daily life in school, it could mean including everybody in the same activities and would therefore question an ability grouping of students. From a social perspective, inclusive education refers to the ambitions of turning students into citizens; bringing them into the political and cultural life of society; bringing them into the labor market; and so on.

Several assumptions are integral of the discourses on inclusion: Inclusion has to be pursued, and exclusion has to be prevented. Thus, it is broadly assumed that there exists a straightforward "good-bad" duality to inclusion and exclusion. This assumption is operating when strategies for inclusive education become elaborated and justified. In this chapter we are going to address this discursive duality-assumption. We find that the processes of inclusion and exclusion are very complex, and there is no unequivocal way of evaluating them.

Although inclusion and exclusion are two words, we think of them as representing one integral concept of inclusion-exclusion. We consider it as an explosive concept, and we will return to this explosiveness in the final section of the chapter. We will consider inclusion-exclusion discourses related to education in general and in mathematics education in particular.

Critique as Uncertainty, pages 95–109.

For exploring the discursive duality assumption, we have found inspiration from outside education: from discussions related to postcolonialism, which in recent years has been explored profoundly. Being an "ism," postcolonialism refers to a theoretical-political position which developed in steps: colonialism, anticolonialism, postcolonialism, and beyond. One can see Frantz Fanon's book (1961) *The Wretched of the Earth* as one of the primary expressions of anticolonialism, while *Orientalism: Western Conceptions of the Orient* by Edward Said (1979) represents the beginning of the postcolonialism. The *Location for Culture* by Homi K. Bhabha (1994) and *A Critique of Postcolonial Reason* by Gayatri C. Spivak (1999) represent the recent developments of this "ism."

We are going to present some cases where the inclusion-exclusion complexity becomes illustrated. The cases are (1) an example from a textbook, (2) the aspirations of some indigenous students in Brazil, and (3) a case of inclusion-exclusion in a special school for visually impaired children and teenagers in Brazil. However, before we get to these examples, we will say a few words about postcolonialism and about (post)colonial discourses.

Postcolonialism

The worldwide colonial system was at its peak just before the First World War. Then there started a slight crackling. However, it was only after the Second World War that the colonial powers lost their firm grip. Many colonies got independence, and the map of the world got many more different colours.

However, many structures of dominance are maintained, although in different organizational formats. Ideological structures became recast in other ways of legitimizing suppression. Discourses, including different forms of stigmatizing, are shifting "grammar." The principal feature of postcolonialism is to address all such forms of suppression and stigmatization.

As we take a look at *A Critique of Postcolonial Reason* by Spivak we get a clear impression of the broad categories that can be applied in such a critical approach. Thus, Spivak (1999) addresses philosophy, literature, history, and culture and identifies along these dimensions the stereotyping discourses as constituting parts of the "postcolonial reason."

With respect to stereotyping Spivak draws on Bhabha (1994), who characterizes it as a limited kind of otherness. Bhabha grounds his stereotyping concept on Said's (1979) book *Orientalism*. Here it is pointed out that the idea about the East was created for and by the West and has nothing to do with the East. In other words, it is nothing more than a representation made by the West for the West. The Oriental subject is a stereotype, in fact a web of stereotypes, as "the stereotype requires, for its successful signification, a continual and repetitive chain of other stereotypes" (Bhabha, 1994, p. 110). With such layers of stereotypes, the West was justifying their control over what they discursively constructed for themselves as noncivilized groups, or at least, not so civilized as them: the colonizers from the West.

However, the stereotyping includes much more than just creating a particular image of the colonized and uses this as a justification for the discrimination. Bhabha emphasizes that stereotyping goes much further than this:

> It is a much more ambivalent text of projection and introjection, metaphoric and metonymic strategies, displacement, over-determination, guilt, aggressivity, the masking and splitting of "official" and phantasmatic knowledges to construct positionalities and opposionalities of the racist discourse. (Bhabha, 1994, p. 117)

What happens is that the colonized population is presented as inferior and, consequently, as carrier of the very justification for being colonized and suppressed. Even in a postcolonial world, populations can be trapped in such circles of interpretation.

Such interpretations are related to social, political, economic, and cultural processes of inclusion and exclusion: The stigmatizing discourse opens for some form of inclusion into the worlds of domination; simultaneously providing the justification for a variety of forms of exclusion. The critical approach of postcolonialism is extremely important today, when stigmatizing discourses make a large part of a range of inclusion-exclusion processes, also in education.

(Post)colonial Discourses

The notion of discourses is extremely broad: It is in need for some clarification. In order to do so we can refer to Nietzsche's conception of "truth" and Foucault's conception of "regime of truth."

Nietzsche emphasizes that there does not exist any "reality" about which "truths" can be claimed to exist. Truth is not about anything. "What, then, is truth?" Nietzsche asks. And he answers,

> A mobile army of metaphors, metonyms, and anthropomorphism—in short, a sum of human relations which have been enhanced, transposed, and embellished poetically and rhetorically, and which after long use seem firm, canonical, and obligatory to people: truths are illusions about which one has forgotten that this is what they are; metaphors which are worn out and without sensuous power; coins which have lost their pictures and now matter only as metal, no longer as coins. (Nietzsche, 2010, pp. 29–30)

Truth is an illusion. It just makes part of a way of metaphoric or rhetoric way of talking. This observation brings us a step closer to the notion of discourse. And most likely, Bhabha had this Nietzsche citation in mind when he characterizes stereotyping as including metaphoric and metonymic strategies.

The further step in clarifying the notion of discourse can be taken with reference to Foucault's observations concerning "regime of truth." Foucault states that

> each society has its regime of truth, its "general politics" of truth—that is, the types of discourse it accepts and makes function as true; the mechanisms and instances

> that enable one to distinguish true and false statements; the means by which each is sanctioned; the techniques and procedures accorded value in the acquisition of truth: that status of those who are charged with saying what counts as true. (Foucault, 2000, p. 131)

A regime of truth can be seen as a discourse.

A paradigmatic example of a colonial discourse is presented by Said (1979) in *Orientalism*. He quotes from a speech in the English parliament in June 1910 made by Arthur James Balfour. The issue addressed was the Egyptian question, Why should we, the English Empire, be troubled with Egypt?[1]

> First of all, look at the facts of the case. Western nations as soon as they emerge into history show the beginnings of those capacities for self-government. . . .You may look through the whole history of the Orientals in what is called, broadly speaking, the East, and you never find traces of self-government.

The problem is self-government. Somebody has to govern the Orientals; they are not able to do so by themselves. To govern is, according to Balfour, a task "we have taken upon ourselves in Egypt and elsewhere." And he emphasizes that this is "the dirty work, the inferior work, of carrying on the necessary labour." Through such formulations Balfour echoes Rudyard Kipling's (in)famous phrase "the white man's burden." Balfour then asks,

> Is it a good thing for these great nations—I admit their greatness—that this absolute government should be exercised by us? I think it is a good thing. I think that experience shows that they have got under it far better government than in the whole history of the world they ever had before, and which not only is a benefit to them, but undoubtedly a benefit to the whole of the civilized West.

Yes, the English are doing a tremendous job: the colonized nations benefit greatly from being governed by the English. And Balfour further claimed, "We are in Egypt not merely for the sake of the Egyptian, though we are there for their sake; we are there also for the sake of Europe in large."

One can imagine the cheering from the members of the Parliament. Did Balfour speak the truth? This sounds like a silly question. But he established a perspective on the Egyptian question. We see this talk as a paradigmatic example of stereotyping, including "projection and introjection, metaphoric and metonymic strategies," as addressed by Bhabha.

Balfour's speech belongs to the colonial times. The speech represents an elaborated form of racism and orientalism from before the First World War. Orientalism and racism have not been vanished today. They have instead been transformed and multiplied into a range of different discourses, including a wide range of stigmatizing expressions.

---

[1] The quotations from Balfour's speech can be found in Said (1979, pp. 32–33).

In their text *Stereotypes of Muslims and Support for the War on Terror*, John Sides and Kimberly Gross (2013) from George Washington University, show some results of their investigation about American's stereotypes of Muslim and Muslim-Americans. They found that

> Negative stereotypes relating to violence and trustworthiness are commonplace and that little distinguishes Muslims from Muslim-Americans in the public's mind. Furthermore, these stereotypes have consequences: those with less favorable views of Muslims are more likely to support several aspects of the War on Terror. (p. 1)

They also found that "Black and Latino respondents tended to view Muslims in the same way as Whites." They summarized their observations along four dimensions: peaceful-violent, trustworthy-untrustworthy, hardworking-lazy, and intelligent-unintelligent. Muslims, Muslim Americans, and Blacks were considered violent, Asian Americans and Whites were peaceful, while Hispanic Americans were in an in-between position. Muslims and Muslim Americans were considered untruthful, Asian Americans and Whites were trustworthy, while Hispanic-Americans and Blacks were in an in-between position. All categories were considered more or less hard working, except Blacks who were lazy. All categories were considered intelligent, although Asian Americans and Whites more than the others. In general, stigmatizing and stereotypes make part of discourses that produce "truths" and "justify" policies, both with respect to inclusion and to exclusion.

The Escalator at the Holborn Tube Station is 156 Feet Long

Let us now make a jump from the general comments on postcolonialism and (post)colonial discourses to mathematics education.

In the article "Western Mathematics: The Secret Weapon of Cultural Imperialism," Alan Bishop (1990) refers to a textbook which contains the following problems:

> If the cricketer scores altogether r runs in x innings, n times not out, his average is $r/(x - n)$. Find his average if he scores 204 runs in 15 innings, 3 times not out.

The escalator at the Holborn tube station is 156 feet long and makes the ascent in 65 seconds. Find the speed in miles per hour.

Years ago, during British colonial times, the textbook was used in schools in Tanzania. The use was recommended by the English colonial education officer. Thus, Bishop observes that in India and in Africa, schools and colleges were organized in such a way that they "mirrored once again their comparable institutions in the 'home' country" (Bishop, 1990, p. 55).

It was a mirroring that included many elements: the school structure; the teaching forms; the dress code including school uniforms; and the textbooks including their ways of the contextualization. From today's perspective of learning, the form of contextualization presented in the textbook does not make much sense.

Naturally, we would assume that a contextualization has to relate to the students' environment and experiences.

However, in a colonial context, the educational system should serve the empire. It should instill "proper values" among the students. It was important that the students who were successful in this colonial educational program would turn into well-adjusted state functionaries. It should be the functionaries who appreciated the British Empire and acknowledged British superiority. In the most direct Foucault-interpretation, colonial education meant disciplining. And as part of this disciplining also the exercises concerning cricket scores and the Holborn escalator served their purposes. The exercises can be seen as making part of an empirical gatekeeping, ensuring that those who demonstrated a readiness to assume the colonizers' perspective got included, while the rest got excluded. From the point of view of running an empire, the colonial officer in Tanzania might well have made relevant recommendations.

Considering a colonial educational system, inclusive education does not appear, at least to us today, to be any straightforward good thing. It certainly meant inclusion into a range of pregiven structures. Concerning *colonial discourses* Bhabha states, "The objective of colonial discourse is to construe the colonized as a population of degenerate types on the basis of racial origin, in order to justify conquest and to establish systems of administration and instruction" (Bhabha, 1994, p. 101).

With small reformulations, we can state that the aim of *colonial education system* is to form a population of degenerate types on the basis of racial origin in order to justify submission and to establish systems of government and administration. Thus, the aim of colonial education in Tanzania was to elaborate a British "regime of truth." Colonial education meant colonial disciplining, and inclusion meant inclusion into the colonial world order.

Colonial times belong to the past, but we find many leftovers from colonial disciplining even today. For instance, inclusion at some places still means preparation for certain working conditions. Many types of production get outsourced to countries that previously functioned as colonies and which now are referred to as Third World countries. It could be a polluting production, which is not welcomed in the home country. However, by outsourcing the production, the company could benefit in different ways: lower salaries for the workers and fewer security measures. For the companies, however, it is important to make the outsourcings to a location with qualified workers, and this demand can easily have implications for the education offered in the region. What would an inclusive education mean in this situation?

Also in the postcolonial times, processes of inclusion can be associated with the most questionable interests and priorities. Inclusion can mean domination, and *inclusive domination* can be put in operation by the educational system and by mathematics education as well. Naturally exclusion can as well be problematic. What we want to emphasize is that *inclusion and exclusion refer to contested processes. It is impossible to assume that inclusion constitutes a straightforward attractive educational endeavor, while exclusion has to be condemned. Inclusion and exclusion processes can assume any kind of qualities and serve very different purposes.*

In the Fields, We Use Division a Lot

The article "Before You Divide, You Have to Add" (Skovsmose, Alrø, & Valero, 2007) presents an interview with some Indian students living in an Indian reserve in the São Paulo State. The reserve is located close to a big city with several universities and faculties. The Indian village is composed of many small houses hiding in the landscape. The village has a recently constructed school building, and in the middle of the village we see a small football field.

Here we do not summarize the whole interview, but concentrate on two students, Patrick and Matheus. The students themselves decided to use these names for the interview. First, the students were asked to present themselves and they said,

Patrick: I'm 17 years old. I was born in the city and used to come here on the weekends. When I was a bit older, I decided to come live here in the village with my grandfather. I couldn't get accustomed to the ways of the city; everything here in the village is calmer. The work is harder, because we have to work in the field. The village is big, and there is plenty to do here. There's a reservoir where we swim, and a football field. I'd never leave the village now. My girlfriend is pregnant, and we're going to live together.

Matheus: My name is Matheus. I'm 16 years old. I've lived here . . . since I was born. My mother used to live in another village, but when she and my father got married, she came to live in his house. . . . We play football every Sunday in the football field. . . . We have lots of parties here, too, and at the parties, we dance forró.

Patrick emphasizes that his real friend is his grandfather: "Since I go to work in the field with him, we talk the whole day long. He tells me that he would like me to get an education. He thinks that I could help our people more if I study." And Patrick continues,

Patrick: But I have other friends too. I have one from the college prep course who is cool. . . . He wants to study how to work with computers. I want to be a nurse.

This is what Patrick wants: to become a nurse. Patrick and Matheus were asked about their opinions regarding going to school and what they were doing. Patrick is joining a course preparing him for entering the high school:

> Patrick : In the college prep course, I learn all the subjects taught in high school. I learned the theorems, [algebraic] equations, roots, how to transform meters into kilometers. I learn a little of everything, since they are the subjects you need for the college entrance exams.

He also tells that when he was younger he liked solving mathematical problems. He seems to have enjoyed mathematics. Matheus is of a different opinion.

> Matheus: I think it's kind of boring, this subject [mathematics]. I think it's because I don't pay very much attention. I don't like the teacher very much, either. She goes too fast; there is hardly enough time. I am learning to find the area of a square or a rectangle, but I think that's geometry.

However, could this (boring) mathematics turn out to be used somewhere?

> Patrick: In the fields, we use division a lot, too. We divide the land according to the seed we are going to plant. We also divide our profits and expenses. We use division for everything, and addition, too, since before you divide, you have to add.
>
> Matheus: I agree with Patrick. In the fields, or there in the vegetable garden, we use division to divide the area to be planted and the seeds.

The students were then asked about the future:

> Patrick: I want to become a nurse to help all my Indian relatives to be more healthy. . . . I think I can even manage to achieve it. But first I have to study for that. I don't want to go live in the city.
>
> Matheus: I don't think I want to live in some other place, no. I like the village. I want to continue working in agriculture.

And mathematics, what sense to make of mathematics with respect to the future: becoming a nurse, working in the fields? Patrick points out the instrumental value of mathematics: "And really, what good are trigonometry, theorems, and roots if they don't get me into the university?" However, for the very profession of nursing he does not consider it relevant. With respect to farming, Matheus makes similar observations: "I don't think that the 'catheti' and the hypotenuse are very useful for the farmer. Maybe for me to pass the college entrance exams I should know it, but on the farm, you don't use any of that."

What could inclusion-exclusion mean for Patrick and Matheus? What direction does inclusion take? Patrick wants to become a nurse. He wants to enter a university. He has no intention of operating within, say, the outlook of natural medicine. He intends access to regular education to become a nurse within the Brazilian education system. And in order to obtain this, he is ready to study mathematics, even though he doubts that mathematics might be relevant for the daily practice of nursing. Patrick wants to work as a nurse in the indigenous community.

Matheus seems less interested in going to school, but he likes to work in the fields. Both Patrick and Matheus emphasize that in fieldwork they are using mathematics, in particular when dividing the crops, profits, and expenses. As Patrick points out, you have to add before you divide. This principle is a statement of solidarity: one has to bring together everything that has been produced by whoever before you divide it among everybody.

Inclusion-exclusion can have different poles of reference—for the Brazilian society or the indigenous society. We do not have to do with a straightforward duality. Some indigenous communities are located in or close to big cities, like São Paulo, and appear as *favelas* (slums). We also find rather isolated indigenous communities, which cultivate traditions and operate without any money system. Still, in these remote communities, one also finds TV sets, young people walking around in shorts and t-shirts with the colors of the Brazilian football team. Some people have cars. Furthermore, we find many in-between communities, as the one in which Patrick and Matheus are living.

So what does inclusion-exclusion mean for them? We have to consider the different poles from which inclusion-exclusion can be addressed. One possibility is to consider inclusion of indigenous students into the Brazilians society, and in particular into the education system. (Let us immediately emphasize that the very expression "Brazilian society" is loaded with presumptions. Thus, the indigenous society we are addressing is Brazilian as well. Anyway, we use this loaded expression in order to simplify some formulations.) We can consider to what extent Patrick becomes properly prepared for entering the university. And in general, to what extend do they get prepared for participating as citizens in the Brazilian society.

We can, however, address the same issues taking departure in the indigenous society. In what sense do they become educated to make part of the Indian way of life? Sure this is an explicit priority for both Patrick and Matheus. Should the Indian schools use the Brazilian textbooks in mathematics so as to facilitate their further advancement in the Brazilian education system? Or should the schools address the procedures for making divisions, which makes a crucial part of the Indigenous welfare system?

Patrick and Matheus know the village; they also know the city. They know young people here and there. They certainly know Indian dances, but they also like the *forró* and the disco. The first years in the school they have studied their Indian language, but certainly Portuguese is their mother tongue as well. They are *operating within different life worlds, and the processes of inclusion and exclusion can be considered with reference to these different worlds. Inclusion and exclusion refer to the processes with complex multiplicities and various poles of possible references.*

I'm Not Blind. Am I?

Inclusion-exclusion not only concerns differences between cultural groupings, it also concerns groups that are different from the stipulated "normality." The following data were produced between August and early December 2011 by the first author. During this period, one classroom was observed and recorded in audio and video, in addition to field notes. These data is a part of the first author's PhD study (with the second author as supervisor). The teacher Fabiana and an ophthalmologist, Virgílio, were interviewed (pseudonyms). Fabiana was working in a special class with eight students, in the third school year, but all of them with an advanced age for this, they were between 12 and 15 years old, while usually at third year, in Brazil, the students are 8 years of age and not more than 9 years old. It was a special class, inside a special school—a school for blind and low-vision students.

As soon as we got there, we met Carlos, a special student, in a special class, in a special school. He was 14 years old in 2011, and he joined the first year in a braille class. Until the year before, Carlos was studying in an "ink class." Let us explain what this means. In the special school where we were working, there are two kinds of classes: one for "legally blind" students, which means that even if they still have some visual residue they are not able to see properly, and they are not able to read as well, so they need to be educated in braille; and another for students with low vision but who are still able to read printed texts using some strategies, like magnification devices and high color contrast. The division between legally blind and low vision is based on parameters accepted all over the world, legitimatized by WHO (World Healthy Organization).[2]

We were interested in Carlos' case firstly because it was a border case, meaning it is a case where it is not clear whether he is blind or low vision, but, specially, because of his reaction about being placed in a braille class. He was continuously repeating, "I'm not blind, I should go back to my ink class, I can see!" But some teachers of the special school did not agree with him; they declared Carlos to be "educationally blind"—he was able to see his friends and many things, so he may not be declared legally blind, but he was not able to study in an ink class accordingly with the teachers. So the school decided that moving him to a braille class was for his best interest.

[2] Is possible to learn more about blindness, its meaning, and its definition in this website http://www.who.int/blindness/en/.

In one occasion during a math class, Fabiana was teaching problem resolution using arithmetic, and the tool used by the blind students in order to do calculations was the soroban,[3] an old but still very much used calculator (especially for blind people). Fabiana was trying to convince Carlos that he should not use his eyes to see the soroban. Instead, he should use his fingers and learn how to do calculations in soroban using only his tactile sense. But Carlos was insisting on using his eyes trying to see the accounts on the soroban, while Fabiana kept repeating: "Forget your eyes, use your fingers"!

Fabiana is a very experienced teacher. She had been working at the special school for the past 18 years (at the time of the interview in 2011); 4 years as a supervisor and 14 years as a teacher in braille classes. Her undergraduation is in Literature and Portuguese Language, but she teaches all the subjects, including mathematics. Fabiana is also blind; she became so when she was a teenager. In our interview about Carlos, Fabiana says,

> About this case, Carlos' case, he has a reasonable visual residue, but the low-vision teachers concluded that it wasn't enough to work with ink. So, this is something that is always on his mind. In the future, if he loses all his sight, he may have some problems, because he did not stimulate his touch for braille, he is still reading braille with his eyes, so, let see what'll happen. We are all done about talking to him, insisting that he should stimulate his touch, read with his fingers instead of his eyes. But now, it's all up to him, we can't just blindfold and obligate him to use his hands, it'll be a violence, yet, he'll need to face this one day.

Fabiana's statement shows her concern about Carlos' case. What is most important, that the student goes to the proper classroom following to the teacher's position based on student's "abilities?" Or to follows the student's desires? We asked Fabiana about the process to decide whether or not all these students should go to a braille classroom.

> The decision is really complex. Even here, a special school where we deal with inclusion all the time, there is a huge prejudice about this. Some teachers believe that low-vision students should stay in ink classes until is not possible anymore, until they lose their sight completely. There are teachers here [who say] that, if I had studied here, probably they would have tried to do with me the same that some doctors did when I was a teenager: forced me to stay at ink, even if I have failed all the time. . . . And it happens that some people, even with a good visual residue as one of my colleagues, also a teacher, can do much better [studying] in braille than in ink, because they have much more resourcefulness in braille than in ink. . . . So, because of this taboo that many people, even here, still have over blindness, it looks like learning braille, symbolically, blinds the person definitely, it's like going to a braille classroom means that the student has failed. Braille is a stigma.

[3] For soroban, look at http://en.wikipedia.org/wiki/Soroban.

Also, the students with disability in a special school can be subjected to prejudices and exclusions. Inclusion is not a straightforward good thing as we said before; it is much more complex than we usually imagine. The stratification of students starts in the regular classes, in "normal" schools, where they become classified as normal or not, as intelligent or not, as capable or not. Then the non-normal students get a different treatment, sometimes even a different school, a special school. Once in the special school, one could assume that exclusions and stratifications are finished, but students continue to be classified and divided in categories, as being blind or with low vision as in this case. As we saw, if a student in a braille class is not completely blind, she or he is nonnormal, then different, and then excluded. Yes, braille can be a stigma.

We are not stating that it is a bad idea to send students to a special school, or that it is a bad idea to separate low vision students from blind students, or to separate students based on any kind of classification for that matter. It all depends on the actual acted-out interpretation of inclusion and exclusion.

We cannot just start researching about how to teach mathematics properly for blind students, deaf students, normal students, low vision, and so on, without thinking why we are doing this. Surely, teaching the contents is urgent and necessary, as all the privileged positions in the society use this knowledge to exclude, but reflections on how the processes of inclusion-exclusion might operate in particular situations are as important as the content matter issues.[4]

The ophthalmologist director of the eye hospital annexed to the special school, Virgílio, made the following comments about Carlos' case:

> I think he should go for low-vision classroom, an ink classroom, and we provide the resources for that. Usually when one has a very small residual vision, she or he will need video magnification to read, and we already have this here at school. And if the visual residue is too low, the student can even lead braille in parallel, but if he or she has some vision residue, so the student should stimulate this. Besides, the video magnification features are getting much more accessible. When we bought our first CCTV,[5] which were the first to use the video magnification resource, it was very costly. Nowadays we already have at least two domestic manufacturers for that, and you still can use an application on a tablet or a smartphone which magnify the letters, a feature that will be cheaper and cheaper over time, as a vision accessory to improve visual performance.

The inclusion and exclusion processes are complex. They depend on so many different factors, including the development of technology.

---

[4] In this sense, researches like Healy and Fernandes (2011), Fernandes and Healy (2007), Marcelly (2010), Sales (2013) are very coherent, because they are dealing with contents and simultaneously reflecting about the very processes of inclusion and exclusion.

[5] A CCTV (closed-circuit television) uses a video camera to project a magnified image onto a TV screen. For individuals with visual impairments, a CCTV can help to accomplish a lot of daily tasks, such as reading, writing, viewing photos, working crossword puzzles, or sewing.

Following the ophthalmologist's position, Fabiana should encourage Carlos to use his visual residue in order to stimulate it. But Fabiana finds that the best way to teach and learn arithmetic and solve problems is to apply the soroban calculation mechanism using touch, and not vision, because in the future, if the student loses his sight completely, he should be at least able to use the soroban. Thus, at the same special school which is a very well conceptualized in the entire Brazil, one meets at least two different positions about how to deal with a case like Carlos'.

The different approaches have particular implications. While we use a regular positional algorithm to do sum, subtraction, division, and multiplication in ink, the soroban system is based on a different kind of system (not based on visual positioning of numbers). Carlos, when doing calculations, was always trying to imagine the position of the numbers on the algorithm, worried about divisors, multiplicands, first parcel, second parcel, rests, and so on, something that does not make much sense in a braille classroom. Thus, when a class is divided into ink and braille, it also becomes divided with respect to mathematical algorithms.

We find several conceptions of inclusion-exclusion living together, blending, founding other hybrid concepts of inclusion-exclusion. *Processes of inclusion and exclusion make part of complex discourses that include metaphors and metonyms; they include stigmatizations, and they include assumptions about what to consider normal and abnormal.* We find discourses, rich in stereotyping presumptions, operating when Balfour presented his view on the Egyptian question. We find as well complex discourses brought in operation when inclusion and exclusion are discussed today with respect to education. Not that the discourses are similar, rather far from this. But we have to deal with discourses blended with presumptions and hybrid concepts of inclusion and exclusion and also when we deal with a particular context of specialized education. We have to deal with different possible regimes of truths.

An Explosive Problem

Let us now try to summarize our observations: *First*, inclusion-exclusion refers to contested processes. It is not possible to assume that inclusion constitutes a straightforward attractive educational endeavour, while exclusion has to be condemned. Inclusion-exclusion processes can assume any kind of qualities, and serve very different purposes and interests. There is no "good-bad" duality that directly can be associated to inclusion-exclusion.

Second, processes of inclusion-exclusion refer to different perspectives. We have to consider a multiplicity of references, or poles. What could be considered inclusion from one pole might mean exclusion considering a different pole. We can refer to economic interests, thus we can think of the economic order of the capitalist society, but also of the economic structure of an indigenous society where one adds before one divides. Poles can refer to different cultural priorities, different neighbourhoods, different life-worlds.

Third, in educational discourses and practices we find contesting definitions of inclusion-exclusion living together, blending, and founding new hybrid conceptions. Inclusion-exclusion makes part of complex discourses, which includes different suggestions for what to consider normal and abnormal. Not even in the most specific case of inclusive education, can one expect to find agreed-upon interpretations of inclusion-exclusion. One can claim that in order to learn calculation in soroban it is necessary to stop looking. However, one can also claim that it is always important to stimulate the visual residue. Or would it be possible to create a hybrid pedagogy, teaching properly the soroban mechanisms and yet stimulating the visual residue?

The summary brings us to emphasize that inclusion-exclusion is an *explosive concept.* By this we understand an open concept without any simplifying definition. If one tries to provide a clarification of such a notion, one comes to use other notions just as open as the notion itself. As an example, one can think of the notion of democracy—if one tries to define what it could mean, one becomes drawn into a spiral of complexities. One comes to use notions like "justice" and "equity" just as hard to clarify as "democracy" itself. Furthermore, explosive concepts operate as nucleus within different contesting discourses. Thus, one can find "democracy" operating in very many radical different discourses.

In the article "Explosive Problems in Mathematics Education" (included as Chapter 19 in this book), Ole tries to characterize principal research problems in mathematics education. The article was written as a response to Richard Barwell's editorial in *For the Learning of Mathematics*, inviting researchers to address what could be considered fundamental problems in mathematics education. In 1984, David Wheeler, the previous editor of *For the Learning of Mathematics*, had provided a collection of such problems, and Barwell wanted to engage people in a new discussion of such problems.

Naturally, an identification of such problems reminds us of David Hilbert's presentation in 1900 of fundamental research problem in mathematics. We want, however, to emphasize that while Hilbert's problems were formulated in terms of relatively stable concepts, principal research problems in mathematics education cannot be formulated this way. And, further, there is no point in trying to provide reformulations of such problems by substituting explosive notions by notions that appear stable. Instead, one should expect research problems in mathematics education to be formulated in terms of explosive concepts. In this sense, we think of basic research problems in mathematics education as being explosive.

We find that the issue of inclusion-exclusion through mathematics education is an example of such an explosive problem. One should not assume that some conceptual clarification and simplification are to be expected. Nor should one try to ignore the problem as being too diffuse. We have to address explosive problems in mathematics education. In particular we have to address inclusion-exclusion problems in all their dramatic complexities.

REFERENCES

Bhabha, H. K. (1994). *The location of culture*. London, UK; New York, NY: Routledge.

Bishop, A. J. (1990). Western mathematics: The secret weapon of cultural imperialism. *Race and Class, 32*(2), 51–65.

Fanon, F. (2004). *The wretched on the earth* (R. Philcox, Trans.). New York, NY: Grove Press. (Original work published 1961)

Fernandes, S. A. A., & Healy, L. (2007). Transição entre o intra e interfigural na construção de conhecimento geométrico por alunos cegos. *Educação Matemática em Pesquisa, 9*(1), 121–153.

Foucault, M. (2000). *Power* (R. Hurley et al., Trans.; J. D. Faubion, Ed.). New York, NY: New Press.

Healy, L., & Fernandes, S. A. A. (2011). The role of gestures in the mathematical practices of those who do not see with their eyes. *Educational Studies in Mathematics, 77*(2/3), 157–174.

Marcelly, L. (2010). *As histórias em quadrinhos adaptadas como recurso para ensinar matemática para alunos cegos e videntes* (Unpublished master's thesis). State University of Sao Paulo, Rio Claro, Brazil.

Nietzsche, F. (2010). On truth and lie in a nonmoral sense. In T. Carman (Ed. &Trans.), *On truth and untruth: Selected writings* (pp. 17–49). New York, NY: Harper Perennial/ Modern Thought. (Original work published 1873)

Said, E. W. (1979). *Orientalism: Western conceptions of the Orient*. New York, NY: Vintage. (Original work published 1978)

Sales, E. (2013). *A visualização no ensino de matemática: uma experiência com alunos surdos* (Unpublished doctoral dissertation). State University of Sao Paulo, Rio Claro, Brazil.

Sides, J., & Gross, K. (2013). Stereotypes of Muslims and support for the war on terror. *Journal of Politics, 75*(3), 583–598. Retrieved from http://home.gwu.edu/~jsides/research.html

Skovsmose, O., Alrø, H., & Valero, P. (with Silvério, A. P., & Scandiuzzi, P. P.) (2007). Before you divide you have to add: Inter-viewing Indian students' foregrounds. In B. Sriraman (Ed.), *International perspectives on social justice in mathematics education.* (pp. 151–167). The Montana Mathematics Enthusiast, Monograph 1. Charlotte, NC: Information Age.

Spivak, G. C. (1999). *A critique of postcolonial reason: Toward a history of the vanishing present*. Cambridge, MA; London, UK: Harvard University Press.

CHAPTER 9

RESEARCHING POSSIBILITIES

What could it mean to research "what is not?" What could it mean to research possibilities? In order to clarify such questions, one could consider what it means to research "what is" and discuss a positivistic and a naturalistic approach. The first tries to uncover a reality; the second to grasp the multiple set of constructed realities. In this way, both approaches pay particular attention to the descriptive qualities of doing research.

Researching possibilities in an educational context is discussed with reference to three different situations. The current situation refers to the circumstances before any educational reorganisation takes place. The imagined situation, far from being a well-defined entity, includes ideas about how things could be organized. The arranged situation refers to educational reorganization restricted by practical, political, and other circumstances. Researching possibilities includes a concern with all three types of situations, and in particular, it addresses the relationship between the situations.

Explorative reasoning originates from an arranged situation. However, the idea is not simply to try to further clarify the arranged situation, but also to gain a better grasp of an imagined situation and, in so doing, to shed new light on the current situation. In this way, an imagined situation could be developed and specified, and eventually become an alternative to what has taken place. Researching pos-

Critique as Uncertainty, pages 111–126.

sibilities means to search for, to construct and to realize possibilities, and in these processes, explorative reasoning is essential.

One can try to *outline* possibilities. One can *talk about* possibilities or *indicate* possibilities. But in what sense can one *research* possibilities?

It is often claimed that research must provide descriptions, and a variety of discourses have evolved around science as a descriptive tool. Some discourses assume the existence of a reality which is waiting to be portrayed. Others claim that any reality is constructed, if not created, and that the task of research is to report as adequately as possible on this multitude of constructed realities. Such discourses do not interpret scientific activities as including a search for alternatives to what can be observed. My concern, however, is to talk about "researching possibilities" as including a search for new possibilities.

The ideas I am going to present have been developed over a long period of time. One crucial event took place at a PhD seminar in Durban in March, 1996.[1] The seminar was part of a South African-Danish research cooperation aiming to establish doctoral studies within the postapartheid situation in South Africa. One dilemma was experienced by the South African doctoral students: On the one hand, they wanted to do research that could meet the proper standards of research quality; on the other hand, they felt uncomfortable researching the current situation, as this was dominated by the "sins of apartheid," which refers to all those consequences of the apartheid regime that could not easily be eliminated when entering a postapartheid period, such as, for instance, the apartheid-determined locations of neighborhoods.[2] Thus, it could be difficult to explore multicultural mathematics classrooms, as the apartheid regime had established complete segregation between "Whites," "Blacks," "colored," and "Indians," which still dominated the context of schooling. At the seminar, we found it relevant to research what was not there, but which could become a new reality.

The ideas instigated at the Durban meeting have been presented and elaborated in very different contexts at conferences and workshops. They have turned out to be useful for several studies both related to, but also outside of, the South Africa-Danish research cooperation.[3] Through such work, the ideas have taken on new dimensions. Together with Marcelo Borba, I have presented some elements of the story (Skovsmose & Borba, 2004), and what follows has its origins in this presentation. My intention, however, is to make a reformulation and an update based on the many contributions to the approach of researching possibilities. I am now speaking of "researching possibilities" and not, as I have previously, about doing "critical research." This change in terminology may avoid some misunderstandings, which I will return to in the concluding remarks.

---

1 See Vithal (2003) for a reference to the meeting and for a substantial further development of the ideas.

2 The expression "sins of apartheid" has been used by Bopape (2002).

3 See, for instance, Bopape (2002); Vithal (2003); Barton, Autagavaia, Poleki, and Alangui (2002); Eriksen (2003); and Pfannkuch (2005).

I want to consider to what extent it makes sense to address not present or past situations, but rather what could be called possibilities and alternatives to a present situation. What could it mean to research something that is not really "in sigh?" In order to clarify this question, I discuss what it could mean to research "what is." A short intermezzo brings me to the ideas of researching "what is not." Then I discuss the process of explorative reasoning, which I believe to be important for researching possibilities (and not just, say, "talking" about possibilities). After pointing out the variety, specificity, and collectivity of explorative reasoning, I will turn to the concluding remarks.

Researching "What Is"

In order to clarify what it could mean to research "what is not," I will comment on what it could mean to research "what is."

In many discourses, it appears a simple truth that research must address what can be observed and collected (or produced) as data. Inventing data would, thus, be out of the question. Imagine a situation in which one has observed a classroom and is preparing a transcription. Choices have to be made: What transcriptions to include in the data material. The transcriptions can be organized in different ways, and this composes part of the interpretation. In many cases, translations must also be done (for instance, when Danish research is presented at an English conference), and translation also involves interpretation. All this is well recognized. However, what if one starts speculating: Well, according to the tape, the teacher asked this question, and the student answered that way. But what would have happened if the teacher had put the question differently? Like this, for instance! What might the student have possibly answered then? One possibility is . . . and the student's possible answer becomes "transcribed" as well. That was in fact an interesting answer! The person doing the transcription continues the dialogue, and in the best Ernest Hemmingway style, he or she produces a fascinating dialogue between teacher and student. This creative transcription continues until it is long enough to be analyzed, at which point the researcher concentrates on studying the constructed dialogue. Such an approach seems to go far beyond what is generally acceptable in research. (Naturally, there are many exceptions, and later I will come to that.) The researcher must address reality, not fiction. In this sense, research methodology often expresses a commitment to providing descriptions. This commitment is acted out in positivistic as well as in naturalistic inquiry. I will comment on both trends.

The ideal of positivism is that research should begin with that which is "positively given." But it is well known that this simple principle leads to huge difficulties, as the "positive given" does not mean any "simple and direct given." Since Francis Bacon, in *The New Organon* from 1620, presented induction as the principal method for obtaining knowledge, it has been emphasised that *idola* could prevent human beings from looking beyond their own preconceptions.[4] A huge methodological apparatus could be involved in eliminating preconceptions and inaccuracies inherent in the particular observations, as well as presumptions introduced by the researcher. The researcher involved could inadvertently adopt a particular perspective (making observations in a particular way, at a particular time, etc.), but a sound methodology should help to bring forward a universal perspective on things. This universality would establish the positive given, from which a sound theorizing could proceed.

A much more sophisticated expression of researching "what is" is found in Yvonna S. Lincoln and Egon G. Guba's *Naturalistic Inquiry* (1985). They make a clear point in addressing what we could call the "myth of the positive given." They simply abandon the positivist assumption that it is possible to get to the positive given by eliminating disturbing factors, such as, for instance, those caused by the presence of the researcher. Lincoln and Guba stress the difference between the naturalistic and the neopositivist paradigm by specifying the notion of reality. According to them, the positivist paradigm may assume the existence of an objective reality which can be investigated. However, positivism could also assume the existence of a reality which may not be possible to uncover through any research but which could be "approximated" through properly designed research. Lincoln and Guba find that both these perspectives are problematic. Instead, they consider the possibility that reality is constructed (in fact they also briefly consider the more radical stand that reality is created). They assume that a multiplicity of ways of looking at things is legitimate, and they doubt that any supreme perspective could provide "objectivity." The naturalistic inquiry of Lincoln and Guba includes a sophisticated interpretation of subjectivism in research methodology. Their point is that reality does not exist before it has been constructed by an actor or created by a participant (p. 87).[5] The aim of naturalistic research, then, is to investigate such constructed realities. They point out that when disturbing subjective factors are eliminated, no positive given remains. In fact, nothing remains, as any research presupposes the presence of someone.

I consider "naturalistic inquiry" to be different from "positivist inquiry" in many ways. Although both forms of inquiry are addressing "what is," positivism talks about "what is positively given," while naturalism talks about "what is constructed." To a positivist, reality hides beneath a multiplicity of interfering factors

[4] See Bacon (1960).

[5] One radical position within positivism, as for instance represented by Ernst Mach, does not assume the existence of an external reality. As sensory impressions are personally based, and as knowledge and science have to build on sensory impressions, this "radical" positivism does not claim the existence of any underlying reality. Such a claim would be an illegitimate metaphysical invention. The only "reality" to consider is the experienced one. This radical position within positivism has been integrated into radical constructivism, particularly as presented by Ernst von Glasersfeld (1995), and Guba and Lincoln disclose some similarities with this radical positivism. So neopositivism, which Lincoln and Guba (1985) contrast with their naturalistic approach, is only one position within positivism.

to be eliminated through the research process. To a naturalist, reality emerges in a multiplicity of forms as part of the research process. Either way, neither of these interpretations grasps what I have in mind when I talk about researching possibilities, as this includes not only a study of "what is" or "what is constructed" but also a study of "what is not" and "what could be constructed." It includes a construction of possibilities.

Intermezzo

In order to facilitate the discussion of researching possibilities, I told at the Durban meeting a fable (see also Vithal, 2003, p. 12), and let me confess, I felt tempted to tell it again:

Let us imagine a country with traditions that are quite different from those we are accustomed to. In this country, so far away, men do the dishwashing. It has always been like this, and nobody has in fact taken notice of this. This is the way things are done.

In this country lives a philosopher, and one day, in a moment of inspiration, he (we imagine that it is a man) formulated the sentence, "Women can help men in the kitchen doing dishwashing." A most surprising combination of words! In fact, the philosopher continued this surprising stream of thought and arrived at the sentence, "Men and women could, on equal terms, help each other with the dishwashing." Grammatically speaking, it had always been possible to formulate such a sequence of words, but it was the first time that words had in fact been combined in this way. The philosopher had formulated a possibility which had never before been expressed.

Not surprisingly, he became enthusiastic, but also worried. What would people think of this? The new combination of words opened up possibilities for new imaginations, but it was far from certain that such imagination could be made real. Many reasons could imply that the imagination had to remain in the world of dreams. Anyway, the philosopher was curious to investigate what it could mean for men and women to cooperate in the kitchen. But from where to get empirical material? There were no cases whatsoever of such a cooperation. Should the philosopher embark on a positivist or a naturalistic approach, data collection would appear impossible. There was nothing to observe.

The philosopher mentioned his idea to different people, men as well as women. What he said was received with great surprise. People shook their heads; nobody could imagine a thing like this. And it was impossible anyway. Women could not enter the kitchen as it was all too obvious, come to think of it, that people who work in the kitchen eventually grow beards.

The philosopher, however, insisted on collecting at least some data that could illuminate his imagination. He tried to persuade some women to get into the kitchen, but they refused. He did not give up and figured out new strategies. He designed some very colorful and stylish gloves for dishwashing, and these gloves, everyone admitted, were very comfortable to wear. In addition, champagne was served in the kitchen. In the end, the philosopher succeeded in getting some women into the kitchen together with the men already there. After some cheers and humorous remarks, some dishes were also washed. The philosopher had started to collect some data.

Researching "What Is Not, but Could Be"

Let me try to be more specific about researching "what is not, but could be." For instance, one could try to research what it would mean to bring computers into a school in a poor neighborhood where students have no access to this technology. How would students experience such learning possibilities? What would it mean for them to have the opportunity to experiment with mathematical programmes? But bringing computers to a school in a poor neighborhood is not a simple thing, so how to research such questions if there are in fact no computers in school? Could one instead bring the students to the university and let them work in the lab of the university? Could such an approach reveal anything about what could happen if computers were in fact available in school?

By the *current situation*, I refer to the situation before any speculations about alternatives have had implications. In the far-away country where it is the custom for the men to wash the dishes, it refers to the situation before the philosopher had formulated any new possibility. It could refer to the situation in a *favela*, where there are no computers in sight.[6] In what sense, however, could we talk about *the* current situation? I have sometimes been asked if this expression presupposes the existence of one and only one reality. I do not think so. It could refer to many realities, although not to any kind of reality. If we imagine an investigation of the teaching and learning of mathematics in the *favela* in question, neither a positivist nor a naturalistic researcher would find any students working with computers to observe. They could make different interpretations of what they see, but this could not bring them to claim that they have collected (or produced) data about students working with computers.

There could be many realities referred to by the term "current situation," but we could nonetheless imagine alternatives. The philosopher formulated the incredible idea that men and women could cooperate in the kitchen. And we could consider what bringing information and communication technology to a *favela* could mean for the students' understanding of mathematics and for their experiences of sociopolitical processes of exclusion and inclusion. We could imagine that students came to work with dynamic geometry. We could imagine their activities organized as group work and that they communicated with students from other parts of the world about some of their results. We could imagine so many things. This brings us to the notion of an *imagined situation*. Certainly, this is not a well-defined entity. One could imagine very many and very many different things, both about dishwashing and computers. There is nothing definite about

[6] In what follows, the word *favela* refers to poor neighborhoods, townships, etc. in Brazil or elsewhere.

an imagined situation. The best one can say is that it is partially grasped by what concepts might be at hand. An imagination is vague, it is flexible, it is partial, and it might be far from any realities. The imagined situation may include educational hopes and aspirations.

Neither the current nor the imagined situation is well-defined, although for rather different reasons. The current situation could refer to many different realities, while the imagined situation could refer to a range of partial and inconsistent speculations—and it could hardly be called a "situation." We are dealing with a multiplicity of hopes, visions, expectations, and such. With this in mind, I continue to talk about the current and the imagined situation. While a current situation, in all its multiplicity, can be addressed by a positivist or a naturalistic approach, the imagined situation, even in its much more profound multiplicity, remains beyond the scope of both positivist and naturalistic inquiries.

To me, the imagined situation is essential for researching possibilities, in other words, for researching "what is not, but could be." But then, how does one look at an imagined situation? What's to look at? It appears only to be available through concepts. What kind of data is to be collected? Maybe it is not possible to "look at" an imagined situation, but we could still do something more specific than just speculating. We could try to realize some elements of the imagined situation. Thus, the philosopher tried to illuminate some of the imagination by bringing women, one way or another, into the kitchen. "Something" could be arranged. For sure, what was possible to arrange was remote from any imagination. If one cannot investigate what it could mean for students in a *favela* to study mathematics with computers simply by bringing computers to the *favela*, then one may be able to bring students to the computers.

In this way, we get to an *arranged situation*, illustrated by the philosopher's success in eventually bringing some women (with gloves) into the kitchen (with champagne). By using the term *arranged*, I want to emphasize that we are dealing with an unusual and a new situation, and also a situation that might be somewhat artificial. In the case of the students from the *favela*, the situation is different from the current situation, where there were no computers in the school, and it certainly differs, as well, from the imagined situation, where the computers were available in the *favela* environment. The arranged situation is a kind of in-between situation. I could have used the notion of experimental situation, but the word "experimental" might give the wrong connation. It might indicate that we are looking at certain variables, and that these are, so to speak, controlled, at least in the sense that they can be enumerated. Connotations of "experimentation" might indicate that one has identified some well-described alternatives and that through the experimentation we try to find out which one is preferable. When I talk about an arranged and not an experimental situation, I admit that things are out of control. There are no control groups, but things go as well as they can. An arranged situation is full of complexities.[7]

Explorative Reasoning

Could one address the arranged situation by a naturalistic (or positivistic) inquiry? Yes, this makes good sense. However, my point is that researching possibilities cannot be reduced to researching arranged situations through a naturalistic (or positivistic) approach. There is something different at stake here. The crucial point in researching possibilities is to research *relationships* between the current, the arranged, and the imagined situations. The different situations, in all their multiplicities and interwoven relationships, have to be addressed when one researches possibilities. This brings me to the notion of *explorative reasoning*, which comes to represent a way of exploring possibilities.

Explorative reasoning has its starting point in an arranged situation.[8] However, the point of explorative reasoning is not simply to specify conclusions about the arranged situation. The point is to also address situations that differ from the one in which the data is produced. Thus, one aim of explorative reasoning is to develop a deeper understanding of the imagined situation. (In a positivistic or naturalistic approach, however, the point of studying an arranged situation is to come to know better the arranged situation itself.) It is through such a process that an imagination becomes substantiated. Explorative reasoning, drawing on what is seen and experienced through the arranged situation, might turn an imagined situation into an assessable alternative. In this way, new possibilities may be created through exploration.[9]

Explorative reasoning means studying imagined situations through the windows that have been opened by an arranged situation, established in light of what is possible given the current situation. The arranged situation contains many limitations and restrictions, some of a political nature, others of a practical nature, while some may simply be accidental. However, the point of explorative reasoning is not to keep analyzing these particular forms of limitations but to see "through" the arranged situation in order to qualify an imagination. So, when the philosopher was analysing the data produced, one point was to look "beyond" the

[7] The ideas I am presenting may have strong similarities with the approach of action research. See, for instance, Atweh, Kemmis, and Weeks (1998).

[8] Deductive reasoning means drawing logical conclusions from premises or axioms. Deduction, as cultivated in mathematics, has been advocated by rationalism as the principal tool in sciences. Induction has been advocated by empiricism as the principal methodological approach in science. Hypothetical-deductive reasoning has been highlighted by critical rationalism (as formulated by Popper), and abduction has been presented by pragmatism (inspired by Peirce), as a principle for identifying a theory. I see explorative reasoning as still another form of reasoning within science.

[9] In doing so, it provides new insight into the current situation as well. Thus, explorative reasoning might draw attention to what obstructions for changes might be present in the current situation, and it might reveal some of the mechanisms through which the current situation regulates imagination.

more accidental aspects of the arranged situation (the gloves and the champagne). The date could be used as a departure for grasping preliminary possibilities indicated by the formulation: “Men and women could, on equal terms, help each other with the dishwashing.” And when students were brought to the university, the point of explorative reasoning was *not* to look at the particularities of what it means to *bring* students to the new learning environment but instead to consider what this environment might provide of new significance for their learning.

A certain situation might represent a necessity when no alternatives have been conceptualized. If we are able to point out alternatives, however, the situation becomes a contingency. Alternatives that we are able to conceptualize *modulate the situation.* A situation that appears to be a necessity can be modulated through the construction of alternatives. The philosopher’s first insight revealed that “men do the dishwashing” is not a necessity but a contingency (although up to that point it had appeared to be a necessity). By researching “what is not, but could be,” one contributes to the modulation of a situation. The important feature of modulation is to conceptualize alternatives to a present situation. Herein lies the basic difference I see between researching possibilities and the positivist as well as the naturalistic approach. Neither of these approaches relates researching to modulation. Researching given or constructed realities is different from modulating realities, and explorative reasoning is an important tool for modulation.

The Variety of Explorative Reasoning

Current, arranged, and imagined situations can be related in different ways. Thus, an imagined situation can illuminate a current situation and invite a critique (in the form of a modulation) of what is taking place. Something could be changed! So, researching possibilities includes establishing an insight of the relationships between the different situations, explorative reasoning being only one such format.

In her study, Kathrine Krageskov Eriksen (2003) pointed out different forms of explorative reasoning. The study addresses the introduction of courses with reflective and ethical dimensions in chemistry education at the University of Copenhagen. However, possibilities for carrying out more profound experimentations with new courses did not exist. Any arranged situation was confined within strict limitations, mainly due to the overall study programs and the dominant conception of what is considered to be proper chemistry education. This meant that the teaching sequence that Eriksen was in fact able to carry out appeared to be rather remote from what could be imagined. Somehow the explorative reasoning had a long way to go.

Eriksen (2003) divided the explorative reasoning into three main parts, the first being *initial explorative reasoning*, which was carried out in a dialogue between students and teachers who participated in the very limited arranged situation. The initial explorative reasoning was based on experiences of participating in the arranged situation. Notions and categories emerged from the summarizing of what had taken place, keeping in mind what could be imagined.

The second part, called *expanded explorative reasoning*, established connections back to the ideas that emerged in the initial explorative reasoning and the first version of the imagined situation, but also forward to new ideas:

> I contemplate the "conclusions" drawn in the process of initial explorative reasoning through the theoretical lenses developed throughout the preceding parts of the study. And, crucially, I attempt to do this in a "forward-looking" way. That is, I am not drawing conclusions on the arranged situation on the background of my theoretical concepts. Rather, I aim to develop these concepts through the experience gathered in the arranged situation. (Eriksen, 2003, pp. 249–250)

So the input to the expanded explorative reasoning is the initial explorative reasoning, as well as the notion and concepts used in formulating the imagined situation. However, this imagination now appears preliminary. It can be reformulated and developed.

Finally, the concepts developed are used in the formulation of new possibilities, which is what characterizes *constructive explorative reasoning*: "Here I use the interplay between the 'forward-looking' evaluations of the arranged situation and the theoretical apparatus which made up the expanded explorative reasoning for developing new theoretical constructs." (Eriksen, 2003, p. 250) Thus, ultimately, the explorative reasoning brings about new concepts, which I prefer to call constructs, and through these constructs, new imagined situations can be formulated. The important point is that explorative reasoning is directed towards the formulation of new constructs through which it is possible to express possibilities and alternatives in a refined way.

The Specificity of Explorative Reasoning

Identifying possibilities means identifying *particular* possibilities in a *particular* situation. So explorative reasoning needs to be specific both with respect to content matter issues (being mathematics, chemistry, or something different) and with respect to the sociopolitical context.[10]

In the section "Researching What Is," I referred to the possibility of turning a transcription into an invented dialogue, not following what is on the tape, but instead what comes to the transcriber's mind. This approach has been described by Morten Blomhøj (2006). He experienced the relevance of doing this when he was observing students working with the program *Geometers SketchPad* (GSP)

[10] Methodological discussions in general education have, to a large extent, set the stage for how to do research in specific educational fields, as well, like mathematics education. I could refer to Andrew Brown and Paul Dowling (1998) as an example (despite the fact that both authors have close connections to mathematics education). The book is an illustrative example of the phenomenon that when methodology is described in general terms, regardless of content-matter issues, then it easily comes to highlight the descriptive qualities of research and not its potential for locating alternatives. The emphasis is put on grasping what is taking pace and not on providing proposals for alternatives.

and studied the teacher-student communication. In one situation, the teacher suggested to a student that he or she drag the corner *B* of a triangle shown on the screen in order to study the relationship:

$$a = \frac{b\sin(A)}{\sin(B)}$$

The student did so, and the conversation continued. Blomhøj found, however, that different mathematical observations could be made if the teacher had suggested that the student drag the corner instead. When one drags *A*, *a* remains the same. This seems clear enough looking at the figure, but it is not that clear looking at the formula. How to interpret the situation? Blomhøj found it useful to illustrate these educational possibilities by transcribing the conversation up to the point where the teacher suggested dragging *B*. But at this point, Blomhøj switched the "transcription" into an invented dialogue, wherein the teachers advised the students to drag *A*. In the end, Blomhøj made this whole dialogue (the original beginning and the invented continuation) the object of an analysis. The reconstructed dialogue was discussed with the teacher, and this turned out to be a powerful way to illuminate educational possibilities. Blomhøj specifies his approach into a well-defined pattern for the construction of episodes. These episodes originate from what has been observed in the classroom, but they are constructed.

The strategy presented by Blomhøj can be seen as a way of trying to be specific about explorative reasoning and about alternatives. Naturally, the approach cannot be confused with fiddling with data, and Blomhøj is careful to describe what is done and what is not done. I see this as one promising way of exploring possibilities. Explorative reasoning can be initial, explorative, or constructive, and the example illustrates how specific it could be. In fact it could include identification of a suggestion for a teacher-student dialogue that has not yet taken place concerning what corner of a particular triangle to drag.

It should also be noted that we might not be dealing with a new technique. Jean-Jacques Rousseau (1979) published *Emile* in 1762 as a novel rich in conversations between a teacher and a student. This book has been investigated in detail without assuming it to be the transcription of a real conservation. It was very clear that it was an invented conversation. Yet this can well reveal educational possibilities.[11] Elizabeth de Freitas (2004) describes how fiction can be used as a methodology in mathematics education.[12] As an example, she presents what she calls a "political text," where Agnes, a mathematics teacher, reflects on her experiences with mathematics as a student and on how she is concerned about her present role as a teacher restricted by the dominant school mathematics tradition. By presenting a story about Agnes' reflections, Freitas finds that the "interior monologue and the life history of Agnes cast a personal light onto specific curriculum content.

---

[11] Lakatos' *Proofs and Refutations* (1976) can be considered in a similar way.

[12] See also Hannula (2003).

The re-storying of seemingly non-narrative knowledge functions subversively in carving out a new space of critical pedagogy." (de Freitas, 2004, p. 271) And de Freitas concludes,

> Fiction, as a methodology, has the potential to defamiliarize, to cross boundaries, to transgress cultural norms. . . . The contingent and particularized aspect of narrative remains intact in the story, but through fiction I am able to draw more heavily on imagination as the informant. Consequently, the story undresses the myth of mathematics as apolitical and objective by dwelling on the counter-images of Agnes. (p. 272)

de Freitas' formulation of political texts provides counterimages. Both the approaches presented by Blomhøj and de Freitas try to establish more specific forms of imagination which could show that what has taken place has not taken place out of necessity. Things could be different. Both the constructed episodes and the counterimages operate as critical remarks to what has taken place. They ensure a modulation of the given. In this way, both approaches facilitate explorative reasoning. They also illustrate the specificity of explorative reasoning. In order to bring explorative reasoning to a "conclusion" by formulating new possibilities, a particular insight is needed. This insight could concern particular mathematical properties (as illustrated by Blomhøj's construction of episodes), but also overall conceptions of mathematics (as illustrated by de Freitas' presentation of counterimages). Such specificity cannot be reached through a general methodological approach.

If one wants to identify possibilities, then not only mathematical but also sociopolitical specificity is important. As pointed out by both Mathume Bopape (2002) and Renuka Vithal (2003), the search for changes and educational possibilities in postapartheid South Africa must continue to confront the "sins of apartheid." Racism is part of daily life, as the "geography" of apartheid is still in place. By this geography, I refer to the locations of the townships; the construction of roads; the locations of White, Indian, Black neighborhoods in a city like Durban; the location of schools; and so on. Laws can be changed overnight, but changing geography involves long-term efforts. And the sins of apartheid concern not only geography, they also concern economy and ownership. They are present in ways of talking and experiencing. So when one seeks educational possibilities through explorative reasoning, one must take into account the sociopolitical reality. Alternatives must always be constructed with reference to a particular context. This is true for South Africa as well as any other place. It is also true for the design of chemistry education at Copenhagen University, where the "sins of the past," for instance, in the form of a professor-centred curriculum design, presently dominate education.[13] Constructing possibilities is not a free enterprise; it is limited only

[13] For particulars addressing the socio-political dimension of research in mathematics education, see Vithal and Valero (2003), Valero (2002), and Valero and Zevenbergen (2004).

by a lack of imagination. It is a laborious process taking place within powerful realities and dominant discourses. Possibilities must be sought while facing these realities and discourses.

Collectivity of Explorative Reasoning

A situation, a teaching-learning practice, could constitute an object of research, and one might assume that a researcher would conduct research *on* teachers and *on* students. Alternatively, research could take place in a collaborative form, meaning that the researcher could do research *with* teachers and *with* students. Data could be produced in a collaborative way. (The naturalistic approach seems to provide good opportunities for doing research *with*, while positivist research would likely consider any collaboration as introducing a possible bias of the "objectivity" of data.) The distinction between researching *on* and researching *with* is important when one tries to research possibilities, as qualities of researching possibilities are grounded in collaboration.

To be more specific, we could consider what collaboration could mean for the quality of explorative reasoning. What could it mean to look "through" the limitations of an arranged situation and try to grasp more clearly what the alternatives could consist of, and what possibilities might be accessible. In the case described by Eriksen (2003), it seems clear that in order to clarify what the introduction of reflective elements in chemistry education could mean, the researcher, together with the students involved in the arranged practice, could try to look beyond the limitations of the situation they have experienced.

If we believe that explorative reasoning should in fact be explorative, there is no point in considering students or teachers' participation as being a disturbing factor. It appears instead to be a constructive factor. It appears obvious that processes of formulating possibilities can be strengthened through the contributions of all participants. In this sense, the change from doing research *on* to doing research *with* appears to be a natural consequence of researching possibilities.

This brings us to the notion of dialogue as part of a research process. This idea has been developed by Diana Stentoft (2006). Here, I want only to point out a principal point, which has to do with the construction of concepts. A dialogic process can help to ensure that the concepts by means of which we explore imaginations and possibilities become shared. Possibilities are always possibilities for somebody, and possibilities that have not yet been recognized are not really possibilities. The remark referred to previously, made by Lincoln and Guba (1985) with respect to realities, seems appropriate with respect to possibilities as well. Realities as well as possibilities must be someone's realities and possibilities, and the notion of multiplicity seems to apply equally well to reality and possibility. This point indicates the importance of searching for possibilities as a shared process, including a sharing of concepts.

Concluding Remarks

As mentioned in the introduction, I have previously talked about "critical research" instead of "researching possibilities." Often, then, I have been asked, Does that mean that all critical research must research possibilities? The answer is "no." We could clearly imagine that research serves a critical function without taking the form of researching possibilities. Thus, in postapartheid South Africa, some descriptive statistics-based research served a critical function in providing numbers and figures which had never been accessible during the apartheid. Could we, however, assume that researching possibilities is in fact critical? Again the answer is "no." We could easily imagine that the exploration of possibilities could serve quite different and also rather questionable interests (companies researching a new possibility for marketing, etc.). I do not assume any identity between critical research and researching possibilities. Then, more openly, we can address to what extent researching possibilities resonates with the critical concerns expressed, for instance, in critical education and critical mathematics education.

A few observations related to the idea of modulation might be helpful. C. Wright Mills (1959) talks about "sociological imagination," which expresses a capacity to separate what is necessary from what is contingent and, therefore, possible to change. A fact could be not only a state of affairs but also an experienced social necessity when it is impossible to imagine the fact not being the case. The existence of an imagination that describes alternatives to an actual situation makes a difference. By this imagination, experienced necessities could be reduced to contingencies—they *could* be different. The power of sociological imagination is to reveal a social given as being open to change. One could claim that such an imagination is important for sociology to become critical.

In Max Horkheimer's classic study, "Traditional and Critical Theory," first published in German in 1937, we also meet the idea that a theory, in order to be critical, must provide modulations and not only descriptions of situations.[14] Horkheimer characterizes Traditional Theory as being in line with a positivist interpretation of theorizing which provides a recollection and systematization of facts. Horkheimer's presentation prepares a description of critical theory as a broader sociopolitical and philosophic enterprise, which ensures an analysis that opens the way for changes.

It has been stressed that research in critical education (and research in critical mathematics education, as well) must seek alternatives to a given educational practice, just as critical theory in general seeks alternatives to a social given. In order to be critical, research must try to modulate educational situations. Therefore, one could expect a certain degree of resonance between what it could mean to research possibilities and what it could mean to do critical research also in education.

---

[14] The essay is also published in Horkheimer (2002).

However, one has to be careful. Many considerations about critical theory were developed well before we became fully aware of the limits of modernity and saw the possibilities of postmodern positions. Thus, the notion of critique, as presented by Horkheimer in "Traditional and Critical Theory," could well be characteriszed as a traditional notion of critique. It reflects the idea that it is possible to elaborate a critique taking the form of a critique-with-a-foundation. And in this case, Horkheimer draws heavily on a Marxist foundation. Later, Jürgen Habermas brought critical theory into a different conceptual framework (see, in particular, Habermas, 1971), but it could be argued that Habermas elaborated on a critique-with-a foundation as well.

I have tried to step outside assumptions about the existence of some kind of reliable foundation for becoming critical. However, this brings me into a much more troublesome situation. The notion for critique becomes a most slippery one. It becomes extremely complicated to characterise what critical research could mean. As a consequence, I prefer not to provoke any additional confusion by talking as though researching possibilities and critical research were identical, although it is my hope that they might overlap in many situations.

NOTE

This chapter was first published in M. Setati, M. Vithal, R. Malcolm, C., & Dhunpath R. (Eds.). (2009). *Researching possibilities in mathematics, science and technology education* (pp. 105–119). New York, NY: Nova Science.

REFERENCES

Atweh, B., Kemmis, S., & Weeks, P. (Eds.). (1998). *Action research in practice: Partnership for social justice in education.* London, UK: Routledge.

Bacon, F. (1960). *The new organon.* Indianapolis, IN: Bobbs-Merrill. (Original work published 1620)

Barton, B., Autagavaia, J., Poleki, A., & Alangui, W. (2002). The mathematics enchantment project: A theoretical approach to research and development. In P. Valero & O. Skovsmose (Eds.), *Proceedings of the Third International Mathematics Education and Society Conference* (pp. 144–153). Copenhagen, Roskilde, and Aalborg: Centre for Research in Learning Mathematics.

Blomhøj, M. (2006). Konstruktion af episoder som forskningsmetode: Udforskning af læringsmuligheder i IT-støttet matematikundervisning [Construction of episodes as research methodology: Researching learning possibilities in an IT-supported mathematics education]. In O. Skovsmose & M. Blomhøj (Eds.), *Kunne det tænkes?—Om matematiklæring* [*Could it be?—About learning mathematics*]. Copenhagen, Denmark: Malling Beck.

Bopape, M. (2002). *Mathematics school based in-service training (SBINSET): A study of factors contributing towards success or failure of SBINSET in the South African school context* (Unpublished doctoral dissertation). Department of Education and Learning, Aalborg University, Aalborg, Denmark.

Brown, A., & Dowling, P. (1998). *Doing research/reading research: A mode of interrogation for education.* London, UK: Falmer.

de Freitas, E. (2004). Plotting intersections along the political axis: The interior voice of dissenting mathematics teachers. *Educational Studies in Mathematics, 55*(1/3), 259–274.

Eriksen, K. K. (2003). *The role of reflectivity in tertiary chemical education* (Unpublished doctoral dissertation). University of Copenhagen, Copenhagen, Denmark.

Habermas, J. (1971). *Knowledge and human interests.* Boston, MA: Beacon. (Original work published 1968)

Hannula, M. S. (2003). Fictionalising experiences—Experiencing through fiction. *For the Learning of Mathematics, 23*(3), 31–37.

Horkheimer, M. (2002). *Critical theory: Selected essays.* New York, NY: Continuum. (Original work published 1968)

Lakatos, I. (1976). *Proofs and refutations: The logic of mathematical discovery* (J. Worrall & E. Zahar, Eds.). Cambridge, UK: Cambridge University Press.

Lincoln, Y. S., & Guba, E. G. (1985). *Naturalistic inquiry.* Beverly Hills, CA: Sage.

Pfannkuch, M. (2005). Probability and statistical inference: How can teachers enable learners to make the connection. In G. A. Jones (Ed), *Exploring probability in school: Challenges for teaching and learning* (pp. 267–294). New York, NY: Springer.

Rousseau, J.-J. (1979). *Emile.* New York, NY: Basic. (Original work published 1762)

Skovsmose, O., & Borba, M. (2004). Research methodology and critical mathematics education. In P. Valero & R. Zevenbergen (Eds.), *Researching the socio-political dimensions of mathematics education: Issues of power in theory and methodology* (pp. 207–226). Dordrecht, The Netherlands: Kluwer.

Stentoft, D. (2006). Research as an act of learning: Exploring student backgrounds through dialogue with research participants. In M. Bosch et al. (Eds.), *Proceedings of the Fourth Congress of the European Society for Research in Mathematics Education* (pp. 1193–1203). Barcelona, Spain: Universitat Ramon Lull. Retrieved from http://www.mathematik.uni-dortmund.de/~erme/CERME4/CERME4_WG10.pdf

Valero, P. (2002). *Reform, democracy, and mathematics education: Towards a socio-political frame for understanding change in the organization of secondary school mathematics* (Unpublished doctoral dissertation). Department of Curriculum Research, Danish University of Education, Copenhagen, Denmark.

Valero, P., & Zevenbergen, R. (Eds.). (2004). *Researching the socio-political dimensions of mathematics education: Issues of power in theory and methodology.* Dordrecht, The Netherlads: Kluwer.

Vithal, R. (2003). *In search of a pedagogy of conflict and dialogue for mathematics education.* Dordrecht, The Netherlands: Kluwer.

Vithal, R., & Valero, P. (2003). Researching mathematics education in situations of social and political conflict. In A. J. Bishop, M. A. Clements, C. Keitel, J. Kilpatrick, & F. K. S. Leung (Eds.), *Second international handbook of mathematics education* (pp. 545–591). Dordrecht, The Netherlands: Kluwer.

von Glasersfeld, E. (1995). *Radical constructivism: A way of knowing and learning.* London, UK: Falmer.

Wright Mills, C. (1959). *The sociological imagination.* New York, NY: Oxford University Press.

PART 3

DEMOCRACY AS A CHALLENGE

CHAPTER 10

GHETTOIZING AND GLOBALIZATION: A CHALLENGE FOR MATHEMATICS EDUCATION

Globalization has got some attractive connotations. It refers to the opening up of borders, and it can include a sense of being together and sharing concerns for each other—as if the globe were turning into one great community. However, globalization may simply refer to the fact that new connections are established between previously unconnected social entities; and that which is happening and done by one group of people may affect, for good or for bad, a completely different group of people, even not being aware of the cause of the effect. Globalization can refer to loss of transparency and can mean a destruction of communities. The concept of globalization contains both positive and negative connotations:

> For some, "globalization" is what we are bound to do if we wish to be happy; for others "globalization" is the cause of our unhappiness. For everybody, though, "glo-

Critique as Uncertainty, pages 129–141.

> balization" is the intractable fate of the world, an irreversible process. (Bauman, 1998, p. 1)

Globalization is not a new phenomenon. We can go back to the discoveries of the 15th and 16th centuries, which bought colonization, slavery, and domination. New political orders and economic priorities were installed, and slaves were shipped around the world in order to secure the production of goods needed by the colonizing powers. Thus, globalization was accompanied by suppression, including racism and religious fanaticism. It was also accompanied by a cultural invasion which interpreted the existing culture as inferior, but much later, maybe, as fascinating and picturesque. The idea was to let "us" provide "them" with the best from the West, simply because "the best *is* from the West."

Globalization has to do with policy, industry, markets, and business. It has to do with cultures and conflict. It also has to do with the construction, codification, and distribution of knowledge which turns into goods for sale. It has to do with education, also mathematics education. In what follows, I shall try to indicate how mathematics education comes to form an intimate but critical part of the processes of globalization.

The Value of Knowledge

The notion of an informational economy refers to the world system, emphasizing that the economic processes within this system can be geared higher by means of information and communication technologies. The informational economy represents a crucial aspect of globalization, which can be illustrated by the extension and movement of "supply trains," which are the the routes leading from raw material to the final commodity. Along the supply chains, we follow the refinements of the raw material into the final product, and at each stage we find the spin-off of profit, basically characterised by the fact that the profit spin-off becomes higher and higher the closer we come to the final product. The direction of the supply chain can, nowadays, be changed according to new priorities. Thus, Albert J. Dunlap, one of the world's leading business managers, claims that "the company belongs to the people who invest in it—not to its employees, suppliers, not the locality in which it is situated" (as cited in Bauman, 1998, p. 6). The meaning of this statement is clear: The company is a freely moving entity, and certainly the big companies can move their supply chains as it wants. The company is not located in a particular physical environment, and it has no obligations towards any particular community. Where the company is placed, for the time being, and what the company is producing, for the time being, are determined by the people to whom the company belongs, and these are the people who invest in the company. Sticking to this economic principle has been highly facilitated by the information and communication technologies, and this is one important aspect of the informational economy. The permanent (possibility of) moving the company, the production, and the capital is a defining element of globalization.

Daniel Bell (1980) suggests a knowledge theory of value which claims that knowledge and information are strategic resources. As capital and labor have been basic in any theory of value, so knowledge and information are placed in the same crucial relationship. However, Bell does not further analyze the notions of knowledge and information. In fact both notions operate as "dummy variables" in his theory of value. To me, it appears surprising that it is not necessary to provide any further specification about the nature of the knowledge and information which may serve as strategic resources. Could there be a particular area of knowledge which is significant to the information society? If we study Bell (1980) and Castells (1996, 1997, 1998), the answer seems to be "no." At least the nature and the content of knowledge and information they have in mind are only being addressed in the most general terms. I find that the knowledge theory of value rests upon some more fundamental (and to me, also dubious) assumptions about the relationship between "knowledge" and "progress," and I shall try to clarify this a bit further.

Double Agents

The value of knowledge can be thought of in optimistic terms—a perspective included in Bell's conception. However, such epistemic optimism has deep roots in the whole project of the Enlightenment, representing the idea that knowledge must be produced and spread in order to illuminate the dark and obscure corners of human life. Knowledge signifies progress.

Progress can concern all aspects of life. We can think of progress in terms of production and material life conditions. We can think of progress in terms of economy. We can think of it as linked to health and wealth. We can also think of progress in terms of knowledge and education. The conception of progress has not dominated all periods in history. Within the religious framework of the Middle Ages, human history was not to be thought of in terms of progress. Instead, life on earth was marked by a struggle to prevent people from sinning in order to ensure life everlasting. All human endeavors had to concentrate on this. Some have argued that the idea of social progress is clearly linked to the whole notion of the Enlightenment. Thus, J. R. Bury (1955), in his classical study *The Idea of Progress*, first published in 1932, states that neither the classical nor the medieval world had any conception of progress. A more nuanced picture is presented by Robert Nisbet in the *History of the Idea of Progress* from 1980. However, Nisbet's main point is not to reduce the relevance of the conception of social progress as a major characteristic of the Enlightenment, but to show that the notion of progress has been presented in many different historic contexts. The overall idea about social progress is that a "utopia" is waiting for humankind, and it makes sense to work towards realizing such a utopia.[1]

[1] Conservatism in philosophical terms is characterized by the idea that any relevant utopia is placed in our past. Social development therefore constitutes a gradual decline. In case we were to try to stay as

I want to characterize the Enlightenment by means of the notion of progress, condensed into the assumption that in case we ensure scientific development, then we ensure progress on a grander scale. Science and scientific progress can be seen as engines for progress in general. Thus, Enlightenment, in my characterization, assumes an intrinsic connection between, on the one hand, the development of knowledge in general and the development of scientific knowledge in particular, and, on the other hand, social, political, economic, and cultural progress.[2]

According to the Enlightenment, one straitjacket for knowledge development has been dogmas. Instead, *a free knowledge-production has to be introduced as the promoter of progress.* It will ensure that humankind is not pushed back into a new medieval period encapsulated in dogmatism. The light of reason will be a safeguard against indoctrination. The problem is not knowledge itself but how to obtain it and how to distribute this universal goodness. *Knowledge represents epistemic welfare.* It is not important to exercise any "control of knowledge." The aim must be the opposite: to eliminate all possible institutions that try to control knowledge. Knowledge must be set free. And this definitely also includes scientific knowledge. As scientific knowledge is a "good" and "attractive" resource, it is not meaningful to discuss whether we shall advance science or not. But this idea can be challenged.

In "Cultural Framing of Mathematics Teaching and Learning," Ubiratan D'Ambrosio makes the following comment:

> In the last 100 years, we have seen enormous advances in our knowledge of nature and in the development of new technologies. . . . And yet, this same century has shown us a despicable human behaviour. Unprecedented means of mass destruction, of insecurity, new terrible diseases, unjustified famine, drug abuse, and moral decay are matched only by an irreversible destruction of the environment. Much of this paradox has to do with an absence of reflections and considerations of values in academics, particularly in the scientific disciplines, both in research and in education. Most of the means to achieve these wonders and also these horrors of science and technology have to do with advances in mathematics. (D'Ambrosio, 1994, p. 443)

---

close as possible to the qualities signified by any utopia, then we have to maintain the status quo. T. S. Eliot's famous poem "The Waste Land"can be, and has been, interpreted in many different ways. One possibility is to consider it as representing a conservative philosophy warning about what it could mean to lose sight of the qualities of the past.

2 Many philosophers and thinkers have presented ideas by means of which I have characterized the Enlightenment. Thus, I do not try to define the Enlightenment as a specific historical period, but as a network of ideas and assumptions. Let me just mention, Francis Bacon (1561–1626), Rene Descartes (1596–1650), Baruch Spinoza (1632–1677), John Locke (1632–1704), Gottfried Wilhelm Leibniz (1646–1716), Christian Wolf (1679–1754), François Voltaire (1694–1778), Denis Diderot (1713–1784), Adam Smith (1723–1790), Immanuel Kant (1724–1804), Nicolas de Condorcet (1743–1794), Henri de Saint-Simon (1760–1825), G. W. F. Hegel (1770–1831), Auguste Comte (1798–1857), John Stuart Mill (1806–1873), Karl Marx (1818–1883), Herbert Spencer (1820–1903), John Dewey (1859–1952), Bertrand Russell (1872–1970), Rudolf Carnap (1891–1970), and Karl Popper (1902–1994).

D'Ambrosio strongly indicates that mathematics is positioned in the nucleus of social development, the essential point being, however, that D'Ambrosio considers both science and mathematics as operating as double agents, serving both "good" and "bad." The assumption of the Enlightenment is questioned.

The notion of paradox refers to a situation where reason seems to have to give up. D'Ambrosio indicates that something is problematic in the way reason operates in social affairs. How could it be that mathematics, this seemingly sublime form of reason and rationality, can be associated with not only wonders, but also horrors? I find that this apparent paradox with respect to knowledge signifies a main challenge to the assumption of the Enlightenment: that there exists an intrinsic connection between scientific progress and sociopolitical progress in general, and to any optimistic formulation of a value theory of knowledge. This paradox also signifies a challenge to mathematics education, which brings us to consider one particular aspect of globalization.

Ghettoizing

If the "value of knowledge" is not a simple and straightforward positive "good," as assumed by the Enlightenment and by Bell (1980), then we should not be surprised if the information economy and the processes of globalization include "wonders" as well as "horrors." To be more specific, talking about globalization also means talking about ghettoizing.

In the *End of Millennium*, Castells defines social exclusion "as the process by which certain individuals and groups are systematically barred from access to positions that would enable them to an autonomous livelihood within the social standards framed by institutions and values in a given context" (Castells, 1998, p. 73).

Castells emphasizes that the "informational, global capitalism" is characterized by "economic development and underdevelopment" and "social inclusion and social exclusion" (see Castells, 1998, p. 82). Bauman makes the following observation: "Globalization divides as much as it unites; it divides as it unites—the causes of division being identical with those which promote the uniformity of the globe." (Bauman, 1998, p. 2) To me, globalization and ghettoizing represent different aspects of (or different perspectives on) the informational society.

Castells makes the following observations which bring him to consider the notion of the Fourth World:

> This widespread, multiform process of social exclusion leads to the constitution of what I call, taking the liberty of a cosmic metaphor, the black holes of informational capitalism. These are the regions of society from which, statistically speaking, there is no escape from the pain and destruction inflicted on the human condition of those who, in one way or another, enter these social landscapes. (Castells, 1998, p. 162)

The Fourth World makes up the black holes in information capitalism:

> The Fourth World comprises large areas of the globe, such as much of Sub-Saharan Africa, and impoverished rural areas of Latin America and Asia. But it is also present in literally every country, and every city, in this new geography of social exclusion. It is formed of American inner-city ghettos, Spanish enclaves of mass youth unemployment, French banlieues warehousing North Africans, Japanese Yoseba quarters, and Asian mega-cities' shanty towns. And it is populated by millions of homeless, incarcerated, prostituted, criminalized, brutalized, stigmatized, sick and illiterate persons. They are the majority in some areas, the minority on others, and a tiny minority in a few privileged contexts. But, everywhere, they are growing in number, and increasingly in visibility, as the selective triage of informational capitalism, and the political breakdown of the welfare state, intensify social exclusion. In the current historical context, the rise of the Fourth World is inseparable from the rise of informational, global capitalism. (Castells, 1998, pp. 164–165)

It might be possible to think of a ghetto as a small community. A ghetto might be reserved for a certain group of people, who stand out from the society in which they live. They might not be welcome in society while they in the ghetto can support each other and live according to their own culture and traditions.[3] It is, however, important to observe that those ghettos, which are growing in the process of globalization, might be very different from "communities." In fact, it can be claimed that globalization means a destruction of communities.

In *Community*, Bauman makes different observations about the new forms of ghettos, which could also be called hyperghettos: "We may say that the prisons are ghettos with walls, while ghettoes are prisons without walls" (Bauman, 2001, p. 121). As part of the positive and attractive characteristics of globalization, moving and traveling have been celebrated. Thus, globalization means to be aware of different cultures, traditions, and places. Globalization means a celebration of being international. Ghettoizing means exactly the opposite. It means to be prevented from moving. Ghettoized people are immobilized people. The point is simply that there is no need for these people. And considering the celebration of globalization, then to be hyperghettoized is a much harder experience than being enclosed in a "classic" ghetto, which at least represents a community and not only a prison. As emphasized by Bauman, "Ghettoes and prisons are two varieties of the strategies for 'tying the undesirable to the ground' of confinement and immobilization." (Bauman, 2001, p. 120)

Does it make sense to confine people? In the network society, the flexibility of the labor force is celebrated, although what is flexible is the supply tail. It is important that the qualifications of the labor force can be developed, as one characteristic of the informational society is the rapid change of demands for labor. Thus, one aspect of globalization is that any scheme of production can spread

[3] The Jewish ghettoes established over centuries within European cities might serve as an example. Isaac B. Singer's description *In My Father's Court* tells about the small life at the time up to the First World War of Krochmalna Street within a Jewish neighborhood of Warsaw. Such a neighborhood might serve well as an illustration of "community," as seen in Bauman (2001).

globewise. So, in case we consider the ghettos as a reservoir of additional labor, then the construction of the modern ghetto seems irrational. The point, however, is that the modern ghetto does not serve as any reservoir. And certainly not as a reservoir for possible consumers who could help to speed up the informational economy. The modern ghetto can be considered a dumping ground for people who have no role to play in the informational economy. There is no need for their labor, nor for their demands; they are disposable people. Bauman refers to Loïc Wacquant for observing that "whereas the ghetto in its classic form acted partly as a protective shield against brutal racial exclusion, the hyper-ghetto has lost its positive role of collective buffer, making it a deadly machinery for naked social relegation" (see Bauman, 2001, p. 122).

In the *Dialectic of Enlightenment*, Horkheimer and Adorno make a related observation:

> On the one hand the growth of economic productivity furnishes the conditions for a world of greater justice; on the other hand it allows the technical apparatus and the social groups which administer us a disproportionate superiority to the rest of the population. The individual is wholly devalued in relation to the economic powers, which at the same time press the control of society over nature to hitherto unsuspected heights. Even though the individual disappears before the apparatus which he serves, the apparatus provides for him as never before. (Horkheimer & Adorno, 2002, p. xiv)

To me, this statement from the preface written in May 1944 appears extremely up-to-date in identifying the double aspect of globalizations. On the one hand, globalization seems to provide advancement (at least for some groups of people); yet on the other hand, the very same processes seem to represent dehumanization and deprivation (at least for some—other groups of people). Globalization divides as it unites.

To think of ghettoizing as an aspect of globalization is also a way of indicating that "progress" is not a simple concept. Knowledge and information may play significant elements in the informational economy, but the output of this economy can only be talked about in terms of both "goods" and "bads." Whenever knowledge is put into operation, we come to face uncertainties. The contradictory situation with respect to knowledge forms part of the informational economy, and of the informational society in general. In particular, we have to think of globalization and ghettoizing as aspects of the same processes of which mathematics education is part.

Constructors, Operators, Consumers, and Disposable

The D'Ambrosio paradox provokes the loss of an illusion in mathematics education. Mathematics education might be doing wonders, and this would be fine. Yet there might also be a downside of mathematics education. Mathematics education cannot be operating from the simple assumption that the activities within it simply represent good and attractive activities. Mathematics education might represent processes with very different social implications. Naturally, this observation represents a challenge to the practice of mathematics education and also to mathematics education as a scientific discipline.

In order to illustrate this point, I shall make some comments on different groups of people who might be involved in or affected by mathematics education: the "constructors," the "operators," the "consumers," and the "disposable." (Naturally, I could comment on other groups as well who benefit from or are affected by programs of mathematics education, the mathematicians for instance.)

One partriclar group is going to maintain and develop further the "apparatus of reason" which is fundamental for any technology. (This apparatus refers to knowledge, techniques, ICT systems, economic resources, management priorities, etc., by means of which technology, in the broadest interpretation of the term, is maintained and developed further.) This group we can call *constructors*, and mathematics will appear as one element to be included in their set of competences. It is the task of universities and other institutions of further and higher education to provide these competencies, and any education of engineers, economists, computer scientists, pharmacists, and such includes mathematics. Any such education must, however, face the question, How is it possible to include mathematical competence in broader technological competence without assuming or promoting the impression that mathematical techniques contain qualities which will ensure (or at least make it more likely) that the quality of activities which are carried out as part of the expertise is improved simply because mathematics is involved? How to avoid giving the impression that mathematics is an unproblematic (albeit difficult) tool with which to operate? The point is to establish the possibility for students to acknowledge uncertainties related to the functioning of the apparatus of reason, also when mathematics appears to be an indispensable part of this apparatus. When sociotechnological actions are resources through an education, an ethical demand becomes part of this education. This applies to any university education where the candidates are qualified to function as constructors. To me, this is a main challenge not only to university science and mathematics education but to all forms of further education which educate "constructors."

We can, however, also see mathematics education as preparation for people who are not going to use what they learn in mathematics lessons in secondary school for pursuing further mathematically dense studies. Many of these people will nevertheless face a working situation where they will operate with mathematics, although often in an implicit way. Mathematics might be available in "packages" (easy to install), which it is important to be able to use, but the details of how the packages function need not be grasped by the person who operates the package. Mathematics does not surface in the situation. Much mathematics education can be seen as preparation for people who are going to operate in contexts where mathematics is incorporated in their job situations. Thus, much mathematics education serves as preparation for people to operate within work processes rich in implicit mathematics. These people we can call the *operators*, and mathematics education will also serve the preparation of operators. One characteristic of the school mathematics tradition is the overwhelming number of exercises which the students have to do. During primary and secondary school, the total number of exercises easily exceeds 10,000. Following any progressive thinking in mathematics education—represented by constructivism, problem-solving approaches, realistic mathematics, and such—this extreme preoccupation with exercises seems pathological. It might, however, appear that a readiness to follow orders and do so in a careful way is "functional" for being an operator. From the perspectives of "adaptability" and "functionality," the mathematics offered in their secondary or tertiary education might appear adequate. However, such adaptability is problematic. Any use of mathematics, be it explicit or implicit, must be accompanied by doubt and uncertainty.

Statements from experts are expressed each and every day on television and in the newspapers. Let us just take one example: Some countries are designated risk countries, referring to the risks possible investors might face when they want to invest in a particular country. In June 2002, Brazil rose to become the second highest risk country in the world, although well surpassed by Argentina. Numbers can be associated with these ratings and experts can be interviewed. So, surely somebody must be listening to all this. I will call these people the *consumers*. Mathematics education also means preparation for consuming. Much consideration has addressed what it could mean to develop a mathematics education, not for a particular job function, but for citizenship. Such citizenship could be passive, but it does make sense to consider how mathematics education could prepare for critical citizenship.

Preparing for citizenship, active or passive, might not be the only function of mathematics education. Not everybody is necessarily in the informational economy, as it is defined through capitalist priorities and interest. Within a capitalist economy, some groups of people seem *disposable*. We have to consider in what sense mathematics education might also prepare some groups for the role of being disposable. It is possible to see the structures of testing, passing or rather failing an examination, as a system which helps to single out those people who are not necessary within the dominant order of the informational economy. Pierre Bourdieu (1996) talks about the designation of a "state nobility" as the group of people who could get access to power and control. Bourdieu claims that mathematics education appears to have a role to play in this context. Success in mathematics might be part of being nominated for the state nobility. I find it important to consider as well that failure in mathematics might be part of nominating people as disposable.

Mathematics Education Is Critical

If we see mathematics education as part of global processes preparing the ground for the informational economy and the informational society in general (and I see mathematics education this way), we must also be aware that the process of globalization "divides as much as it unites."[4] In fact, according to Bauman, "it divides as it unites," and further, "the causes of division being identical with those which promote the uniformity of the globe." This is an essential reminder to mathematics education. We should not be surprised in case this education divides as much as it unites. In case we see mathematics education as part of the universal processes of globalization, then we should also see it as part of the universal processes of making exclusions.

I find the role of mathematics education in the further formation of the informational society to be critical. To me, this means two things. First, I find that mathematics education plays a *significant* role in sociopolitical processes. This is indicated by the fact that mathematics education can be seen as part of the foundation of the technological society and that it can be seen as a cultural invasion, and so on. I do not claim that I am in a position to do any clear analysis of the nature of this role. I do not see the development of the informational society, including the processes of globalization and ghettoizing, as any predetermined process. I find that several factors play a role in this formation, one of them being mathematical based technologies, another being mathematics education. Secondly, I find that mathematics education, in its very many forms, plays an *undetermined* role (or a possible double role). The sociopolitical functions of mathematics education cannot simply be associated with attractive features. Conversely, they are not social mechanisms which can be associated with problematic social roles.

In saying that the role of mathematics education is *critical*, I mean its role of being significant as well as being undetermined. The functioning of a mathematics education can go "both ways." Here, I use "critical" in a similar way, as when we talk about a patient's condition as being critical. This means that he or she may survive, but also that nothing can be taken for granted. His or her condition is simply critical.[5]

To claim that the sociopolitical role of mathematics education is critical means to claim that alternatives are possible. I do not relate mathematics education to any optimistic position claiming the existence of an intrinsic connection between mathematics education and, say, democratic values. Nor do I claim that mathematics education *per se* will serve antidemocratic interests. Instead, I simply claim that *the sociopolitical role of mathematics education is critical as it is significant and undetermined.* Mathematics education can be education for democracy, but it can also be education pointing in a completely different direction. It can become education for adaptability and functionality. No actual functions of mathematics education represent the essence of this education. There is no such essence.

---

[4] See Castells (1999) for comments on the notions of "information society" and "informational society."

[5] The discussion of the critical position of mathematics education is based on Skovsmose and Valero (2001), where the relationship between mathematics education and democracy is presented as being critical. See also Skovsmose and Valero (2002a, 2002b) and Valero and Skovsmose (Eds.) (2002).

Critical Mathematics Education

The critical nature of mathematics education causes a great uncertainty. Naturally, it is possible to try to ignore this uncertainty. This can, for instance, be done by assuming that mathematics education can somehow be "determined" to serve some attractive social functions when organized in, say, a national curriculum crowned by some attractive aims and objectives. But I find this to be an illusion. The actual function of mathematics education is not determined by some overall guiding principles put at the top of the curriculum. The function might depend on many different details of the context in which the curriculum is acted out. The first thing one can do is to acknowledge the critical nature of mathematics education and to acknowledge the great uncertainty related to this subject. To acknowledge this uncertainty is a characteristic of critical mathematics education.

Critical mathematics education is not to be understood as a special branch of mathematics education. It cannot be identified with a certain classroom methodology. It cannot be constituted by a specific curriculum. Instead, I see critical mathematics education as defined in terms of some concerns emerging from realizing the critical nature of mathematics education. If no intrinsic relationship exists between mathematics education and some attractive sociopolitical development, then the relationship has to be acted out with reference to the particular context. Here I shall briefly refer to two issues, which could give some idea about what concerns I have in mind.[6]

First, as mathematics education is not assumed to possess any essence, critical mathematics education is concerned about the different possible roles which mathematics education could play in a particular sociopolitical setting. Critical mathematics education is concerned about how mathematics education might be stratifying, selecting, determining, and justifying inclusions and exclusions. By a borderline school, I understand a school from where students can find access to the informational society as well as experience a route to the Fourth World.[7] Critical mathematics education must be concerned about what is happening in such borderline schools. What kind of opportunities do they provide for the students? How is the economic situation of borderline schools? It is important to consider mathematics education from the perspective of globalization, including all the attractive features which a globalization might include. But it is just as important to consider what mathematics education could mean for the excluded. Thus, many studies seem to reveal that introducing computers in the classroom provides sig-

---

[6] In another presentation, I try to express the concerns of critical mathematics education by addressing: (1) what realism with respect to mathematics could mean; (2) how learning can be dialogic; (3) how knowledge can mean action; (4) how reflections can be public; (5) how learners can be "realized," and also realized as being noisy; (6) how conflicts can set the scene; and (7) how mathemacy can mean hope. Such enumeration of issues might appear unsystematic, but I am not searching for systematization. I only look for a conceptual sensitivity which can be fruitful for critical mathematics education.

[7] For a discussion of borderline school, see Chapter 2.

nificant new learning opportunities. What does that observation mean when we consider well-researched schools? What does the same observation mean when we consider schools in poor areas of the world where there might be no possibilities to get access to any computer nor to electricity? Both questions have the same significance for critical mathematics education.

Secondly, critical mathematics education is concerned with the nature of those competences which mathematics education might support. The content of mathematics education concerns some forms of knowledge that play a significant role in the further formation of the informational society. The notion of *mathemacy* signifies competencies related to mathematics, similar to the notion of literacy, as developed by Paolo Freire. The task of Freire was not simply to teach illiterate people to read and write (still, one should keep in mind that Freire's programme was extremely efficient in making people literate, in the traditional sense of "literate"). For Freire, reading and writing could also mean reading a situation (and not just a text); a sociopolitical situation becomes open to interpretations and to critique. In this sense, Freire expanded the program of literacy into a support for the development of critical citizenship, implying that people not only need to see themselves as affected by political processes. Critical citizenship also includes participation. Mathemacy refers to different competences. One such ability is to deal with mathematical notions, a second ability is to apply such notions in different contexts, a third ability is to reflect on such applications. This reflective component is crucial for the competence of mathemacy. More generally, critical mathematics education is concerned with the development of the competence of mathemacy in such a way that it provides empowerment similar to the empowerment expressed by literacy. One direct meaning of empowerment simply refers to the possibility for the individual to go beyond the limitations which a sociopolitical situation has imposed on a group of people. It could be any group of people, and also people belonging to the Fourth World. More generally, mathemacy means a support for critical citizenship, whatever group of people we might have in mind (constructor, operator, consumer, disposable people). On this background, it becomes important to consider the question, In what way is it possible to establish mathematical learning which might support the development of mathemacy? It is in no way a given that this question can be answered in any clear and satisfactory way, but it remains a major concern of critical mathematics education.

NOTE

A preliminary version of this paper was published in 2003 in the *Proceedings for XI Inter-American Conference on Mathematics Education*, July 13–17, Bumenau, Brazil (on disk). A Portuguese version of the paper has been published in *Revista Zetetiké*, *13*(24), 113–142.

REFERENCES

Bauman, Z. (1998). *Globalization: The human consequences*. Cambridge, UK: Polity.

Bauman, Z. (2001). *Community: Seeking safety in an insecure world.* Cambridge, UK: Polity.

Bell, D. (1980). The social framework of the information society. In T. Forrester (Ed.), *The microelectronics revolution* (pp. 500–549). Oxford, UK: Blackwell.

Bourdieu, P. (with de Saint Martin, M.) (1996). *The state nobility: Elite schools in the field of power*. Cambridge, UK: Polity.

Bury, J. B. (1955). *The idea of progress: An inquiry into its origin and growth* (with and Introduction by Charles A. Bead). New York, NY: Dover. (Original work published 1932)

Castells, M. (1996). *The information age: Economy, society and culture. Volume I, The rise of the network society*. Oxford, UK: Blackwell.

Castells, M. (1997). *The information age: Economy, society and culture. Volume II, The power of identity*. Oxford, UK: Blackwell.

Castells, M. (1998). *The information age: Economy, society and culture. Volume III, End of millennium*. Oxford, UK: Blackwell.

Castells, M. (1999). Flows, networks, and identities: A critical theory of the informational society. In M. Castells, R. Flecha, P. Freire, H. Giroux, D. Macedo, & P. Wilis, *Critical education in the new information age* (pp. 37–64). Lanham, MD: Rowman and Littlefield.

D'Ambrosio, U. (1994). Cultural framing of mathematics teaching and learning. In R. Biehler, R. W. Scholz, R. Strässer, & B. Winkelmann (Eds.), *Didactics of mathematics as a scientific discipline* (pp. 443–455). Dordrecht, The Netherlands: Kluwer Academic.

Horkheimer, M., & Adorno. T. W. (2002). *Dialectic of Enlightenment* (J. Cumming, Trans.). New York, NY: Continuum. (Original work published 1944)

Nisbet, R. A. (1980). *History of the idea of progress*. New York, NY: Basic.

Skovsmose, O., & Valero, P. (2001). Breaking political neutrality: The critical engagement of mathematics education with democracy. In B. Atweh, H. Forgasz, & B. Nebres (Eds.), *Sociocultural research on mathematics education* (pp. 37–55). Mahwah, NJ: Lawrence Erlbaum.

Skovsmose, O., & Valero, P. (2002a). Mathematics education in a world apart—Where we are all together. In P. Valero & O. Skovsmose (Eds.), *Proceedings of the Third International Mathematics Education and Society Conference* (pp. 1–9). Centre for Research in Learning Mathematics. Copenhagen, Roskilde, Aalborg: Danish University of Education, Roskilde University Centre, Aalborg University.

Skovsmose, O., & Valero, P. (2002b). Democratic access to powerful mathematical ideas. In L. English (Ed.), *Handbook of international research in mathematics education* (pp. 383–407). Mahwah, NJ: Lawrence Erlbaum.

Valero, P., & Skovsmose, O. (Eds.) (2002). *Proceedings of the Third International Mathematics Education and Society Conference*. Centre for Research in Learning Mathematics, Danish University of Education, Roskilde University and Aalborg University.

CHAPTER 11

LINKING MATHEMATICS EDUCATION AND DEMOCRACY

Citizenship, Mathematical Archaeology, Mathemacy, and Deliberative Interaction

In *Reflexive Modernization*, Ulrich Beck, Anthony Giddens, and Scott Lash discuss social development in terms of modernity, individualization, posttradition, reflexion, risk society, and several other constructs from recent social theory. In separate chapters, each author presents an analysis; and in *The Reinvention of Politics*, Beck interprets the industrial society as an expression of modernity. However, the very success of industrialization brings with it a development towards a different kind of modernity: "This new stage, in which progress can turn into self-destruction, in which one kind of modernisation undercuts and changes another, is what I call the stage of reflexive modernisation" (Beck, Giddens, & Lash, 1994, p. 2). This development is not decided upon or monitored by any democratic institution. It is a social, "subconscious" process. In this way reflexive modernization brings us into a risk society (Beck, 1992).

If we study the index of *Reflexive Modernization*, we do not find any reference to "mathematics." However, Beck's chapter contains the following observation:

Critique as Uncertainty, pages 143–159.

"Risks flaunt and boast with mathematics" (Beck et al., 1994, p. 9). To me, this comment is crucial, but in *Reflexive Modernization* it is left as an insignificant side remark. A discussion of mathematics and mathematics education with respect to democracy does not appear to be considered significant for a deeper understanding of today's societies. Instead, works like *The Theory of Communicative Action I-II* by Jürgen Habermas (1984/1987) and *The Constitution of Society* by Anthony Giddens (1984) seem to subscribe to the idea that mathematics is a harmless phenomenon.

I shall maintain that the opposite is the case: An analysis of mathematics, mathematics education, and democracy is essential to an interpretation of social development and to a discussion of, for instance, reflexive modernization. I shall return to this point in my conclusion. In what follows, however, I shall first of all concentrate on issues which relate mathematics education and the problems of democracy.

For this discussion, my involvement in a mathematics education project in South Africa has meant considerable inspiration to me.[1] In a significant way, the development in this country has raised issues about democracy and education.

Democracy

The linguistic roots of "democracy" are found in Greek. *Demos* means "people," and *cratos* means "rule"; so, literally speaking, "democracy" means "ruled by people." This conceptual clarification, however, raises two fundamental questions: What does ruling mean? And, Who are the people? Ruling can refer to legislation, but it can also refer to the actual process of carrying out decisions. In ancient Greece, the answer to the latter question was "free men"; slaves and women were not included.

Two extremes in the interpretation of democracy can be illustrated by the ideas of Joseph A. Schumpeter and Jean-Jacques Rousseau. In *Capitalism, Socialism and Democracy*, published in 1943, Schumpeter suggests an interpretation of democracy which is clear and simple. To him the role of the people is to produce a government, and he defines the democratic methods as "that institutional arrangement for arriving at political decisions in which individuals acquire the power to decide by means of a competitive struggle for the people's vote" (Schumpeter, 1987, p. 269). This definition makes the American campaign for presidency an exemplary case of the democratic method. This method does not include people being involved in the actual discussion of political matters. The democratic method takes the form of a "competitive struggle," and the actual content of people's democratic activities is to cast their votes. According to Schumpeter, this elitist interpretation of democracy is important, because it is an illusion to assume

[1] This project involves the following institutions: University of Durban-Westville, Aalborg University, and the Royal Danish School of Educational Studies.

that "the people" can actually take part in a rational process of deliberation and decision-making.

The opposite idea is exemplified by Rousseau's (1968) definition of democracy, put forward in *The Social Contract*, first published in 1762. Rousseau discusses the notion of sovereignty: Who is the sovereign? In a monarchy the answer is the King. But who, then, is the sovereign in a democracy? Rousseau's answer is simple and straightforward: The people! According to Rousseau, it should be possible for the people actually to participate in the ruling, and therefore his interpretation exemplifies direct democracy, while Schumpeter's definition represents the extreme of indirect democracy. In *The Social Contract*, Rousseau analyses conditions for democracy, and he takes the view that this form of governing can only exist in small communities. Democracy is not for "normal" societies.

In recent discussions, democracy certainly refers to formal procedures for electing—being of a government or of any other kind of ruling body. But it has also been emphasized that democracy is subject to the fulfillment of certain conditions: a fair distribution of social services and goods, equal opportunities and obligations for every member of society; opportunities for citizens to participate in discussions and decision-making.

The interpretation of democracy I want to suggest does not follow the line set out by Schumpeter. Instead, it is inspired by the notion of direct democracy. To me, democracy concerns not only questions of governing a state but of governing institutions of any kind. Furthermore, "ruling" presupposes "citizenship" involving a variety of forms of "participation." Thus, "democracy" comes to mean "a way of life." This brings us to the notion of *deliberative democracy*, which "refers to the idea that legitimate lawmaking issues from the public deliberation of citizens. . . . [It] presents an ideal of political autonomy based on practical reasoning of citizens" (Bohman & Rehg, 1997, p. ix). From this perspective, it certainly makes new sense to explore the relationship between democracy and education.[2]

Democracy and Education

The Second World War raised the questions, How could Nazi barbarism emerge in the heart of Europe? Did education have a responsibility in not trying to prevent the development of authoritarianism? The claim of critical education is that fundamental undemocratic developments must be challenged by education! In fact, the development of "critical education" can be seen as an educational attempt to provide a new foundation for education for citizenship.[3]

[2] Paola Valero has drawn my attention to the notion of deliberative democracy as being of particular relevance to the discussion of mathematics education. See also the paper by Valero in part 2 of these analyses in *Zentralblatt für Didaktik der Mathematik* 1999/1. For a general discussion of "deliberative democracy," see Bohman and Rehg (1997).

[3] As part of the early discussion of critical education, inspired by Critical Theory, the notion of *politische Bildung* was crucial. However, politische Bildung cannot adequately be translated to "political education" as the German word Bildung has other connotations than "education." *Politische Bildung*

However, already in *Democracy and Education* from 1916, John Dewey integrates a broad perspective on democracy with an interpretation of education. He elaborates this perspective in subsequent writings. According to Dewey's pragmatic interpretation of science, nothing can be taken for granted. Science must refuse any form of dogmatism. Every question must be investigated with fresh eyes, and such a process, particularly as developed in empirical sciences, represents a process of inquiry. Basically, this process is initiated by a problem and based on experience. A research process can, however, also be interpreted as a process of learning.

> It is a cardinal precept of the newer school of education that the beginning of instruction shall be made with the experience learners already have; that this experience and the capacities that have been developed during the course provide the starting point for all further learning. (Dewey, 1963, p. 74)

In his introduction to John Dewey's *On Education*, Reginald D. Archambault summarises Dewey's point of view concerning democracy, the scientific method, and learning in the following way:

> Dewey saw democracy as the political manifestation of scientific method, with its combination of purposiveness and objectivity, freedom and discipline, individual speculation and public verification. . . . The aim of education is the development of reflective, creative, responsible thoughts. Hence, Dewey's whole conception of science, and its methods and its aims is directly relevant to education. (Dewey, 1974, p. xvii–xviii)

The scientific method bridges over education and democracy. Education which organises itself in line with an inquiry process becomes "education for democracy."

That education has a social role to play can also be observed "negatively." During the apartheid period of South Africa, Fundamental Pedagogics was developed in order to help "justify" the apartheid system. The dictatorship found it essential to involve education in such a task, and Fundamental Pedagogics became education for nondemocracy. An important step to be taken in the education of the new South Africa is precisely to redevelop education as part of a democratic endeavor. Still, we have to keep in mind that whatever undemocratic aspects of global development we might refer to, the ultimate in barbarism against democratic life took place in central Europe.

Dewey makes two claims: (1) Education and democracy form aspects of the same discussion; and (2) the scientific method provides the important bridge between education and democracy. To me, Dewey's first claim is as actual as ever,

---

refers to a wide range of issues concerning, for instance, "empowerment," "sociological imagination," and "autonomy." An important work in this area is Negt (1964). See also Paffrath (1987). For a recent discussion of education and democracy, see, for instance, Giroux (1989) and Young (1989).

certainly illustrated by the development in South Africa. However, I find Dewey's second claim problematic; also when we take a look at mathematics education.

Democracy and Mathematics Education

If mathematics is interpreted as language, the speech act theory of language (see Austin, 1971, and Searle, 1969) will raise the question: What can be done by means of mathematics? Mathematics can be interpreted not only as a descriptive tool but also as a source for decision-making and action. This brings into focus the notion of "symbolic power," discussed by Pierre Bourdieu (1991), and the theme of "knowledge and power." Michel Foucault has concentrated much of his work on unmasking the interplay between "knowledge" and "power" (see, for instance, Foucault, 1980, and Kelly, 1994). I suggest that these more general discussions should be specified with mathematics in mind. A careful study of knowledge (in terms of mathematics) and power is still waiting to be carried out. Mathematics as a possible codification of power is not spelled out by the recent philosophy of mathematics, being neoempirical or constructionist. While waiting for such a clarification, we can, however, consider the thesis of the *formatting power of mathematics*: Social phenomena are structured and eventually constituted by mathematics.[4]

The claim is not that this thesis is true, but that the thesis expresses a possibility which is important to consider when mathematics and mathematics education are investigated from a social and political point of view. Nor is the claim that the thesis is simple. Naturally, it does not make sense to claim that mathematics per se has a formatting power. The thesis concerns mathematics in context. Social, political, and economic interests can be pursued by means of the powerful language of mathematics. In this way, the thesis of the formatting power becomes a thesis of the existence of interplay between mathematics as a source of technological decision-making and action and other sources of social development.

The possibility of such an interpretation of mathematics has, as already mentioned, not been part of the general sociological discussion; here mathematics appears "harmless" and insignificant. To Dewey, the scientific method as such represents a unification of research, learning, and democratisation, and therefore it certainly has a social role to play. However, Dewey's interpretation of this method and his whole perspective of science do not allow him to observe a possible double role of science. Dewey's interpretation of "education and democracy" is based on complete reliance on the scientific method as a faultless supporter of democratic thinking. This is a problematic assumption, certainly also when mathematics is considered. We have, therefore, to look for other ways of bridging over mathematics education and democracy than by means of scientific methodology.

[4] For a discussion of the formatting power of mathematics see Skovsmose (1994). See also Keitel (1989, 1993), Keitel, Kotzmann, and Skovsmose (1993), and the discussion of the "prescriptive use of mathematics" in Davis and Hersh (1988).

Mathematics education might serve very different social functions, and fundamental questions have to be raised concerning the actual social organization of mathematics education.[5] As already emphasized, it is not the actual truth of the thesis of the formatting power of mathematics which makes it relevant to the discussion of mathematics education. The relevance is produced by the challenge: Does mathematics education produce critical readers of the formatting? Or does mathematics education prepare a general acceptance of the formatting, independent of the critical nature of the actual formatting?

Essential functions in the technological society depend on how competence in mathematics is distributed by means of the educational system. Mathematics education can serve as a "blind" instrument for providing the mathematical competence in a form that "accommodates" the present technological development. I call a social or political act "accommodating" insofar as it does not challenge any aspects of the predominant distribution of power or any power relationships. The structure of the educational system can make sure that the mathematical competence is distributed in such a way that, for instance, the adequate number of people needed in developing the information technology in fact receives sufficient mathematical competence.

Mathematics education can also make sure that the "inverse competence" is in place and distributed in an "accommodated" way, meaning that a sufficient number of people come to understand that mathematics is not their business. Excluding a certain number of people from competence can also be "useful" as a potential group of critics is eliminated. In *Mathematics by All*, John Volmink writes:

> Mathematics is not only an impenetrable mystery to many, but has also, more than any other subject, been cast in the role as an "objective" judge, in order to decide who in the society "can" and "cannot." It therefore served as the gate keeper to participation in the decision-making processes of society. To deny some access to participate in mathematics is then also to determine, a priori, who will move ahead and who will stay behind. (Volmink, 1994, pp. 51–52)

In other words, awareness of the politics of mathematics education leads to a discussion of conditions for equity and democratic life. Much research in mathematics education has, however, ignored this discussion. For instance, in *Mathematics Education as a Research Domain: A Search for Identity* (Sierpinska & Kilpatrick, 1998) we only find one reference to democracy. This, however, is a relevant one. In his chapter *A Postmodern Perspective*, Paul Ernest enumerates 10 interdisciplinary themes which are a shared concern for both science and mathematics education, one of them being: "Critical citizenship, epistemological empowerment and enhanced democracy through science and mathematics for all" (Ernest, 1998, p. 82).

[5] An essential step in developing the politics of mathematics education is taken by Mellin-Olsen (1987). See also Fasheh (1982), Niss (1994), Noddings (1993), Powell and Frankenstein (Eds.) (1997), Tate (1996) and Woodrow (1997).

"Our Community"

What I have tried so far is to indicate that it is possible, from a general perspective, to relate mathematics education and a discussion of democracy. However, what could such a linkage mean from the perspective of classroom practice? With reference to one example of educational practice, I shall discuss the following four issues:

- Citizenship
- Mathematical archaeology
- Mathemacy
- Deliberative interaction

My intention is to illustrate how these issues, on the one hand, concern classroom practice in mathematics education, and, on the other hand, concern democracy.

The project, "Our Community," took place in a comprehensive school in the village of Hinnerup in Denmark.[6] Fifteen students, about 16 years old, were involved in the project, which was planned, organized, and carried out by two teachers, Jørgen Boll and Jergen Vognsen. The students were all in the 10th form, which is the last form before they leave secondary school. The intention of the project was to give the students an idea of some of the conditions and possibilities, not only for living and working in a local community but also for influencing its social and political life.

In Denmark, it is quite normal for students to practise different sorts of jobs for a short period of time as part of their school education. Such a one-week trainee service introduced the project, "Our Community," and the students worked as porters, teachers, environmental supervisors, unskilled laborers, librarians, technical assistants, kindergarten assistants, and such. Besides participating in the jobs, the students had to clarify some general questions, formulated by the teachers, about their workplace, like, How is the workplace organized? Who makes the decisions? The idea was that the students should not only experience the actual nature of the work but also get an impression of how the workplace was organized.

After the trainee service, the students, working in groups, concentrated on selected problems of Hinnerup. Two of them were Studying the population forecast for Hinnerup; and making suggestions with respect to the economic situation of the local Music School.

Population forecasting: The students had gathered statistics specifying the number of people in the different age groups of the Hinnerup District. Based on these statistics, they could produce their own forecast and compare it with the available official forecast. It became apparent that some decrease in the group of young people would take place and that not every school in the Hinnerup District could expect to get a 10th form in the future. More general implications of the

---

[6] A full description of the project "Our Community" is found in Skovsmose (1994, pp. 141–154).

forecast concerned the program of house and school building and the capacity of kindergartens and rest homes.

The financing of the Music School: This school gives lessons to students interested in practicing music. The Music School is organized as a private school, which means that the expenses are not covered by the government but by the parents of the students who take lessons. However, in many other towns in Denmark, the local authorities support the music schools financially. The students had gotten the necessary information about the economy of the Music School, and they made different budgets and experimented with alternatives forms of financing. The results of the calculations were organized as recommendations actually handed over to the head of the Music School. The students' investigations, however, continued. It was calculated what it would mean, in terms of increased taxes, if the authorities decided to support the Music School. It was discussed whether local organizations, other than the Music School, could also argue for the right to receive financial support. In this way, emphasis was placed on the fact that it does not make sense in a democracy to take care of an isolated case without discussing whether a particular decision has general consequences or not. It was also estimated what more extensive general financial support to different cultural activities in Hinnerup would mean for the level of taxes.When the project was about to finish, the mayor visited the class. He was a former teacher, and as the Hinnerup District is small, his visit could be organized without many difficulties. Different topics of particular interest to young people in Hinnerup were raised and discussed. It was obvious that during the project work the students had gained much information relevant to such a discussion; they had more ideas about structures of decision-making in a local community and about economic possibilities and limitations.

Citizenship

The project "Our Community" shows that mathematics education for democracy could simply mean *learning about the local community* of which you are a member. Such learning may provide citizenship. Learning for democracy does not only mean learning about constitutions, rules for election, and such, it also means being involved in democratic processes.

"Our Community" relates to the work of Marilyn Frankenstein (1989), who shows how a study of statistics and figures, concerning, for instance, unemployment, may develop an understanding of the society of which the students are members. She tries to show how relearning of mathematics can support people in developing a critical competence in living in a society, in which many decisions are made with reference to mathematics or attempts are made to justify them with reference to certain calculations, reliable or not.

That this is also the case in a small community is illustrated by "Our Community." The students worked with population forecasting and observed how such figures could be used in justifying decisions about house building programs and the planning of schools and kindergartens. The students also had the opportunity to get an impression of the reliability of such figures, as, for instance, they were based on certain expectations of people moving in or out of the district. The students saw how the organization of the Music School and the kind of activities which it could offer depend on the policy of payment. They could make experiments, using spreadsheets, with alternative forms of payment, including new assumptions about the number of students, the amount of fees, support from the district, and so on. They could use mathematical tools to investigate alternatives, and in this way get an impression of the basis upon which actual decisions were made. The idea was to make mathematics, as a language of power, accessible to critical discourse.

The aim of "Our Community" was not to make the students accept social facts. Citizenship does not only imply being ready to live in and to face the "output" from authorities, it also means providing an "input" to authority, a "talking back" to authority. Education for citizenship, therefore, also presupposes participation. In the project, this was experienced in the most direct way, as the students in fact met with a person in power. That this person happened to be the mayor is naturally sheer coincidence, but the general idea is clear: Learning by doing makes sense, also when we talk about being engaged in a local community.

To me, citizenship means being involved in such activities as illustrated by "Our Community." But citizenship can also be analysed in more theoretical terms, one of these is "empowerment," another is *Mündigkeit*. The German word *Mündigkeit* has a double meaning. The term relates to legal theory referring to a person coming of age and thus acquiring an adult's rights and responsibilities in society. However, the term also has an informal meaning: that of having the capacity to speak for oneself. This notion was developed as part of a theoretical frame of critical education in order to contrast the idea that education should produce "followers" (see, for instance, Adorno, 1971). During the 1930s, too many followers were produced by European education. The present project shows that *Mündigkeit* can be given a specific interpretation, such as the students being able to participate in political discussions taking place in a local community. And, most important, *Mündigkeit* also includes competence in investigating decisions with mathematically formulated arguments.[7]

Mathematical Archaeology

Social structuration,[8] at the macro- or at microlevel, often takes place with reference to mathematics. Such an interplay between knowledge and power is

[7] I must emphasize that I am not using the example, "Our Community," to make any empirically based claim about what the students involved in this projects actually did learn. I only use the references to the project to illustrate how a notion like citizenship can relate to both a discussion of democracy and a discussion of mathematics education. This is also the case for the next notions I want to discuss.

[8] According to Giddens: "The basic domain of study of the social sciences, according to the theory of structuration, is neither the experience of the individual actor, nor the existence of any form of societal totality, but social practices ordered across space and time" (Giddens, 1984, p. 2). And further,

referred to by the thesis of the *formatting power of mathematics*. As mathematics is basically an "invisible" part of social structuration, we need strong analytical tools to capture the role of mathematics.

This leads to the notion of *mathematical archaeology*. By this activity, I understand the process of excavating mathematics which might be encapsulated in certain political arguments, technologies, or administrative routines. As part of the ethnomathematical research program, many instances of mathematical archaeology have been carried out. Ethnomathematical studies have emphasized that mathematics in many forms is present in "traditional" societies, and that mathematical competences are frozen in routines, techniques, and handicrafts, in all kinds of ordinary life. (For an overview, see Gerdes, 1996, and Powell & Frankenstein, 1997.) However, to me, this does not seem to be a phenomenon that is specific to "traditional" societies. Different forms of mathematics are frozen in different societies, although the actual role of the frozen mathematics may be very different. A distinction can be made between a mathematical archaeology which tries to identify mathematics which has been "frozen" as part of a technological design, a procedure for decision-making, and such, and an archaeology which searches for mathematical patterns in activities not originally "formatted" by mathematics. The last type of study has been a main focus for ethnomathematical research.

When mathematics exercises a formatting power, citizenship presupposes that this power is excavated, and a mathematical archaeology becomes a useful tool. This activity is illustrated by the project, "Our Community," both with respect to the population forecast and the budget for the Music School. The reliability of arguments for certain decisions concerning house-building programmes became accessible to critical investigation only after mathematical archaeology had made it clear to the students what kind of mathematics-based argumentation was supposed to support the decisions.

Mathemacy

The notion of "literacy" has been developed by Paulo Freire to mean much more than just being able to read and write. Literacy also includes competence in interpreting social life. The notion of *mathemacy* can be developed in a similar way to mean more that an ability to calculate.[9]

In order to do this, it is important to pay attention to the notion of reflection. When a calculation, based on a mathematical task, is carried out, it is possible to reflect on the actual result. Such a reflection has mathematical concepts and algo-

---

"Analysing the structuration of social systems means studying the modes in which such systems, grounded in the knowledgeable activities of situated actors who draw upon rules and resources in the diversity of action contexts, are produced and reproduced in interaction" (Giddens, 1984, p. 25).

[9] For a discussion of mathemacy, see D'Ambrosio (1985). See also Skovsmose (1994) and Skovsmose and Nielsen (1996).

rithms as its objects. A question guiding such a *mathematics-oriented reflection* can be, Are the calculations made correctly?

When setting up a mathematical model, in order to solve a nonmathematical problem, we face many difficulties. Thus, any modelling process presupposes that certain simplifications are established. This means paying attention to certain aspects of "reality" and neglecting others. This is, for instance, what takes place when a population model is used for the purpose of forecasting. Reflections referring to a modeling process are of a different kind than mathematics-oriented reflections. While this kind of reflection has mathematics calculations as its object, a *model-oriented reflection* has the relationship between mathematics and an extra-mathematical reality as its object. A question guiding a model-oriented reflection can be, Is the output of the modeling process reliable? Model-oriented reflections are carried out with an interest in improving the model, that is, with a technological interest. Such reflections concern the validity of the model.

A modeling can, however, also be considered from a more general perspective and with a different interest. The guiding questions could be, What is the actual purpose of carrying out the modeling? What, in fact, is the political and social function of applying mathematics to a certain situation? A reflection guided by such questions can be called a *context-oriented reflection*. "Context" is here understood as a political, social, or cultural context. The mathematical archaeology, carried out as part of the project, "Our Community," illustrates how context-oriented reflections become possible, when the latent functions of the model become unearthed. Such reflections try to address the issue of "mathematics and power."[10]

Beck distinguishes between two processes: reflexion and reflection (see Beck et al., 1994, pp. 5–8). As mentioned above, reflexion refers to a social process which, so to speak, takes place outside the democratic institutions of society. This is the process by means of which the successful industrial society produces an input to itself. This effective production means precisely that a new social form is created: a risk society. According to the thesis of the formatting power of mathematics, this reflexitivity is also an expression of a mathematical activity. Beck emphasizes that reflection is different from reflexion. Reflection means reconsidering what is taking place. Reflection is a mental, a conscious, or a theoretical activity. In my terms, it is a critical activity and a process of grasping basic processes of social development. If reflexive modernization has mathematics as constituent, then reflections with respect to mathematics become of particular importance. This brings the notion of reflection into focus in an attempt to redefine mathemacy so that this competence gets a critical potential. Tentatively, mathemacy can be understood as a kind of competence in which mathematics-oriented, model-oriented, and context-oriented reflections are brought together as an epistemic unit.

---

[10] In Christiansen, Nielsen and Skovsmose (1997) (in Danish) a distinction is made between different kinds of reflections, similar to the one made between mathematics-oriented, model-oriented, and context-oriented reflections.

I want to constitute mathemacy in parallel with literacy in such a way that mathemacy becomes a kind of competence for acting in the world structured by mathematics. In "Our Community," I saw students expressing this kind of competence when they were engaged in community problems and, in fact, were able to understand some of the principles in decision making and to challenge some of the decisions actually made. The notion of citizenship, then, becomes refined in terms of mathemacy.

A different kind of reflection, however, must also be considered. A study made by Mathume Bopape (1997) of children's attitudes towards mathematics in schools in South Africa shows an interesting phenomenon. Even though the children were considered "bad" students by the teachers, and even though their school was located in a problematic area, the children expressed a very positive attitude towards learning mathematics. They simply considered it important to develop mathematical skills. Some of them referred to the fact that they might be able to help their parents in solving problems having to do with money and business. The children expressed a remarkable responsibility for learning mathematics. This can also be seen as a manifestation of a reflection related to mathematics, although not a mathematics-oriented, model-oriented, or context-oriented reflection. This different kind of reflection could be termed a lifeworld-oriented reflection. Also, this kind of reflection forms part of the competence I call "mathemacy."

There is no point in trying to classify all different kinds of reflections. A variety of reflections do exist, and this grand family of reflections becomes an important aspect of mathemacy as a constituent of citizenship.

Deliberative Interaction

It is not difficult to find examples of a teacher-student communication which is ritualized into fixed procedures: The teacher asks a question; the students raise their hands; the teacher asks a particular student; this student gives an answer; the teacher corrects the student if necessary; the teacher asks a new question.

Classroom absolutism refers to the phenomenon that communication between students and teacher is structured by the assumptions that mathematics (school mathematics) can be organized around exercises and questions which have one and only one correct answer, and that, ultimately, it is the teacher' s job to make sure that mistakes are eliminated from the classroom (see Alrø & Skovsmose, 1996a). This absolutism can coagulate into a ritualized form of communication (which might appear to be efficient if, in fact, the job of the mathematics teacher is to eliminate mistakes).

Giddens's chapter in *Reflexive Modernization* has the title *Living in a Post-Traditional Society*. Here Giddens discusses the notion of tradition, and he defines a "formulaic notion of truth" (see Beck et al., 1994, p. 63). Such truths have guardians. Considering mathematcs education, we can reinterpret classroom absolutism in these terms, and a stereotype of communication is exercised when the guard makes sure that the students reproduce the formulaic truths. To me, classroom absolutism brings along a form of communication which is hostile to the development of a democratic culture in the classroom. Education for democracy cannot be based on stereotypes of teaching-learning practices dominated by guardians and formulaic truths. It is important to make possible an interaction in the classroom which supports dialogue and negotiation. *Deliberative interaction* should be a possibility.

With reference to Habermas, Joshua Cohen describes the "ideal deliberative procedure" in terms of freedom, reason, equality, and consensus. This procedure captures "the notion of justification through public argument and reasoning among equal citizens" (Cohen, 1997, p. 72). Deliberative interaction can be defined as an interaction which resonates with an ideal deliberative procedure.

It is essential for the teacher to be aware of the student's good reasons in order to escape the paradigm of classroom absolutism. Even if an answer from a student might be different from what the teacher expects, the answer might be adequate when seen in relation to the student's preunderstanding. An awareness of this can be a useful starting point for a dialogue between teacher and student. In the article "The Student's Good Reasons" (1996b), Helle Alrø and I describe an *inquiry cooperation model*. This model is developed with reference to a teacher-student communication. The model includes different elements: active listening, where the teacher and the student *get in contact*; a process of *discovering* and *identifying* the student's good reasons for thinking the way he or she does; a process of the student *thinking aloud* and, in this way, puts his or her ideas and reasons forward in the dialogue; a *reformulation* by the teacher in order to make sure that he or she understands what the student says; a challenge of the student' s good reasons; a *negotiation* of the teacher and student perspectives; a shared evaluation of the (possible) good reason. The inquiry cooperation model refers to a pattern of communication where the teacher and the student meet in a shared process of coming to understand each other. In this way, the inquiry cooperation model exemplifies deliberative interaction.

A learning theory could take the form of a dialogical epistemology which could bring together two discussions. First, it could help to unite different forms of reflection and in this way provide a basis for analyzing how mathemacy may be developed. Second, it could be a theory of democratic life in a classroom and therefore suggest how certain forms of communication might support the development of *Mündigkeit* and citizenship. This brings the notion of deliberative interaction to the heart of a learning theory of mathematics, and in this way we may come to understand the relationship between reflections (mathematics-oriented, model-oriented, context- oriented, lifeworld-oriented), mathemacy, citizenship, and *Mündigkeit*.

Classroom communication does take place in a school setting which puts limitations to the structure of the communication. The students and the teacher are easily assigned particular roles, and the interaction between these actors becomes predescribed by a didactical contract. In order to make it possible for communication in the classroom to become a dialogue, it might be essential to frame the school situation in new ways. "Our Community" can illustrate an effort in providing a different frame for learning. The students were actually getting a trainee training. They were learning about community by participating in the community life. They were, for a while, released from the roles of students. Back in school, they were facing "real" numbers, and they were interacting with real authority. Deliberative interaction becomes a real possibility.

Conclusion

Citizenship, *mathematical archaeology*, *mathemacy*, and *deliberative interaction* are all relevant issues when the relationship between mathematics education and democracy is considered. The first issue concentrates on the learner as a (new) member of society; the second on the social functions of mathematics and on how to get to grips with mathematics in use; the third refers to an integrated kind of competence; the fourth considers the classroom as a microsociety and deals with the nature of the teaching-learning process. As I have tried to illustrate, these four issues concern both mathematics education and democracy. In this way, I try to confirm Dewey's claim that it is important to link the discussion of education and democracy, but I also try to demonstrate that this linkage can never be established only in terms of a scientific method.

Let me now return to the question, Why is it important to discuss mathematics, mathematics education, and democracy in order to get a deeper understanding of today's societies? According to the thesis of the formatting power, mathematics is an essential instrument when technological authority is exercised. Mathematics is part of technological empowerment, and, as already was indicated, mathematics may play a double role in society, being a means for technological and economic development as well as a source of technological catastrophes. Industrializ,ation is linked to technologies integrating mathematics, and as industrialization links with modernity so does mathematics. As the risk society can be associated with the further development of mathematically based technologies, the notion of reflexive modernization can be analysed in terms of the formatting power of mathematics. Therefore, the discussion of citizenship, mathematical archaeology, mathemacy, and deliberative interaction becomes important to a discussion of modernity, reflexivity, risk society, and other concepts from recent sociology.

Mathematics and mathematics education are put in the middle of social development, being within or outside the reach of democratic institutions, but nevertheless essential to a discussion of democracy. In *Towards a Philosophy of Critical Mathematics Education*, I describe the problem of democracy in a highly technological society referring to the difficulty of matching a critical competence which can match the actual social and technological development (Skovsmose, 1994, p. 40). This difficulty is also touched upon by Lash in his chapter "Reflexivity" and it doubles in *Reflexive Modernization*. Here he writes,

> Citizenship's rights in simple modernity, featuring equality before the law, political rights and social rights of the welfare state, have been transformed into reflexive modernity's rights of access to the information and communication structures. Reflexive modernity's new lower class, which is increasingly in many respects effectively an underclass, is deprived on both obligations and rights of what now is no longer social but predominantly cultural citizenship. (Beck et al., 1994, p. 133)

The question is, Can mathematics education provide an adequate response to this situation? Mathematics education is part of the development of reflexive modernity, and, as remarked by Volmink, mathematics education also serves as a gatekeeper: Who does and who does not get access to the information and communication structures? What role does mathematics education play in the production of "reflexive modernity's new lower class?" What other roles could mathematics education play?

NOTE

This paper was first published in 1998 in *Zentralblatt für Didaktik der Mathematik 98*(6), 195–203,.

REFERENCES

Adorno, T. W. (1971). *Erziehung zur Mündigkeit* [Education for authority]. Frankfurt am Main, Germany: Suhrkamp.

Alrø, H., & Skovsmose, O. (1996a). On the right track. *For the Learning of Mathematics, 16*(1), 2–9, 22.

Alrø, H., & Skovsmose, O. (1996b): The students' good reasons. *For the Learning of Mathematics, 16*(3), 31–38.

Austin, J. L. (1971). *How to do things with words*. Oxford, UK: Oxford University Press.

Beck, U. (1992). *Risk society.* London, UK: Sage. (Original work published 1986)

Beck, U., Giddens, A., & Lash, S. (1994). *Reflexive modernization.* Cambridge, UK: Polity.

Bohman, J., & Regh, W. (Eds.). (1997). *Deliberative democracy: Essays on reason and politics*. Cambridge, MA: MIT Press.

Bopape, M. (1997). *The geometry worksheet in the air!* (Research Report, 9). Copenhagen: Royal Danish School of Educational Studies, Department of Mathematics, Physics, Chemistry and Informatics.

Bourdieu, P. (1991). *Language and symbolic power* (J. B. Thompson, Ed. & Introduction). Cambridge, UK: Polity.

Christiansen, I. M., Nielsen, L., & Skovsmose. O. (1997). Ny mening til begrebet refleksion i matematikundervisningen [New meaning to the concpet of reflection in mathematics education]. In J. C. Jacobsen (Ed.), *Refleksive læreprocesser* (pp. 173–190) [Reflective processes of learning]. Copenhagen, Denmark: Politisk Revy.

Cohen, J. (1997). Deliberative and democratic legitimacy. In J. Bohman & W. Regh (Eds.), *Deliberative democracy: Essays on reason and politics* (pp. 67–91). Cambridge, MA: MIT Press.

D'Ambrosio, U. (1985). *Socio-cultural bases for mathematics education.* Campinas, Brazil: UNICAMP.

Davis, P. J., & Hersh, R. (1988). *Descartes' dream: The world according to mathematics.* London, UK: Penguin.

Dewey, J. (1963). *Experience and education.* New York, NY: Macmillan. (Original work published 1938)

Dewey, J. (1966). *Democracy and education.* New York, NY: Free Press. (Original work published 1916)

Dewey, J. (1974). *On education* (R. D. Archambault, Ed. & Introduction). Chicago, IL: University of Chicago Press.

Ernest, P. (1998). A postmodern perspective on research in mathematics education. In A. Sierpinska & J. Kilpatrick (Eds.), *Mathematics education as a research domain: A search for identity* (pp. 71–85). Dordrecht, The Netherlands: Kluwer Academic.

Fasheh, M. (1982). Mathematics, culture and authority. *For the Learning of Mathematics, 3*(2), 2–8.

Foucault, M. (1980). *Power/knowledge: Selected interviews and other writings 1972–1977* (C. Gordon, Ed.; C. Gordon, J. Marshall, J. Mepham, & K. Sober, Trans.). Brigthon, UK; New York, NY: Harvester.Pantheon. (Original work published 1972)

Frankenstein, M. (1989). *Relearning mathematics: A different third r—Radical maths.* London, UK: Free Association.

Gerdes, P. (1996). Ethnomathematics and mathematics education. In A. Bishop, M. A. Clements, C. Keitel, J. Kilpatrick, & C. Laborde (Eds.), *International handbook of mathematics education* (pp. 909–943). Dordrecht, The Netherlands: Kluwer Academic.

Giddens, A. (1984). *The constitution of society: Outline of the theory of structuration.* Cambridge, UK: Polity.

Giroux, H. A. (1989). *Schooling for democracy: Critical pedagogy in modern age.* London, UK: Routledge.

Habermas, J. (1984/1987). *The theory of communicative action,* I-II. London and Cambridge, UK: Heinemann/Polity.

Keitel, C. (1989). Mathematics and technology. *For the Learning of Mathematics, 9*(1), 7–13.

Keitel, C. (1993). Implicit mathematical models in social practice and explicit mathematics teaching by applications. In J. de Lange, I. Huntley, C. Keitel, & M. Niss (Eds.), *Innovation in maths education by modelling and applications* (pp. 19–30). Chichester, UK: Ellis Horwood.

Keitel, C., Kotzmann, E., & Skovsmose, O. (1993). Beyond the tunnel-vision: Analysing the relationship between mathematics, society and technology. In C. Keitel & K. Ruthven (Eds.), *Learning from computers: Mathematics education and technology* (pp. 243–279). Berlin, Germany: Springer.

Kelly, M. (Ed.). (1994). *Critique and power: Recasting the Foucault/Habermas debate.* Cambridge, MA: MIT Press.

Mellin-Olsen, S. (1987). *The politics of mathematics education.* Dordrecht, The Netherlands: Reidel Publishing.

Negt, O. (1964). *Soziologische Phantasie und exemplarisches Lernen* [Sociological imagination and exemplary learning]. Frankfurt am Main, Germany: Europäische Verlagsanstalt.

Niss, M. (1994). Mathematics in society. In R. Biehler, R. W. Scholz, R. Sträßer, & B. Winkelmann (Eds.), *Didactics of mathematics as a scientific discipline* (pp. 367–378). Dordrecht, The Netherlands: Kluwer Academic.

Noddings, N. (1993). Politicizing the mathematics classroom. In S. Restivo, J. P. van Bendegem, & R. Fischer (Eds.), *Math worlds: Philosophical and social studies of mathematics and mathematics education* (pp. 150–161). Albany: State University of New York Press.

Paffrath, F. H. (Ed.). (1987). *Kritische Theorie und Pädagogik der Gegenwart* [Critical theory and todays education] (Research Report, 11). Weinheim, Germany: Deutscher Studien Verlag.

Powell, A. B., & Frankenstein, M. (Eds.). (1997). *Ethnomathematics: Challenging Eurocentrism in mathematics education.* Albany: State University of New York Press.

Rousseau, J.-J. (1968). *The social contract.* London, UK: Penguin (Original work published 1762)

Schumpeter, J. A. (1987). *Capitalism, socialism and democracy.* London, UK: Unwin Paperbacks. (Original work published 1943)

Searle, J. (1969). *Speech acts.* Cambridge, UK: Cambridge University Press.

Sierpinska, A., & J. Kilpatrick, J. (Eds.) (1998). *Mathematics education as a research domain: A search for identity.* Dordrecht, The Netherlands: Kluwer Academic.

Skovsmose, O. (1994). *Towards a philosophy of critical mathematical education.* Dordrecht, The Netherlands: Kluwer Academic.

Skovsmose, O., & Nielsen, L. (1996). Critical mathematics education. In A. Bishop, M. A. Clements, C. Keitel, J. Kilpatrick, & C. Laborde (Eds.), *International handbook of mathematics education* (pp. 1257–1288). Dordrecht, The Netherlands: Kluwer Academic.

Tate, W. F. (1996). Mathematizing and the democracy: The need for an education that is multicultural and social reconstructionist. In C. A. Grant & M. L. Gómez (Eds.), *Making schooling multicultural: Campus and classroom* (pp. 185–201). Englewood Cliffs, NJ: Prentice Hall.

Volmink, J. (1994). Mathematics by all. In S. Lerman (Ed.), *Cultural perspectives on the mathematics classroom* (pp. 51–68). Dordrecht, The Netherlands: Kluwer Academic.

Woodrow, D. (1997). Democratic education: Does it exist—Especially for mathematics education? *For the Learning of Mathematics, 17*(3), 11–16.

Young, R. (1989). *A critical theory of education.* New York, NY: Havester Wheatsheaf.

CHAPTER 12

DEMOCRATIC COMPETENCE AND REFLECTIVE KNOWING IN MATHEMATICS

In *Schooling for Democracy*, Henry Giroux emphasizes that school needs to be defended as a service that educates students to be critical citizens who can challenge and believe that their actions will make differences in society. Students therefore have to be introduced to forms of knowledge "that provide them with the conviction and opportunity to fight for a quality of life in which all human beings benefit" (Giroux, 1989, p. 214). As part of this enterprise, the discussion of literacy has played a major role, especially provoked by the work of Paulo Freire, who developed the political dimension of education from this term.

Antonio Gramsci has indicated that literacy is a double-edged sword.[1] Literacy is a necessary condition in today's society for informing people about their obligations and for people to be used in essential work processes. However, literacy can also be wielded for the purpose of empowerment because it can be a means to organize and reorganize interpretations of social institutions, traditions, and proposals for political reforms. Literacy is not just a competence having to do with

[1] See the remarks about Gramsci in Giroux (1989, p. 147).

Critique as Uncertainty, pages 161–179.

ability to read and write, an ability which can be simply tested and controled, it possesses also a critical dimension. Giroux (1989) formulates it like this:

> Literacy as a radical construct had to be rooted in a spirit of critique and a project of possibility that enabled people to participate in the understanding and transformation of their society. As both the mastery of specific skills and particular forms of knowledge, literacy had to become a precondition for social and cultural emancipation. (p. 148)

Later he underlines:

> Literacy is not just related to the poor or to the inability of subordinated groups to read and write adequately; it is also fundamentally related to forms of political and ideological ignorance that function as a refusal to know the limits and political consequences of one's view of the world. . . . What is important to recognize here is the need to reconstitute a radical view of literacy that revolves around the importance of naming and transforming those ideological and social conditions that undermine the possibility for forms of community and public life organized around the imperatives of a critical democracy. (p. 151)

The question we shall discuss in this paper is, Could "mathemacy" be substituted for "literacy" in the previous formulations? By mathemacy, we may initially understand an ability in calculating and using mathematical and formal techniques; later we may hope to provide the concept with a more differentiated content. Could mathemacy be conceived as a double-edged sword? Could mathemacy also be wielded for the purpose of empowerment? Could mathemacy help people to reorganize their views about social institutions, traditions, and possibilities in political actions? If we reformulate another of Giroux's formulations, it becomes, mathemacy as a radical construct has to be rooted in a spirit of critique and project of possibilities that enables people to participate in the understanding and transformation of their society and, therefore, mathemacy becomes a precondition for social and cultural emancipation. Could this be more than a half-empty assertion? What might be the meaning of mathemacy if it is to fit into this formulation? Freire developed literacy to include more than simple reading and writing abilities, but what sort of extension of mathemacy is needed? Is mathemacy also related to forms of political and ideological ignorance that function as a refusal to know the limits and political consequences of one's view of the world? Could mathemacy be involved in a project of naming and transforming those ideological and social conditions that undermine the possible forms of community and public life organized around the imperative of a radical democracy?[2]

It would naturally be too simple to assume as axiomatic that mathemacy has a similar role to play in society as literacy. It may be the case, but differences and similarities have to be analyzed. The intention in what follows is not to "prove"

[2] See also Frankenstein (1987, 1989).

that mathematics education has to be developed along some specific lines. The intention is more limited. It is to discuss the possibility of providing meaning for such developments. Giroux talks about critical democracy, several educators have reflected on this term, and I find that a more careful look at ideas and assumption related to democracy can indicate some answers to the problem raised. Although several other lines of analysis are relevant, I shall, in this paper, concentrate on the concept of democracy and try to connect it with mathematics education.

Democracy and Education

There is a broad agreement that democracy is a most attractive feature of society, but at the same time widespread disagreement about what democracy might mean.[3] The concept of democracy refers to a bouquet of different ideas, hopes, and utopias. Therefore, although it is impossible to pinpoint any simple definition of democracy, we could try to grasp the concept by outlining ideas which relate to democracy.[4]

Democracy refers to at least the following four aspects: (1) Formal procedures for electing a government and for the government to carry out its ruling. (2) A fair distribution of social services and goods in society, such as welfare, education, hospitals, and so on. Consequently, a substantial part of the theoretical analysis of democratic ideas is concerned with the types of goods and facilities which have to be distributed in a fair way. And what is the interpretation of "fair"? (3) Equal opportunities, rights, and obligations for every member of society. There must be no differences in opportunities based on differences in social background, sex, or color of skin. Everybody must be treated equally according to the law, and similarly, everybody must obey the law. But what does "equal opportunity" mean? According to the liberal and idealistic tradition, it means the unrestricted possibility of trying to do whatever a person wants (legally) to do; while the materialistic tradition has underlined that it is not enough to decrease the number of restrictions, society must actually provide the conditions for everybody to be able to pursue his or her interests. In this way every discussion about democracy becomes a discussion of freedom. (4) The possibility for and the ability of citizens to participate in discussion and evaluation of the conditions for and consequences of the governing which takes place: a "democratic life" is presupposed.

In other words, democracy refers to *formal* conditions concerning algorithms for election, *material* conditions concerning distribution, *ethical* conditions con-

[3] In *Models of Democracy*, David Held writes, "Nearly everyone today says they are democrats no matter whether their views are on the left, centre or right. Political regimes of all kinds in, for instance, Western Europe, the Eastern bloc and Latin America claim to be democracies. Yet, what each of these regimes says and does is radically different. Democracy seems to bestow an 'aura of legitimacy' on modern political life: rules, laws, policies and decisions appear justified and appropriate when they are 'democratic.'" (Held, 1987, p. 1)

[4] A classical discussion in analytical philosophy of democracy is found in Benn and Peters (1959).

cerning equality, and finally conditions concerning the *possibility for participation* and re-action.

It is important to discuss all these aspects in relation to education, and in fact the second and the third have been discussed in great detail.[5] A fair distribution of social services implies that in a democratic society every child and adolescent must have equal access to schooling and to learning. This naturally leads to a discussion about equality. What does equality in education mean? Obviously, children and students seem to receive very different sorts of education, even in the same society, and even in societies supposed to be democratic. How is this possible? Some find it tempting to maintain that only children with the same abilities can be treated equally. This may be convenient for some practical purposes: it seems easier to teach a group of children who are at about the same level. But in what sense is this in accord with democratic ideals? Several investigations have indicated that in some countries, working-class children receive less from schooling than other children.[6] It is also well documented that differences in achievement correlate with gender, at least in some subjects (in some countries). School seems to serve a reproduction of social structures, including the division of labor, the distribution of power between the individual and the state and between social groups, and finally it seems to reproduce traditional cultural values. What does that mean for our interpretation of education by democratic standards? To be in accordance with the ideals of democracy, schools have to react to the different ways in which society reproduces itself, and it must try to counterbalance some of these reproductive forces to provide an equal distribution of what schooling may offer, including opportunities in further education and vocational life.

Education has to do with the content as well as the distribution of acquired competences. That I am not going to pay attention to the distribution of competences does not mean that this is of minor importance. However, I find the questions relating to content and to aspect (4) above—democracy as involving a sort of participation—sufficiently complicated, and this brings us back to questions about the interpretation of mathemacy. What sort of competence, if any, important for participating in a democracy, can be supported by developing mathemacy? What is the nature of such a competence in a highly technological society? Can mathematics education be of value in providing a foundation for children's and adolescents' later participation in a democratic life as critical citizens? Does it in fact make sense to relate the discussion of the content of mathematics education to the discussion of the nature of democracy?[7] We must take a closer look at the idea of democracy.

---

[5] See, for instance, Purvis and Hales (1983). In Peters (1966) is found a discussion of democracy and education and also a discussion of the meaning of equality and freedom in relation to education.

[6] See, for instance, Bowles (1983).

[7] See also Chapter 11.

Democratic Competence

In *The Social Contract*, published in 1762, Jean-Jacques Rousseau makes a classification of different types of government, and he writes,

> The sovereign may put the government in the hands of the whole people, or of the greater part of the people, so that there are more citizen-magistrates than there are ordinary private citizens. This form of government is known as democracy. (Rousseau, 1968, p. 110)

This definition closely follows the ancient conception of democracy stressing the importance of participation, but Rousseau takes a step further to express explicitly the idea of *direct democracy*. Everybody (or most of the people) should actually be able to participate in the ruling. If we conceive this to be a genuine definition of democracy, it is obvious that democracy is impossible in most modern societies. The concept of democracy gets a limited range of applications; only very small and homogeneous societies can be democratic.

If we give up the idea of direct democracy and try to find a more feasible interpretation, applicable to a wider range of societies, we face the problem of delegation of sovereignty: How is it possible to combine democracy with the necessity of selecting a small group of people to actually do the ruling? This question always follows *representative democracy*. The necessity for the delegation of sovereignty is also implied by the fact that ruling presupposes specific qualifications that are not of a common nature, particularly not if we have a complex modern society or organization in mind. The people in charge must have specific knowledge about the domain to be ruled. Perhaps specialized education is needed. The people in charge must possess a competence that includes information and knowledge. We shall accept the delegation of sovereignty as a necessity in what follows. But how is it possible to control the people in charge? This problem accompanies every attempt to extend democracy beyond exclusive direct democracy.[8]

We have to make a distinction between the competence which the people in charge must possess if they are to be able to take well-founded decisions and act in an appropriate way, and the competence which is presupposed if people are to be able to judge the results and consequences of the ruling. We shall distinguish between *competence in ruling* and a *democratic competence*. Democratic competence is to be ascribed to the majority; it must be presupposed in a functioning representative democracy. Democratic competence is the basis of knowledge and understanding which is necessary if the delegation of sovereignty is to be subjected to any sort of control. It is a condition for participation and reaction. Interpretations of this competence vary between two extremes. One sees demo-

[8] Democracy need not be linked to the question of governing a country or a community. In what follows, we shall have a more general approach in mind, meaning that democracy is seen as a characteristic of governing some sort of organization, be it a large or small community or a private or public organization. However, we shall still refer to these as "society."

cratic competence as a natural ability of human beings, the other sees it as an acquired ability. In what we could call the classical and idealistic interpretation of democracy the basic idea is that while the ruling-competence of the people in charge is of a special nature, the judging-competence is natural (but perhaps only as a potential capacity because only a certain attitude will stress the importance of a democratic way of social control, and a democratic attitude is not of a common nature).

It is my assumption that we cannot presuppose that a democratic competence automatically exists. It has to be developed. However, I do not find the fundamental reason for this in the "limitation" of the (although limited) human nature but in the demands which the complexity of society makes on democratic competence. To participate in democratic life in a highly technological society seems a challenging task. But by which institutions in society has a democratic competence to be developed? It is a basic assumption that this question makes sense. Does this imply that we assume the existence of some sort of evaluative authority? But can it make sense to accept a delegation of sovereignty but not to accept that people are sovereign in their evaluation of the result of this delegation?

Before getting into that discussion, I shall mention a particular way of "solving" the problem of the extension of democratic competence. The classical interpretations of democracy draw attention to the question of how to deal with ruling. The actual election of the people in charge has not been the important issue. Naturally this election has to be democratic, and formal rules for election have to be specified and obeyed. But election has been seen as subordinate to the real concern of democracy: the appropriate way of ruling. However, it is possible to turn this upside down and rephrase democratic life as primarily concerned with election and not with ruling. What must be democratic, then, is the election of the government. This makes the formal condition important in a discussion of democracy. The other conditions for democracy become insignificant, and no attention need be paid to the establishment of a democratic competence. That is a possible "solution" to our problem about the nature and development of democratic competence.

Such an interpretation of democracy is suggested by Joseph A. Schumpeter (1987), who conceives the election of the government to be the primary concern of a democracy. Democracy, then, becomes a formal characteristic; it does not have to do with the actual questions to be dealt with in governing. In fact, Schumpeter maintains that producing a government practically amounts to deciding "who the leading man shall be" (p. 273). Schumpeter's interpretation is provocatively simple, but let me point out one of its "advantages": It has great descriptive value. It explicates how the concept of democracy is normally used when countries describe themselves as democratic.

This interpretation neglects every concern about the nonformal conditions for democracy, such as a fair distribution of goods and equal opportunities. Even the basic idea that democracy should be a way of keeping power in the hands of the people is eliminated. Also it is obvious that if we take Schumpeter's interpretation literally, we need not care any more about the existence and nature of democratic competence. Democracy does not presuppose participation and has nothing to do with decision making or with criticism and the evaluation of decisions and proposals put forward by the government. Democracy has to do only with the production of a government. It is the procedures and algorithms for election which are democratic or not. That is the most simple and radical solution of the problem of delegation of sovereignty. The only input from the people to the democratic life is their votes. Beside that, they have only to receive.

A corollary of Schumpeter's interpretation is that education need not pay any attention to democratic life. To do so simply expresses a misunderstanding not only of education but also of the nature of democracy. Considerations about the potential of literacy as a support of a critical democracy become irrelevant. Several trends in (mathematics) education ignore the analysis of democratic demands as part of the guidelines for the curriculum, although without an explicit acceptance of Schumpeter's interpretation. However, such ignoring may express a Schumpeter-like interpretation of democracy.

The concept of democracy to which I subscribe does not have the great descriptive value of Schumpeter's. It is much closer to the classical interpretation which regards democracy as a characteristic of governing, although the actual production of a government should not be ignored. I shall not return to the utopia of direct democracy, so the delegation of sovereignty has to be dealt with. Therefore, democracy also characterizes the ways of participating in discussions and criticism of the actual ruling. A democracy must make room for a critical citizenship, which is the actual performance of a critical competence.

So the problem of democratic competence is back on the agenda. I see the development of this competence as one fundamental condition for democratic life. We have, therefore, to characterize the possible content of democratic competence in relation to the main questions which concern the society in question. *The content of the democratic competence depends on the nature of the problems which face the society. And my focus is on the highly technological society.*

The Problem of Democracy in a Highly Technological Society

Humanity is surrounded by technology. Society and technology are integrated, and technology has become the dominant feature of civilization. The relationship between types of technologies and sciences varies greatly. Manual tools are developed independently of science. The development of steam engineering took place in parallel with the theoretical understanding of thermodynamics, while information technology is totally based on developments in mathematics. Mathematics is the logical underpinning of information processing, and mathematical thinking is the foundation for the actual applications of information technology as well. In fact, every application of a computer can be seen as an application of a simple or complex mathematical model. Therefore, from a logical point of view, information technology need not be interpreted as a *new* way of formal manipulation; but it is an enormous extension of these manipulations. The effect of computers is that the applications of formal methods have colonised all areas of life. This is what characterizes the information society.

In a society based on the use of manual tools, a classical interpretation of democratic competence remains plausible; no specific technological knowledge seems to be needed to evaluate the acts and decisions of the people in charge. Quite the contrary is the case in a highly technological society, where all sorts of social, economic, and political decisions also concern technology. On the face of it, only a limited group of people is able to manage this complexity. In fact, this competence seems to presuppose a certain amount of technological knowledge. How can anybody evaluate decisions which must take into consideration consequences of technological enterprises without a fair amount of technological knowledge?[9]

Democracy may be destroyed by a dictatorship which obstructs formal democratic procedures. That has been seen often and is normally conceived as *the* problem of democracy. It is what some countries accuse some other countries of not solving. But unless we accept Schumpeter's interpretation, the algorithm for the election of the people in charge is only one aspect of democracy. Democracy can be undermined in other ways than by just neglecting the rules of election. Democracy refers not only to formal but also to material and ethical conditions and to possibilities for participating and reaction. In particular, democracy can be destroyed if a critical citizenship cannot be brought to life. When society changes rapidly, the main principles in the mechanisms of the development of society become hidden and difficult to identify. How is it possible to evaluate decisions taken by the people in charge if neither the conditions for, nor the implications of, their decisions are visible? How can anyone other than experts control experts? Will the conditions for critical citizenship be eroded by social and technological development itself? This phenomenon we shall call the *problem of democracy in a highly technological society* (keeping in mind that this naturally is far from the only problem which faces a democracy).

In the information society, the ability to collect, systematize, and use information seems to be the vehicle for social development and, simultaneously, it becomes a source of power. In the article "The Social Framework of the Information Society," Daniel Bell maintains that a knowledge elite has power within intellectual institutions but only influence in the larger world where policy is made. Bell finds that even if political questions become more and more mixed up with technical issues, the knowledge elites have no power to take decisions, although they can define problems, initiate questions, and provide the technical bases for answers. Bell concludes that the political power belongs, inevitably, to the politician rather than to the scientist or economist: "In this sense, the idea that the

[9] In my terminology, the concept "technological knowledge" refers to the competence needed in developing and applying a technology.The concept will de discussed further in the section, "Reflective Knowledge."

knowledge elite will become a new power elite seems to me to be exaggerated." (Bell, 1980, p. 542)[10]

However, even if this argument may be sound, it does not solve our problem about the existence of democratic life. Let us assume that Bell is right: The elite with knowledge may have power within intellectual institutions but only influence in politics. But the problem of democracy raised by the development of the highly technological society is not just that of the influence or the power of a knowledge elite placed outside the arena for political decisions. The problem is not just whether or not technological development reduces politicians to marionettes giving voice to the consequences of prefabricated technological calculations. The problem concerns the relationship between, on the one hand, the people in charge (the elected politicians) and the technological elite and, on the other hand, the people who are affected by the ruling. So even if the knowledge elite is only able to influence while the politicians remain with the power, the conditions and the argumentation for the decisions to be made may go beyond the reach of ordinary people.

Therefore, if the condition for critical citizenship is important in a democracy, we have to return to the problem that the grounds for the decisions taken by the authorities may be inaccessible to people other than the technician and the people in charge. Technological development may *erode* part of the nonformal conditions for democracy, leaving behind only an algorithm for election. That erosion is a real threat to democracy in a highly technological society. But is it possible to secure a critical citizenship in a highly technological society? To find a positive answer to this question is equivalent to conceiving democratic life as possible (also) in the future (without being forced to return to Schumpeter's definition). We shall not try to turn the clock back. We cannot propose abandoning technological environments. We cannot expect the conditions for critical citizenship to decline. The problem is to develop a general critical competence which can actually match the social and technological development. This constitutes the problem of democracy in a highly technological society. To analyze this further, I shall take a look at the position of mathematics in society. This formal science can be seen as a structuring part of "expertocracy."

The Formatting Power of Mathematics

Technology today can be assumed to be characterized by the dominance of formal methods, so we shall take a look at the role of mathematics in society. The thesis which we shall discuss says that mathematics makes a real intervention in reality, not only in the sense that new insight may change interpretations but also in the sense that mathematics colonizes part of reality and rearranges it. The thesis

[10] A little later, Bell concludes, "The fear that a knowledge elite could become the technocratic rulers of the society is quite far-fetched and expresses more an ideological thrust by radical groups against the growing influence of technical personnel in decision making." (Bell, 1980, p. 543).

is that *mathematics is formatting our society*. This thesis of the formatting power of mathematics does not imply that mathematics itself cannot be seen as a social construct (which I conceive it to be) and interpreted as colonized by economical and cultural interests. What is emphasized by the thesis of formatting is that this social construct, although formal, is able to do something to reality, and I concentrate on this aspect because of my interest in the conditions for democracy.

We shall distinguish between two different kinds of theoretical constructs, *thinking abstractions* and *realised abstractions*.[11] Thinking abstraction is used to facilitate reasoning, and this type of abstraction may be exemplified by mathematical concepts and mathematical modeling. Reasoning about economical development may be helped by using a concept like the National Product, defined as a mathematical function. Thinking abstractions only exist as mental models or as images exist. Their existence is similar to the existence of a character in a novel. However, realised abstractions have a different ontological status. They are taken for granted. We normally do not question whether what we are dealing with is a realized abstraction. Perhaps we do not even have the possibility for identifying such abstractions. We live with realized abstractions. Ways of calculating taxes, child benefits, salaries, strategies of production, and such are not just thinking models, they have a real influence on our lives.[12] Exchange values of goods in the form of money are real, they are not just models for expressing the degree of usefulness of some goods or for expressing the time needed for their production. Money systems become real states of affairs, and even the National Product becomes real; it has reached a different status from being just a mathematical summary of calculations based on the values of some parameters. The National Product enters the political and economical discussion as an independent object and as a real figure.

Every society and every culture have developed a realm of realized abstractions. But from what source? Where do they come from? They must be brought into existence by some creative act. Going back in history, we may be able to trace some realized abstractions in ideological and religious structures and metaphysical systems; but habits become norms and standards and rules. Realized abstractions have to be taken into consideration as part of our lives. They are not any longer just models for our thinking. In that sense, thinking abstractions become realized abstractions. Further, the main source for realized abstractions is different in today's highly technological societies. Mathematics and formal sciences have become a new bubbling source for the invention of rules and structures. Formal sciences not only create ways of describing and handling problems, they also become the main source for the reconstruction of reality.

We can formalize a language or a part of a language, and mathematics can be interpreted as such a type of formalization. But it is not only possible to formalize

[11] Ideas described in this section are developed in cooperation with Christine Keitel and Ernst Kotzmann as part of the work in the BACOMET-group. See also Keitel (1989).

[12] See the discussion of the prescriptive function of applied mathematics in Davis and Hersh (1988).

a language: Actions and routines, that is, way of behaving, can also be formalized. In this case, the result of a process of formalization is not a new language but new structures for management perhaps in the shape of "manuals," that is, descriptions of how to behave in a prescribed algorithmic way. This phenomenon can be illustrated by "scientific management" as developed by F. W. Taylor (1947). The basic idea is that complex work processes have to be decomposed into their atomic components. Then each component must be investigated in order to find out the best way of handling the operations, and the "proper time" for its execution must be measured. Then the atomic components have to be sequenced to define the work processes for each worker, and the total of the worker's algorithmic behavior will make up a new "megamachine." Taylor described how he investigated specific work processes and, as an example, he gives the story of the engagement of the "perfect" worker named Schmidt, who never asked questions but just followed the prescribed algorithms. Schmidt thus became the first "Taylorized" man.[13]

Formalization of language and formalization of actions are closely connected. The application of a formalized language in order to describe phenomena makes it easier to "see" formal structures, and by doing so, a first step in adapting reality to our image of reality is taken. A formal description highlights some aspects and ignores others. If, for instance, the object of our description is work processes or economic transformations, and our intention is to do further systematisation and Taylorizing, a formal description will facilitate our steps. It makes it easier to work out new algorithms of behavior. We witness a transmutation of thinking abstraction into realized abstraction, caused by the transformation of a formalized language into a formalization of routines. We create a semantics for our formal description by inventing algorithms and routines, that is, sorts of behavior referred to by the formal language. Mathematics intervenes in reality by creating a "second nature" around us, by giving not only descriptions of phenomena, but also by giving models for changed behavior. We not only "see" according to mathematics, we also "do" according to mathematics. Mathematical structures come to play a role in social life in the same fundamental way that ideological structures organise reality. Schmidt is no longer the only Taylorized man.

Reflective Knowledge

I find that the problem of democracy in a highly technological society has to be seen in the perspective of the thesis of the formatting power of mathematics. If mathematics intervenes in reality, one of the principles for organizing work processes, economical management, and such is created by a source hiding behind the scene of politics. But if mathematics has a special role to play, it becomes natural to suppose that mathematics education must be put into focus, and that

[13] See Taylor (1947), as well as Chapter 15 for a more detailed description of scientific management and the related process of robotting.

brings our discussion back to the concept of mathemacy. To specify the content of this concept, some distinctions may be useful.

Let us as an example look at the problem of motoring: Too many (private) cars cause pollution, too much gas is burned into the atmosphere, and so on. This form of transport carries some serious risks (of an ecological nature, for instance) which we are going to face in a not-too-distant future. The way to confront these emerging problems is not to develop the drivers' driving skills, that is, their abilities in maneuvering a car in the traffic, nor is it to give drivers more information about mechanics—how the car is actually constructed, how the brakes works, how it could be repaired, and so on. Naturally, it is useful both to be able to repair a car and to drive it in a better way, but this is not a satisfactory answer to the problem of motoring. To face this problem and to react to it in an adequate way, we have to develop a better understanding of "motoring," seen as the complex phenomenon of organizing transport and traffic in general. What are the economic and ecological consequences of "motoring?" What social and political actions are needed and which seem to be possible? We have to learn *about* motoring to address such questions. Obviously, to learn about motoring is not itself a solution to the problems caused by motoring, but it is the epistemological step to be taken to get hold of the problem itself. Knowledge at a metalevel must be developed if our actions are not to degenerate into measures of desperation. To stick to the improvement of driving skills would be to adopt an ostrich-like policy.

This can be stated in a more precise way. Let us call the knowledge necessary for developing and using technology *technological knowledge* and, as an example, we may include both the knowledge necessary for driving a car and the know-how necessary for repairing and constructing it. Driving skills are not of the same type as knowledge about motoring. The latter is an example of meta-knowledge, and we call it *reflective knowledge*.[14] The fundamental thesis relating technological and reflective knowledge is that technological knowledge itself is unable to predict and analyse the results of its own production; reflections are needed. The competence in constructing cars is not sufficient for the evaluation of the social consequences of car production. Improved road ability does not produce a better understanding of "motoring." Technological knowledge is born short-sighted. Reflective knowledge must be based on a wider horizon of interpretations and preunderstandings. It has to grasp the situation in which technological knowledge is at work, but no simple steps lead from technological to reflective knowledge. Technological and reflective knowledge constitute two different types of knowledge, but not two independent types. It may be important to master some technological insight to support reflections. To be able to understand and discuss the social implications of the pollution caused by motoring, we have to know about the basic principles and conditions related to the constructions of cars; but it is

[14] See also Skovsmose (1989a, 1989b) as well as Chpater 18.

not necessary to master all aspects of these constructions. If that were the case, democracy in a highly technological society would become impossible.

Reflective knowledge cannot be analyzed into technological components. Even if we collect every bit of technological information, we shall not be able to build up reflections from these parts alone. Technological knowledge does not address a self-critique, nor a specification of the alternative trends in technological development, so reflective knowledge does not have its epistemological basis in technological problems but in the technological way of handling such problems. While technological knowledge aims at solving such problems, the object for reflective knowledge is a suggested technological solution to some (technological) problems.

It may be necessary to make a further distinction between technological and *mathematical knowledge*, which refers to the competence normally understood as mathematical skills, including competences in reproducing mathematical thoughts, theorems, and proofs, as well as mastering a variety of algorithms. These competences differ from abilities in model building, that is, the ability in applying mathematics in the pursuit of technological aims.

Now we can reformulate (a part of) the problem of democracy in a highly technological society. The question is whether the development of reflective knowledge can take place without presupposing a complete development of mathematical and technological knowledge. If it is possible to show that the instruments for identifying and criticizing the use of formal methods in society do not push us completely into the formal sciences themselves, then we may hope to find a solution to the problem of democracy. In that case, technological development need not necessarily erode the conditions for democratic life, leaving behind only Schumpeter's algorithm.

Which institutions in society could take over the job of developing democratic competence? It cannot be assumed that it can be done in a straightforward way, but one answer given is that education must be in charge. I take in the assumption that education plays a specific role in developing democratic competence, and this raises a set of new aims for education. Traditionally, a major concern of education has been to prepare pupils and students for later participation in work processes in society, but different trends in education have stressed that education must also prepare for dealing with the aspects of social life outside the sphere of work, including political and cultural aspects. In short, one of the aims for education must be to prepare for a critical citizenship. The pursuit of such aims was a strong trend in German education after World War II. Also in Scandinavian countries such aims have been put on top of the agenda, underlined by the the use of the German term *Allgemeinbildung* (general education), meaning that education must focus on more than the conditions for work possibilities. Education must prepare pupils and students for (political) life in society. We are now back to the idea, also stated by Giroux (1989), that education has a specific obligation with regard to democracy, but now we are able to say more about the importance and nature of mathemacy. The idea which I have tried to provide with meaning (not to prove) is that *if mathemacy has a role to play in an education—similar to but not identical with the role of literacy—in trying to develop a democratic competence, then mathemacy must be seen as composed of different competences: mathematical technological and reflective. And especially, reflective knowledge has to be developed to provide mathemacy with a radicalized power.* Reflection on the application of formal methods is one important element in the identification of the conditions for social life and therefore a part of a democratic competence. That means that the guiding principles for mathematics education have to be raised to a metalevel; they are no longer to be found in pure mathematics, neither in any epistemological theory which focuses on the development of mathematical knowledge as such.[15] That means that the whole nature of the discussion of mathematics education has to be changed. The focus must be on the *functions* of the applications of mathematics in society and not just on modeling as such.[16] The discussion of the content of mathematics education has to be guided by the question of whether or not it will be able to clarify the actual function of formal methods in today's societies.

Reflective Knowing in Small Steps

Does it make sense to try to develop reflective knowledge as part of an educational task? Has such an intention any educational meaning? One implication seems to be a criticism of all those epistemologies which concentrate on the development of mathematical knowledge in itself. The principal example is found in the genetic epistemology of Jean Piaget, who concentrated on the nature of the development of mathematical knowledge. The main idea in Piaget's constructivism is that immanent schemes of operations can become objects for abstraction and made into explicit logicomathematical patterns of thought. In this way, the architecture of mathematics is constructed. This preoccupation with mathematical knowledge is characteristic also in the further development of constructivism. Reflective knowledge has as its object the use of mathematics, and therefore it becomes important to step outside the cathedral of formal knowledge to take a more general view of this construction.

Let us try to take some steps in *reflective knowing*, but now without any guarantee that the steps lead into the center of a democratic competence nor into the

[15] See also Davis (1989).

[16] A similar conclusion has also been pointed out by Mogens Niss: "It is of democratic importance, to the individual as well as to society at large, that any citizen is provided with instruments for understanding the role of mathematics [in society]. Anyone not in position of such instruments becomes 'victim' of social processes in which mathematics is a component. So, the purpose of mathematics education should be to enable students to realize, understand, judge, utilize and also to perform the application of mathematics in society, in particular in situations which are of significance to their private, social and professional life" (Niss, 1983, p. 248). Of special importance is also the discussion in Ernest (1991) of the public educator program.

concept of reflective knowledge in its general formulation related to the evaluation of technologies in society. We are only talking about steps put in some analytical order, not about the steps which actually may be taken by children and students (when I talk about the first, the second, etc., step, it just has to be read as different steps). We may hope that this may provide a bit more educational meaning to "mathemacy."

A *first* group of questions formulated by pupils, students, and teachers about their work in the mathematical classroom could be, Have we done the calculations right? Have we followed the algorithm accurately? Are there different ways of controlling the calculation? Such questions are all addressing the mathematical aspects of the problem-solving process, and any attempt to answer these questions immediately moves us back into the area of mathematics. Still, they may be seen as rudimentary steps in making reflections about what has been done. In school, all metareflections seem to be concentrated on questions of this type. The dominance of such questions supports also the widespread true-false ideology which is incorporated in much of school mathematics. This ideology tells that any answer to a problem or exercise must either be right or wrong. These are the only possibilities when we do mathematics.

A *second* step has to be taken, and we could ask questions like, Have we done the right calculation? Is it possible to choose between different algorithms? Is the algorithm reliable in all circumstances? Is it sound? More generally we could ask, Have we used an appropriate algorithm? Is the algorithm reliable for pursuing our aims?

Metareflections need not be confined to the correctness and consistency of the methods used; we could take a *third* step. Reflections could have to do with the reliability of the solution in the specific context. Even if the calculations are done correctly and the accountability of the techniques is established, it need not be the case that the result should be trusted. We could ask questions like, Even if we have calculated in the right way and used algorithms in a consistent way, do we then find a result which we can actually use? Are the results reliable for the purpose we have in mind? This third group of questions starts attacking the true-false dichotomy and takes into consideration the context of the use of mathematics. The questions have to do with means and aims. In this case, we are looking at the technological aspect, while at the first two steps we have been addressing the mathematical tools. If it is possible for students and children to raise questions of this type in school, it is important that mathematics is contextualized in a way that they see the value in such investigations.

This turns us towards a *fourth* step in developing reflective knowing. We could raise questions like, Is it appropriate to use a formal technique at all? Do we in fact need mathematics? Is it important to introduce a formal method? Could we find out without mathematics? Is the result based on a mathematical calculation more or less reliable than intuitive interpretations of the situation in question? These questions draw attention to the fact that formal techniques and mathematics need not be necessary tools for obtaining the technological aim. In some cases, an intuitive way of handling a specific problem may be preferred. It is an important experience for the children that they sometimes are able to find out without mathematics. The fourth group of questions attacks that variant of the true-false ideology which tells that formal methods must be preferred. Formal methods may reach further in some situations, but they do not always work to give an appropriate answer. By contrasting formal techniques with intuitive ones it becomes possible to see formalization as only one possible way of handling a problem, and this experience is important in developing reflective knowing. The question, sometimes raised after a successful contextualization, is where has the mathematics gone? It sometimes seems to disappear in the middle of a project. But this could also be interpreted as a positive condition for reflections on the necessity of formal tools.

We take a *fifth* step towards reflective knowing when we look for broader consequences of the use of specific techniques for solving a problem. In steps three and four, the reflections concentrate on the technological aims of the task, but now we look outside the original aim for our action and we try to find the general implications of pursuing such a task with formal means. How does the application of an algorithm affect our conception of a part of the world? This is the question addressing the formatting power of mathematics. But how do we handle this question in an educational context? In taking our fifth step, we must not forget that we still try to be settled in a classroom. However, before coming back to this question, we shall take a step even further.

The *sixth* step is to try to reflect upon the way we have reflected upon the use of mathematics. Reflective knowing must address its own status. That rounds off our way towards reflective knowing.

Let me summarize the six steps by key questions, keeping in mind that the steps are logically ordered but not more. (1) Have we used the algorithm in the right way? (2) Have we used the right algorithm? (3) Can we rely on the result from this algorithm? (4) Could we do without formal calculations? (5) How does the actual use of an algorithm (appropriate or not) affect a specific context? (6) Could we have performed the evaluation in a other way? Questions (1) and (2) focus on the mathematical tool; questions (3) and (4) on the relation between the tool and the task; question (5) on the general effect of pursuing the aim with the chosen tool; and (6) looks at the way we have been looking at (1) to (5).

By the fifth step, we definitely seem to have left the classroom. Does it make sense to try to illustrate how mathematics is formatting society? Could it make sense at all to introduce the idea of the formatting power of mathematics in elementary education? Let us try to imagine that children/students are involved in a project about child benefits, and let us imagine that, as a subproject, we try to look at more fundamental questions about child benefits. The actual payment of child benefits in Denmark has varied according to different sets of rules, so let us try to participate in these considerations. Let a specific amount of money—M—be fixed. How might we distribute that amount between a number—F—of families?

Depending on the age of the children/students involved in the project, the questions can be specified in different ways. The number of families could be so small that a description of each family could be given: income of the parents, number of children, jobs, age, and such. The children/students could then be asked to suggest reasonable ways of distributing child benefits. One answer could naturally be $cb = M/F$, but at least only families with children should receive child benefits. And the amount of money given to each family must naturally depend on the number of children in the family. And what about the income of the family? What does it mean if a parent is living alone with the children? And so on.

The important question is, of course, what sort of information about the families is to be used. At one extreme, only a very little; at the other, the whole situation of the family could be taken into account. To design a formula in the first case is not so difficult, but to imagine a design in the second becomes nearly impossible. A design of an algorithm presupposes that some simplifications have to be made, and that could be realized by the children/students during a project. They could get an experience of the difficulties raised by bringing more and more aspects into a formal system. The discussions will inevitably change from what is reasonable and fair to what is possible relative to the technological tool. And to get an impression and understanding of that sort of switchover is quite fundamental to an understanding of the role of formal methods in today's societies.

When it is decided that some sort of formal methods are to be used, then it is also decided that only a limited set of factors can be taken into account. The children/students could come to realise some of the basic features of the formatting power of mathematics and also that in some cases, no alternative to the process of formatting exists, although alternative ways of formatting exist. Once an algorithm for the distribution of money is fixed, a little part of social life is also fixed. The formatting then becomes hard frozen reality for lots of people. These considerations indicate what it could mean to illustrate, at a rather basic level, that mathematics is formatting society. However, I am not asserting that this may imply pedagogical success. I have just tried to indicate in what direction we may find some educational meaning to the idea that mathemacy is of importance in today's technological society as part of a general democratic competence.

Conclusions

Driving skill demonstrates only technological competence, but it is crucial to know *about* motoring and its consequences. Such metareflections are important to society, and that presents us with an educational task, in fact, environmental education is becoming more and more widespread. In the same sense, the guiding principles for mathematics education have not any longer to be found exclusively in pure mathematics nor in applied mathematics but in a broader perspective aiming at reflective knowing. It is possible to connect empowerment and mathematics education to the degree that mathematics education may help in clarifying the formatting role of formal methods in society. Empowerment is not connected to any isolated ability in carrying out mathematical calculations as such, but to an understanding of how mathematics is applied and functioning. To be able to carry out democratic obligations and rights, it is necessary to be able to understand the main principles in the mechanisms of the development of society. We have to know about the risk structures which accompany social development (and some of these risk structures are in fact created by formalization).[17] We must be able to see the constructive as well as the destructive forces linked to technological development. Especially, we must be able to understand what mathematics does to society. That provides (some) sense to the formulation: Mathemacy as a radical construct has to be rooted in a spirit of critique and project of possibilities that enables people to participate in the understanding and transformation of their society.

I see this as a condition for developing the dimension of participation in democracy. If people are to be not only receivers of information and instructions but also able to criticize, evaluate, understand, that is, to provide an input to the democratic institutions, then they must get an understanding of some of basic structuring principles in society. And ideological structure is no longer the only structure of importance—formal structures play a role hand-in-hand with ideological formattings. That is the reason that I find that mathemacy and literacy have similar roles to play. I am not saying that these are the only formatting powers in society (my discussion has not touched on the economy), but the intention has been to see parallels between literacy and mathemacy. Perhaps it is a recent historical fact that mathematics education can play a critical role having to do with the nature of the formattings of today's societies—"mathemacy" can now become a critical power. That gives (analytical) *meaning* to the project of relating mathematics education and democratic development, although the project may not be realizable. It is by no means obvious that education can be turned into a strong, progressive social force. However it *is* possible.[18]

NOTE

This paper was first published in 1992 in *For the Learning of Mathematics 2*(2), 2–11.

---

[17] See Booss-Bavnbek (1991).

[18] It has to be noticed that the line of analysis in this paper is far from the only one possible. For instance, although reflective knowledge can be related to reflections about the use of mathematics in society, it can also, in an educational context, be related to reflections about what is happening in mathematics classrooms. Further, I have analysed the technological development as a threat to democratic development, but it is obvious also that information technology has a democratic possibility as it can facilitate the distribution of information. However, this argument does not exclude the dangers built into the technological development.

REFERENCES

Bell, D. (1980). The social framework of the information society. In T. Forester (Ed.), *The microelectronics revolution. The complete guide to the new technology and its impact on society* (pp. 500–549). Oxford, UK: Blackwell.

Benn, S. I., & Peters, R. S. (1959). *Social principles and the democratic state*. London, UK: Allen and Unwin.

Booss-Bavnbek, B. (1991). Against ill-founded, irresponsible modelling. In M. Niss, W. Blum, & I. Huntley (Eds.), *Teaching of mathematical modelling and applications* (pp. 70–82). Chichester, UK: Ellis Horwood.

Bowles, S. (1983). Unequal education and the reproduction of the social division of labour. In B. Cosin & M. Hales (Eds.), *Education, policy and society* (pp. 27–50). London, UK: Routledge and Kegan Paul.

Davis, P. J. (1989). Applied mathematics as social contract. In C. Keitel, P. Damerow, A. Bishop, & P. Gerdes (Eds.), *Mathematics, education and society* (pp. 24–27). Paris, France: UNESCO.

Davis, P. J., & Hersh, R. (1988). *Descartes' dream*: *The world according to mathematics*. London, UK: Penguin.

Ernest, P. (1991). *The philosophy of mathematics education*. Brighton, UK: Falmer.

Frankenstein, M. (1987). Critical mathematics education: An application of Paulo Freire's epistemology. In I. Shor (Ed.), *Freire for the classroom: A sourcebook for liberatory teaching* (pp. 180–210). Portsmouth, NH: Heinemann Educational.

Frankenstein, M. (1989). *Relearning mathematics: A different third r—Radical maths*. London, UK: Free Association.

Giroux, H. A. (1989). *Schooling for democracy: Critical pedagogy in the modern age*. London, UK: Routledge.

Held, D. (1987). *Models for democracy*. Cambridge, UK: Polity.

Keitel, C. (1989). Mathematics and technology. *For the Learning of Mathematics*, *9*, 7–13.

Niss, M. (1983). Considerations and experiences concerning integrated courses in mathematics and other subjects. In M. Zweng, T. Green, J. Kilpatrick, H. Pollak, & M. Suydam (Eds.), *Proceedings of the Fourth International Congress on Mathematical Education* (pp. 247–249). Boston, MA: Birkhäuser.

Peters, R. S. (1966). *Ethics and education*. London, UK: George Allen and Unwin.

Purvis, J., & Hales, M. (Eds.). (1983). *Achievement and inequality in education*. London, UK: Routledge and Kegan Paul.

Rousseau, J.-J. (1968). *The social contract*. London, UK: Penguin. (Original work published 1762)

Schumpeter, J. A. (1987). *Capitalism, socialism and democracy*. London, UK: Unwin Paperbacks. (Original work published 1943)

Skovsmose, O. (1989a). Models and reflective knowledge. *Zentralblatt fur Didaktik der Mathematik*, *21*(1), 3–8.

Skovsmose, O. (1989b). Towards a philosophy of an applied oriented mathematical education. In W. Blum, J. S. Berry, R. Biehler, I. D. Huntley, G. Kaiser-Messmer, & L. Profke (Eds.), *Applications and modelling in learning and teaching mathematics* (pp. 110–114). Chichester, UK: Ellis Horwood.

Taylor, F. W. (1947). *Scientific management*. New York, NY: Harper & Brothers.

CHAPTER 13

MATHEMATICS EDUCATION AND DEMOCRACY

Democracy not only characterizes institutional structures of society concerning distributions of rights and obligations. Democracy has also to do with existence of a competence in society, and it is some of these noninstitutional aspects of democracy we want to discuss in relation to mathematics education.

We concentrate upon democratic problems in a highly technological society, that is, a society based on a total integration of, for instance, information technology. This integration seems to imply that decisions and discussions about change and development always must be related to a technological insight. That situation brings about what I conceive as the problem of democracy in a highly technological society. That means that I, as an example, have the situation in a country like Denmark in mind.

Several times it has been underlined that mathematics education has a political dimension (see Mellin-Olsen, 1987), and this thesis could, naturally, become further specified. We could argue that mathematics education, in a traditional setting, will favor a certain group of students; that mathematics education will make a strong stratification of the students; or that mathematics education will serve as an introduction to an ideology characterized by rationalism and objectivism (see Bishop, 1988).

Critique as Uncertainty, pages 181–196.

The main problems are to what degree is mathematics education involved in the process of building up (or reducing) a democratic competence in society? Is it possible to develop the content and form of mathematics education in such a way that it may serve as a tool of democratization? Or has mathematics education, perhaps because of its formal and abstract nature, nothing to do with such questions? Or is the situation even worse: Are undemocratic tendencies served by introducing the pupils and students to unrelated bits of knowledge putting the teacher (and the book) in a special role of authority?

The Social Argument of Democratization

It is possible to divide these general remarks into two types of arguments relating mathematics education and democratization. The first one is, for example, stated in Bjørneboe and Nissen (1984), and in Niss (1984). In short, we summarize it as the social argument of democratization. The next argument, the pedagogical argument of democratization, will be summarized in the following section.

The social argument tries to identify a relevant subject matter of (mathematics) education via reflections on possibilities for building up and improving democratic institutions and democratic capabilities in society by improving the content of the education. The social argument of democratization is composed of three statements:

Mathematics has a very extensive range of applications. Mathematics is applied in economics (macroeconomics and microeconomics), in industrial planning, in different forms of management, in marketing, as well as in all traditional fields of application in technology. However, often it is difficult in both primary and secondary school teaching to present illustrative examples of real applications; too often examples show pseudoapplications. Real applications of mathematics are normally "hidden," although extensive and important.

Because of its applications, mathematics has a "society-shaping" function. Mathematics constitutes an integrated and unique part of society. Mathematics cannot be replaced by any other tool serving similar functions. It is impossible to imagine a development of society of the type we know without technology playing a major role and with mathematics playing a dominant role in the formation of technology. So, mathematics has important implications for development and organization of society, although those implications are difficult to identify.

To make it possible to carry out democratic obligations and rights, it is necessary to be able to understand the main principles in the mechanisms of the development of society, although they may be hidden and difficult to identify. Especially, we must be able to understand the functions of applications of mathematics. For instance, we must understand how decisions (economical, political, etc.) are influenced by mathematical model building processes.

The social argument of democratization highlights the applications of mathematics, and the importance of mathematical model building activity is in fact often emphasized in educational literature. The basic idea in what we may call the pragmatic trend in mathematics education is that it is most important for students to learn about model building, and the best way of learning this is to build models. The pragmatic trend builds upon a philosophical assumption about mathematics, stating that an essential feature of mathematics is its usefulness (quite contrary to structuralistic and formalistic philosophy, stating that the essential feature of mathematics is its "logical architecture"). Next, the pragmatic trend incorporates the assumption that the best way of learning is by doing; and especially, the best way of learning model building is to practice a model building activity.

However, my thesis is that fundamental problems concerning the application of mathematics are not visible from inside the modeling process. That means, it is not possible to develop a critical attitude towards the application of mathematics solely by improving the modeling capability of students. Fundamental problems concerning modeling cannot be formulated in the conceptual framework that students develop through practical experiences. That means the knowledge mentioned in the social argument of democratization (understanding the functions of the application of mathematics) is not normally developed in a pragmatic educational process. Therefore, an educational practice aiming at democratization by improving the students' possibilities for criticizing model building activities cannot be solely pragmatic. It is not sufficient to become a model builder. (Later in this chapter, we will try to investigate the epistemological background for this conclusion by relating and developing the concepts of democratic competence and critical and reflective knowledge.)

However, educational practice can anticipate some of these difficulties, and let me briefly sketch the content of *Beskæftigelsesmodellen i SMEC III*, written in 1982 by Mogens Niss and Kirsten Hermann, a sort of mathematical textbook with upper secondary school in mind. The intention of the book is in accordance with the social argument of democratization. The authors find it important for students to learn about real model building activities, not primarily to increase the motivation of the students (motivation seen as something emotional) nor to serve as an entrance to a piece of mathematical theory, but primarily to give students opportunities to investigate different details in a model which in fact has important social implications.

The "Simulation Model of the Economic Council" (SMEC) has been used by Danish economists when advising the government and politicians concerning economic policy and its possible consequences; and SMEC III is the third version. For instance, SMEC relates the number of unemployed to different economic factors, some of which are possible to influence by political decisions. That means that SMEC can be used in forecasting possible consequences of different political strategies, and SMEC has in fact had an important role to play in setting economic policy in Denmark.

It is hard to understand how the economic factors are interrelated and difficult to see how these relationships are built into the mathematical model, but this is explained in the textbook.

However, to develop a more critical attitude towards this model building, we have not only to understand the mathematical construction of the model; we have also to know about its assumptions. We must be able to point out which economical ideas are hiding behind the curtain of mathematical formulas.

It is impossible to start model building without assumptions. A special choice must be made about how to conceive the economic reality. Our conception of reality must be so structured that specific patterns could be identified: We have to select elements from reality which are to be conceived as important, and we have to decide which relationships among these elements are important. These two fundamental selections constitute an interpretation of "reality." A model is not a model of "reality" as such. It is a model of a conceptual system, created by a specific interpretation, based on a more or less elaborated theoretical framework and based on some specific interests.

To identify the "society-shaping" function of SMEC, we have to point out the models' selections of elements from reality and how it interrelates these elements. In SMEC, the Cobb-Douglas function of production plays a crucial role. According to this, we have $Y = (L, I)$, that is, the national product Y is a function of the labour force L and capital investment I.

If we are going to discuss SMEC—and not only to investigate the relationships of its different equations—we must ask questions like, What assumptions have made it possible to create the conceptual system which forms the initial step in the modeling process? Could the assumptions be verified in any independent way? What theoretical frameworks are in accordance with these assumptions? What alternative theoretical frameworks do exist? And especially, What theoretical analysis supports the assumptions built into the Cobb-Douglas function of production? To develop a critical knowledge related to SMEC, we have to ask questions that lead behind the assumptions in the Cobb-Douglas function. This is a necessary condition for understanding the political function of the use of the model; we have to trace its roots back into neoclassical economy. This could indicate some of the political functions of the use of the model.

The textbook *Beskæftigelsesmodellen i SMEC III* is of great help in the formulation of these questions. However, the use of such a textbook is not without problems. It is in fact very difficult for the students to understand the details in the text. The number of simplifications cannot be too great, because the intention of the textbook is to present a real model.

Let us summarize the main features of a teaching-learning material like *Beskæftigelsesmodellen i SMEC III*, that is, teaching-learning material which tries to be in accordance with the social argument of democratization:

- The material has to do with a real mathematical model.
- The model has to do with important social activities in society.
- The material does in fact develop an understanding of the mathematical content of the model, but this more technical knowledge is not the goal. It

is to develop an insight into the assumptions integrated into the model and by this to develop an understanding of processes (for instance processes of decisions) in society.

Teaching-learning materials characterized like this we call *empowering teaching-learning materials*. However, the material need not take the shape of a specific textbook, and more generally we could talk about *empowering teaching-learning situations*.

The Pedagogical Argument of Democratization

The social argument of democratization characterises one aspect of the relationship between mathematics education and democratization. A second aspect could be summarized in the *pedagogical argument of democratization*. Here, the term "pedagogical" must be understood in a broad way. It must be interpreted in relation to the term "social" in the previous argument; but now we look into the educational process (while the social argument looked outside the process). The argument consists of the following statements:

Students receive a variety of impressions during an educational process. Much has to do with the curriculum in question, although often the students do not learn what the teachers expect them to learn. There exists a very big gap between the matter taught and the matter learned. Other parts of what is learned have not much to do with the (official) curriculum but much in common with the structure of the educational process and with the traditions and rituals connected to the subject matter in question.

Also, mathematics education has its "hidden curriculum." Often it is stipulated that mathematics education serves important functions in relation to the students' general epistemological development. It is emphasized that mathematical studies tend to improve the students' abilities to structure and to solve logical problems. However, the rituals of mathematics education take another direction. Students learn (also) to follow explicitly stated prescriptions: "Solve the equation . . . ," "Find the length of . . . , " "Calculate the value of . . . ," and so on. This has not much in common with real processes of investigations or creative ways of structuring problems. It has much more in common with instructions and regulations with which lots of people in routine-like work processes are confronted. Mathematics education socialises (also) into quite another direction than that optimistically presumed in "official" statements about the potential epistemological functions of mathematics education. It serves, too, a function by introducing the new labor force to lots of routine work in the technological society. Education, and especially mathematics education, implants an attitude towards technology. The students learn that some people are able to manage technological problems and that some people are not. And consequently, the "incapable" students learn to become servile towards technological questions and to become servile towards those who can manage.

The possibilities for carrying out democratic obligations and rights are not only related to formal institutionalised democratic structures but also to an individually consolidated democratic attitude. Democratic actions at the macrolevel must be anticipated at the microlevel. That means we cannot expect a development of a democratic attitude if the school system does not contain democratic activities as the main element. If a democratic attitude is to be developed through mathematics education, the rituals of that education cannot contain fundamentally undemocratic features. The dialogue between teacher and students has to play an important role.

The pedagogical argument stresses that mathematics education socializes in quite other directions than expected. Some reasons could be related to the structuralistic movement in mathematics education characterized by the following assumptions: The essence of mathematics can be determined by crystallizing fundamental concepts through a logical analysis of existing mathematical theories, and these fundamental concepts can be conveyed to the learner by means of suitable concretizations in accordance with the epistemological powers of the child. Therefore, basic to structuralism is the idea that the students' knowledge has to be built up in accordance with structures and contents identified independently of the students. And one of the undemocratic features of education is implied by the exclusion of students from curriculum planning.

The possibility of an experience-based mathematics education, an education including the total experience of students both in relation to curriculum planning and in relation to subject matter content, concerns the possible relationships between our ordinary language and constructed mathematical concepts. Are they familiar to each other? Does a close connection exist between language games (to use a term coined by Ludwig Wittgenstein) carrying a mathematical competence and the games of ordinary language? One possibly could be stated as the *thesis of familiarity*. A very smooth and continuous transition exists between ordinary language and the conceptual structures of (school) mathematics.

Opposite to the thesis of familiarity we have the *thesis of dichotomy*: Ordinary language and the language of mathematics constitute two quite different and independent language games. Mathematical concepts are created in a special context, and educational planning is forced to relate the two language games to each other.

The thesis of dichotomy was emphasized by structuralism. The fundamental act of curriculum planning was, via investigations of mathematical structures, to identify fundamental concepts, and on this basis to elaborate a detailed curriculum. Jean Dieudonné, who strongly influenced the reform programs in the 1960s, has pointed out

> If the people responsible for drawing up school curricula may be persuaded to consult professional mathematicians in order to understand the relevance of their decisions to science as it is practiced in the university and beyond, we may yet witness one day some sensible teaching of mathematics from kindergarten to graduate school. (Dieudonné, 1973, p. 19)

Much progressive work in mathematics education is now related to the thesis of familiarity. The intention is to place the children, their interest, their work, and their experimentation in the center of the educational practice and to eliminate undesirable aspects of the hidden curriculum. We find examples in materials developed by IOWO (Instituut voor de Ontwikkeling van het Wiskunde Onderwijs). The examples of educational practice presented in *Five Years IOWO* (1976) show the openness implied by the thesis of familiarity. The basic educational process becomes mathematising, leading from the ordinary language games towards the more regular conceptual structures of the language of mathematics. No specific road has to be planned, because of the familiarity of the language games. I find this thesis crucial in case we want to draw educational implications from the pedagogical argument of democratization.

Further reflections related to the thesis of familiarity are found in the ideas of ethnomathematics. Especially Ubiratan D'Ambrosio has tried to develop the theoretical foundations of ethnomathematics. A social force behind the program is found in the educational mistakes witnessed when introducing "new math" in Third World countries. An idea about a mathematics education adapted to the cultural situation of the country grew up. D'Ambrosio characterizes the situation like this:

> While Western countries have science, technology and modern development and the very concept of progress implicit in their historical evolution, Third World countries have played a subsidiary role in this evolution, and transfer has been allowed or stimulated from Western powers to Third World countries in the measure they benefit. . . . All the rationale of the colonial enterprise, and the apparently distinct discourse of independence and development, are pieces of a game which rules have been and still are dictated by the developed world. (D'Ambrosio, 1984, p. 32)

An important observation is that the well-structured mathematical curriculum could become an obstruction to learning activities. The structure of the curriculum perhaps incorporates undemocratic features as stated in the pedagogical argument about democratization. D'Ambrosio states it like this:

> The "learned" matheracy eliminates the so-called "spontaneous" matheracy. An individual who manages perfectly well numbers, operations, geometric forms and notions, when facing a completely new and formal approach to the same facts and needs creates a psychological blockade which grows as a barrier between the different modes of numerical and geometrical thought. (D'Ambrosio, 1985, p. 472)

The main ideas guiding the ethnomathematical project are (1) It is possible to identify a fundamental but hidden mathematical competence in all different cultural settings. This competence manifests itself in different forms, for instance in craftsmanlike abilities. (2) This hidden mathematical competence could be made explicit as ethnomathematics. (3) It is possible to develop mathematics education based on preestablished ethnomathematical competence. This is underlined by D'Ambrosio in the following:

> We have to learn their language, their logic, their history and their evolution, their science and their technology in order to be aware of their motives and ultimate goals. . . . But at the same time the mathematics in schools shall be such that it facilitates knowledge, understanding, incorporation and compatibilisation of known and current popular practice into the curriculum. In other words, recognition and incorporation of ethnomathematics into the curriculum. (D'Ambrosio, 1984, p. 32)

The ethnomathematical approach incorporates the thesis that a continuous transition exists between ordinary language and conceptual structures of mathematics. Further, this approach stresses that traditional mathematics education often socializes in ways not presupposed in official educational planning; psychological blockades could be established. If we want to break down the hidden curriculum of mathematics education and prevent students from taking up a servile attitude towards technological questions, an educational strategy could perhaps be developed from the ethnomathematical approach characterized by the thesis of familiarity and by opening the teaching-learning situation.

More specific ideas related to the educational practice could, as mentioned, be learned from the work of IOWO, and inspiration from IOWO is found in a great variety of school materials, some produced in Denmark. The intention is not to fix the learning situation, but to create opportunities for the students to make plans for their study activities. A nice example for primary school teaching is found in *Flyv med!* by Peter Bollerslev et al. (1987). The material is to be used for a shorter period, 2–3 weeks. The book has a short introduction to teachers, students, and parents stressing that it is not necessary to build paper airplanes in mathematical lessons, but it is possible to do so. It will face us with a great variety of practical and fundamental mathematical questions. We have to start the production of paper airplanes—lines and geometrical figures on paper. Is it possible to make new constructions? We have to test and to improve our models. We have to systematise the testing results. In the whole process the students have the possibility for making decisions and for being involved in the process of educational planning (although at the microlevel).

A crucial point is the refusal to use prestructured and ready-made materials as developed in connection to the structuralistic movement. Instead, one tries to create situations which could facilitate a mathematisation. The idea of concretization—to be understood as an elementarization of abstract mathematical concepts—is rejected in favor of a mathematization. And we find an important asymmetry between these two sorts of activities. To concretize is to give more abstract terms a concrete interpretation and in this way to make them more comprehensible. The activity of concretization is reserved for the planners of the curriculum (compare the remark of Dieudonné), and as such this activity is removed from the educational process. To mathematise means, in principle, to formulate, to criticize, and to develop ways of understanding. Both students and teachers must be involved in the control of this process, which, therefore, could take the form of a more democratic one.

One of the assumptions underlying these ideas is precisely the thesis of familiarity, which is based on the pedagogical argument of democratization leading to the idea of teaching-learning materials like *Flyv med!* characterized as:

- The material has to do with a topic of subjective relevance for the students.
- The material initiates a variety of activities, not prestructured and fully fixed.
- Several decisions have to be taken when involved in the teaching-learning process, and the decisions normally necessitate a discussion between teacher and students.

Teaching-learning materials and situations characterized like this we call *open teaching-learning materials and open situations*.

A Contradiction

An implication of the pedagogical argument of democratization is that we have to develop open situations in the educational process, that is, situations which could take different directions depending on the results of discussions between students and students, and between students and teacher. Opening the situation means to create possibilities for educational decisions to be taken in the classroom. Students must have the possibility to form the educational process if not to become adapted to unquestionable rituals of mathematics education. Another version of this implication is that we have to develop open teaching-learning materials, which could be used in a variety of situations. Open materials must not presuppose a specific teacher-students relationship or lay down implicit teaching-learning programs.

However, we must not forget about the social argument of democratization. The implication of this is different from the above-mentioned implications. According to the social argument, students have to develop not only pragmatic knowledge about how to use mathematics and how to build up (simple) models but primarily knowledge about the preconditions for the model construction, and this knowledge must be aimed at an understanding of the social functions of applications of "grown up" mathematical models.

An implication of the social argument of democratization is that we have to develop empowering teaching-learning situations and materials, that is, situations and materials which in fact give information about real mathematical models and their functions. If a critical distance to mathematical model building has to be created, if a "curriculum critique" (in German, this key term is *Fachkritik*) should be realized in education, and if we try to ask questions like "Who uses mathematical models? In which situations are they used? What knowledge-constituting interests are connected with the subject matter in question?" then our teaching-learning materials must include potential for the establishment of a critical distance.

And now we have reached what I conceive as a main problem in today's mathematics education. What relationships do in fact exist between, on the one hand, a mathematics education aiming at developing the experiences of the students by opening the situation, by incorporating the decisions of the students into the educational process, and by all the time relating the language of mathematics to ordinary language games and, on the other hand, a mathematics education trying to develop not only a pragmatic but also a critical attitude towards the use of mathematical models?

The problem is that the social and the pedagogical arguments, although both related to the concept of democratization, are pointing in two different directions. Will it become possible to create materials and situations that are at the same time open and empowering? We cannot be sure that our educational enterprise is without contradictions. We face one problem when we succeed in developing open materials (like *Flyv med*): Will it be possible to build up any critical knowledge when occupied in this project? And we face another problem when we succeed in developing empowering materials (like *Beskæftigelsesmodellen i SMEC III*): Will it be possible to avoid too much prestructuring of the situation, too much lecturing, to build up that complicated stock of factual knowledge obviously needed to understand the functions of a real model? Open material could result in open and democratic educational situations, but no empowerment is guaranteed; and empowering material could result in critical understanding, but no openness is guaranteed.

This contradiction has practical implications, but we will try to look at it from an epistemological point of view. Both the social and the pedagogical argument presuppose a concept of democratization, and up till now we have used this concept without any further specification. Likewise, we have talked about a democratic competence without analyzing the concept.

Democratic Competence

Democracy characterizes a means of social control. An unsophisticated interpretation of "democracy" states that the "people" must be "ruling." However, already in ancient Greece the concept needed further interpretation: Who are the "people"? Not only the rich people, but definitely not the slaves, were an answer. And what is the specific meaning of "rule?"

If the "people" means everybody, adult women and men, and "ruling" means actually participating in the act of taking decisions, we have the idea of "direct democracy." In the continuous discussion about democracy, this idea often is underlined as the ultimate aim, but at the same time as a utopia. For instance, Rousseau was arguing in favor of direct democracy.

If we give up the idea of direct democracy and try to find a more feasible interpretation, we will face the problem of "transformation of sovereignty": How is it possible to combine democracy with the necessity of selecting a small group of people to actually do the ruling? How is it possible to control the "people in charge?" The necessity of the transformation of sovereignty is implied by the fact that ruling presupposes specific qualifications that are not of common nature. The people in charge must have a specific knowledge about their ruling subject. Perhaps a specific education is needed. The people in charge must possess a competence, including information and knowledge.

We have to make a distinction between the competence which the people in charge must possess if they are to be able to take well-founded decisions and act in an appropriate way, and the competence which is presupposed if we have to judge if the results and consequences of the ruling are acceptable. This makes obvious what is *the basic assumption in the classical interpretation of democracy.* While the ruling-competence of the people in charge is of special nature, the judging-competence is of common nature. The latter we also call democratic competence. In other words, democratic competence is a common capacity of human beings, but perhaps only a potential capacity because only a certain attitude will stress the importance of a democratic way of social control.

An epistemological way of arguing for the basic assumption behind the classical interpretation of democracy is the following: The judgment of an act carried out by the people in charge is a moral or ethical judgment, and ethics need no foundation in facts. An ethical statement is not implied by any conjunction of factual statements. Norms do not rest on any state of affairs, neither is an understanding of norms restricted to any small group possessing certain qualifications. To argue in favor of a normative statement, only rationality is needed, and rationality is of a common nature.

This classical interpretation of democratic competence we may also call rationalistic or idealistic. This has much in common with classical rationalism in philosophy that refuses every sort of dogmatism. We need not believe any statement unless we have realized that it has to be so. No statement need to be conceived as true because of what authority tells. The only authority we have to believe is our own capacity of thinking.

In opposition to the classical interpretation, we put a *materialistic interpretation of democracy*, which maintains that democratic competence in no way rests upon the inner nature of man. It is not a part of a common rationality. Instead, the basic assumptions are that democratic competence has to be developed. The competence is closely related to a democratic attitude, but they are not identical. The development of a democratic competence presupposes an attitude, but in addition, lots of knowledge and information about the domain of the democratic processes has to be developed. The content of knowledge and information built into a democratic competence is not anything fixed. It will change in accordance with the development of society. That means that democratic competence is a socially developed characteristic of the competence which people to be ruled must possess so they can be able to judge the acts of the people in charge. This competence will vary in accordance with the structures in society.

A specific variant of the materialistic interpretation sees the problem of developing a democratic competence as an educational problem. This point of view was especially stressed in German pedagogy after World War II.

Before we go into more details about the educational variant, it must be underlined that another change in the concept has taken place. Originally, democracy has to do with ruling a society. More generally, democracy describes a sort of social control of an organization. The organization could be large or small; the people in charge could be a government as well as the leaders of a minor company. However, the problems of judging the acts of the people in charge will be the same, from a logical point of view.

I agree with the educational variant of the materialistic interpretation of democratic competence, but it needs further specifications. I find that a certain school subject plays an important role in this context, namely, mathematics. (This assumption underlies what previously has been stated in the social and pedagogical argument of democratization.)

The argumentation for the specific role of mathematics needs to be developed. The materialistic interpretation stresses that the content of a democratic competence is determined by the structure of the "domain of the ruling." The content of the competence depends on the content of the fundamental characteristics of the organization in question. For a determination of this characteristic, the concept of technology plays a crucial role. Society and technology are integrated, and technology becomes the dominating feature of civilization. This idea deals with power and power relations integrated into technological structures. The consequence is that all sorts of decisions concerning society or organizations in society also concern technology.

The relationship between types of technologies and sciences varies very much. Manual tools are developed quite independently of science. The development of steam engineering takes place parallel with the theoretical understanding of thermodynamics questions, while the information technology is totally based on development in mathematics. Mathematics is of absolute importance for the development of today's technology. That mathematics constitutes a unique part of today's technological knowledge is especially stressed in *Mathematik als Technologie* edited by Jürgen Maass and Wolfgang Schlöglmann (1989).

A society based on advanced technology faces a specific problem of democracy. If a society is based on manual tools, the idealistic interpretation of democratic competence becomes plausible; no specific technological knowledge is needed to evaluate the acts and decisions of the people in charge. Quite the contrary occurs in a highly technological society. The content of democratic competence is rapidly changing towards a tremendous complexity. On the face of it, only a limited group of people seem to be able to manage this complexity. In fact, this competence seems to presuppose a certain amount of technological knowledge, including mathematics. The consequences seem to be that only a limited group of people can obtain a democratic competence and then become able to evaluate the actions of the people in charge. This is the problem of democracy in a highly technology society.

Epistemology

Our analysis of the problem of democracy seems to imply that technological knowledge has to be developed at all levels in the educational system. And we have to enlarge mathematics education as an integrated part of technology. In a highly technological society, mathematical competence seems to constitute a main part of democratic competence.

However, from an epistemological point of view, this conclusion is problematic. First of all, we have to make a distinction between technological knowledge and reflective knowledge. It is my thesis that the type of knowledge we have to apply in developing technology is different from the knowledge necessary in analyzing and evaluating technological constructions. Making this distinction, I disagree with an epistemology operating with one sort of (scientific) knowledge as supposed by logical positivism. Also, I disagree with the assumption that a distinction can be made between making descriptions and prescriptions, founding descriptions on facts and prescriptions on emotions and private options. Descriptive knowledge must be incorporated into a well-based normative statement, and, in contrast, a pure descriptive statement is an illusion. The distinction between technological knowledge and reflective knowledge does not recapitulate this classical distinction.

Next, it is my thesis that democratic competence to a great extent is based on reflective knowledge. That means that although technology plays a crucial role in the formation of society, it is not the technological knowledge as such which constitutes the democratic competence. Therefore, the main problems are How are technological knowledge and reflective knowledge interrelated, although different? And how is it possible to develop reflective knowledge?

In fact, the integration of mathematics into technology makes an additional distinction necessary. We have to do with three different types of knowledge related to a process of mathematical modeling:

- Mathematical knowledge itself.
- Technological knowledge, which in this context is knowledge about how to build and how to use a mathematical model. Also, we would call it pragmatic knowledge.
- Reflective knowledge, to be interpreted as a more general conceptual framework, or metaknowledge, for discussing the nature of models and the criteria used in their constructions, applications, and evaluations.

The conceptual framework for discussing and criticizing a mathematical model cannot be reduced to the framework gained from a modeling process. In short, reflective knowledge cannot be reduced to technological knowledge. It has a different nature. Reflective knowledge does not have its epistemological basis in technological and pragmatic knowledge. Neither can technological knowledge be reduced to mathematical knowledge, an idea which is expressed in an educational context as when you learn mathematics you also learn how to apply it, an assumption strongly criticized by the pragmatic trend in mathematics education.

When we make a distinction between different types of knowledge, a basis for the distinction making must be specified. One possibility is to look at the *object for the knowledge*. The empiricistic doctrine, that all knowledge is of the same nature, rests on the idea that all sorts of knowledge has to do with the same sorts of objects, namely, sense data.

Instead of the object, we could use the *process of the development of knowledge* in an attempt to argue for the existence of different types of knowledge. Looking at this process, we can make a distinction between a *monological* and a *dialogical* interpretation. A classical monological theory of epistemology is empiricism. To obtain knowledge is seen as an individual process; it is not necessary to interact with others. The source of knowledge is the senses. Likewise, the rationalistic epistemology is monological. Knowledge is produced by "individual thinkers"—the thinking must be clear and (mono)logical.

A modem variant of monologism is the genetic theory of epistemology, developed by Piaget, combining the empiristic and the rationalistic approach into a specific theory of action. Reflections on the results of the individual's operation on objects move the individual towards mathematical insight. Mathematics gets an empirical base in the operations on objects and a rationalistic base in the reflections on these operations.

Activity theory I see as a big step towards a dialogical epistemology, first of all stressing that we have to do something to obtain knowledge and do it in a social context. Activity is not an isolated act.

In a dialogical epistemology, the dialogue and the discussion come to play a crucial role. The main idea is simple: My knowledge is inadequate, it can be improved. But you are in the same situation. To improve our understanding, to move in the direction of more knowledge, we depend on each other. I cannot tell you anything true; neither can you tell me anything. But if we interact in a dialogical relationship, we will be able to move in the direction of more knowledge. The condition for obtaining knowledge is not that we get additional true information but that we interact in a unique way, characterized as a dialogical relationship.

I do not find a dialogical epistemology to be of general validity. Some sorts of knowledge are not primarily based on dialogue, instead a strong interaction with nature could be needed, but some sorts are. And what I want to maintain is the dialogical nature of reflective knowledge.

The dialogical element in mathematical and technological knowledge is more casual. In fact, in traditional mathematical teaching, the students are "informed" by the teacher. Often, no features of an authentic dialogue are to be identified. This will be underlined by the following remarks of Paulo Freire:

> Through dialogue, the teacher-of-the-students and the students-of-the-teacher cease to exist and a new term emerges: teacher-student with students-teachers. The teacher is no longer merely the-one-who-teaches, but one who is himself taught in dialogue with students, who in their turn while being taught also teach. They become jointly responsible for a process in which all grow. (Freire, 1972, p. 53)

To summarize my main points, reflective knowledge is characterized by its complex object and by its dialogical nature.

This analysis separates reflective knowledge from mathematical and technological knowledge, and, for instance, it becomes obvious that pragmatism in mathematics education will be insufficient to establish a critical distance. New features of the educational process have to be developed (because of the dialogical nature of the processes behind reflective knowledge), and also a new content has to be developed (because of the special nature of the object for reflective knowledge).

Turning back to the social argument of democratization, we find it problematic. It stresses, in the right way, that democratic competence cannot be constituted by certain (democratic) attitudes alone. But no attention is paid to the dialogical aspect of knowledge production. The pedagogical argument stresses that the way of production is important, opening the teaching-learning situation to make room for a dialogue is necessary. However, no way of specifying the content of the dialogue is outlined, and reflective knowledge is not created automatically in an open dialogue. Reflective knowledge is not without an object.

It is now possible to describe the epistemological task. We have to bring about an epistemological theory integrating an analysis of the dialogical way of production of knowledge with an analysis of the complexity of the object of reflective knowledge into an educational theory. And in parallel with this, open and empowering teaching-learning materials have to be developed to get as much educational experience as possible to guide the development of theory.

NOTE

This paper was first published in 1990 in *Educational Studies in Mathematics*, *21*(2), 109–128. The original title of the publication was "Mathematical Education and Democracy," however I prefer the term "mathematics eduation."

REFERENCES

Bishop, A. (1988). *Mathematical enculturation. A cultural perspective on mathematics education.* Dordrecht, The Netherlands: Kluwer Academic.

Bjørneboe, J., & Nissen, G. (1984). Den taberproducerende matematikundervisning [The looser producing mathematics education].*Uddannelse*, *8*, 469–479.

Bollerslev, P., Harbo, V., Hartz, V., Olesen, P., & Petersen, L. Ø. (1987). *Flyv med!* [*Flying*] Copenhagen, Denmark: Gyldendal.

D'Ambrosio, U. (1984). Socio-cultural bases for mathematics education [draft], UNICAMP, Universidade Estadual de Campinas.

D'Ambrosio, U. (1985). Mathematic education in a cultural setting. *International Journal of Mathematical Education in Science and Technology, 16*, 469–477.

Dieudonné, J. A. (1973). Should we teach "modern" mathematics? *American Scientists, 61*, 16–19.

Freire, P. (1972). *Pedagogy of the oppressed.* Harmondsworth, UK: Penguin.

IOWO. (1976) Five Years. IOWO. *Educational Studies in Mathematics, 7*(3).

Maass, J., & Schlöglmann, W. (Eds.). (1989). *Mathematik als technologie?* [*Mathematics as technology*]. Weinheim, Germany: Deutscher Studien Verlag.

Mellin-Olsen, S. (1987). *The politics of mathematics education.* Dordrecht, The Netherlands: Reidel.

Niss, M. (1984). Kritisk matematikundervisning—nødvendig men vanskelig [Critical mathematics education—Necessary but diifcult]. *Unge Pcedagoger, 4*, 21–29.

Niss, M., & Hermann, K. (1982). *Beskæftigelsesmodellen i SMEC III* [The employment model in SMEC III]. Copenhagen, Denmark: Nyt Nordisk Forlag Arnold Busck.

PART 4

MATHEMATICS AND POWER

CHAPTER 14

MATHEMATICS AS DISCOURSE

Statements can be revealing; postulates can be challenging; questions can be amusing. Statements include acting: One can express doubt, superiority, uncertainty, anger, affection, and such. One can do a variety of things through speech-acting. However, there are limits, as the grammar of language guides this acting. Here, grammar has to be understood broadly, not only as a set of rules for formulating correct sentences, but also as a deep-grammar, which embodies a worldview. Thus, what can be said depends on assumptions, ideas, and presumptions that might be engraved in the structure of language. A language provides formats for what can be expressed and thought.

When one wants to keep all such language complexities in mind, one may talk about discourses. There are political discourses, racist discourses, and religious discourses. There are discourses when one is dealing with used cars, when doctors talk with patients; when teachers talk about students. Discourse shapes actions, and simultaneously, it may include a range of questionable assumptions. A discourse cannot be assumed to provide a neutral representation of a domain in question. It represents a way of talking about it, a way of perceiving it, and a way of acting with respect to it. In fact, a discourse is part of the constitution of the domain itself. A discourse not only constitutes a worldview, it constitutes the world as well. A discourse annihilates the distinction between the discursive and nondiscursive, as discursive structures turn real.

Critique as Uncertainty, pages 199–214.

Mathematics

In the following, I want to show that it makes sense to consider mathematics a discourse. Naturally, this does not imply that mathematics needs to be only a discourse. But whatever else mathematics might be, we now concentrate on seeing it as discourse.

Mathematics can refer to school mathematics, engineering mathematics, applied mathematics, everyday mathematics, academic mathematics, and various forms of ethnomathematics. Any such form of mathematics can be seen as discourse. However, in the following, I will concentrate on mathematical modeling as applied in engineering, economics, management, business, and so on, meaning that I will concentrate on mathematics as part of technology, having a broad interpretation of technology in mind.

Together with Keiko Yasukawa and Ole Ravn, I have worked out examples illuminating the formatting power of mathematics and mathematics in action. We have discussed how mathematical models provide strategies for overbooking, and for companies' pricing policies and marketing in general; how mathematics is crucial for modern cryptography, and for defining "trust" in general; how mathematics provides interpretations of the ecological state of the world; how mathematics composes part of the processing of knowledge and information; how mathematics has become a defining element of modern warfare. Such examples bring us towards an interpretation of mathematics as discourse.[1]

In the following three sections, I will explore such an interpretation of mathematics by showing that mathematical modeling does not represent, but *re-present*; that modeling is not objective, but that it invents its objects; and that it is not descriptive, but performative. These three sections are partially overlapping; in fact, one can see them as making the same point in slightly different ways. Regardless, they serve as preparation for characterizing mathematics-based discursive acts, which embody the crucial elements in an interpretation of mathematics as discourse. This brings us, in the final section, to address symbolic power as exercised through mathematics.

Modeling Does Not Represent, It Re-Presents

A certain terminology has been developed with respect to mathematical modeling, often supported by a schematic illustration including four elements. The first element is reality, repeatedly depicted as a nice cloud. This reality includes a problem that needs to be addressed. The next element in the illustration is the mathematical model through which the cloudy reality is represented. This model is often depicted as a squared box, indicating that the model includes a range of

[1] See, for instance, Christensen (now Ravn), Skovsmose and Yasukawa (2009); Skovsmose and Yasukawa (2009); and Skovsmose and Ravn (2011). My previous writings about the issues are, for instance, found in Skovsmose (2005, 2009, 2010a).

simplifications. The third element in the figure is squared as well, and it represents the calculations based on the mathematical model: this is the identified mathematical solution to the problem. The fourth and final element is a bit cloudy as well; it refers to the real solution to the problem based on an interpretation of the mathematically calculated solutions. Naturally, modifications of the model are necessary, and the modeling process could include many loops.

This schematic illustration can be called the *canonical portrayal of mathematical modeling*. I am not going to say that this is basically wrong. But I find that this portrayal provides a simplified perspective on the modeling process. Let me point out one particular feature of this simplification related to the notion of representation.

The notion of representation includes a range of philosophical assumptions, some of which can be related to Descartes' outlook and to his rationalist approach in general. Descartes claimed that the world was composed of two different substances, namely, *res cogitans* and *res extensa. Res extensa* stands for the material world, while *res cogitans* represents the human spirit. Thinking and knowledge belongs to *res cogicans*, while *res extensa* remains passive. How then to think about reality? How to formulate knowledge about reality? According to Descartes, a representation is necessary. Elements of the material world need to be represented by symbols in *res cogitans* in order to be part of intellectual processes and rational thinking about reality. Knowledge about reality depends on representations.

Out of this matter-mind dualism developed a particular understanding of language according to which language is a flexible medium for representation. Descartes was not addressing language in any explicit way. This was to come later. But still we can see an understanding of language as emerging from his dualism. The two basic assumptions of this understanding are (1) language and reality are two independent entities; thus reality is not affected by being represented by language. (2) An affinity can be established between reality and language, as language can provide a more or less accurate representation of reality. These two assumptions have accompanied rationalism and its further ramifications; they have played a crucial role in analytic philosophy (which we will return to later); and they are assumed in much research, particularly within the natural sciences. For short, I choose to refer to these two assumptions as defining for the *analytic duality-interpretation of language*. According to this, language first of all provides descriptions.

This duality-interpretation also composes part of the canonical portrayal of mathematical modeling. Reality is passive and given, while the model is flexible and can be accommodated to reality. Naturally, the canonical portrayal does not assume that a model provides any 1 to 1 representation of reality: it provides a partial representation that can be improved through several iterative loops. Thus, the canonical portrayal assumes a model-reality affinity that at least can be gradually improved during a modeling process.

My suggestion is not to avoid using the term representation, but to use it without assuming any form of affinity. Thus, I suggest interpreting representation as *re-presentation*, which means presenting in a new way. I find that a re-presentation is without epistemic control. It could take any form and serve any purposes. We may lose sight of any form of affinity through a re-presentation. When we make a mathematical model of something, we provide a re-presentation. Naturally, one can still talk about mathematical modeling, but I find it important not to include any canonical assumption about the existence of a degree of affinity between model and reality.

Modeling Is Not Objective, It Invents Objects

Traditionally, a statement has been considered objective if its truth value is independent of any particular opinions and interests but only depends on some observational facts. The aim of science has been claimed to be objective. Although it may not be possible to obtain complete objectivity, it remains a principle aim of any scientific investigation to try to eliminate any form of subjectivity. Through such elimination, for instance with respect to the particular conditions of the observer, one is assumed to approximate objectivity. This was the explicit aim of positivism, but has remained an aim in research reaching far beyond classic positivism.

The concern was how to be objective. How to address reality as it is, and not as it appears. How to depict the objects as they really are. The aim was to look beyond appearance and to grasp reality. This idea was emphasized by, among others, Descartes and Galilei.[2] As an illustration of his point, Galilei discussed the concept of heat, and he emphasized that the experience of heat, for instance when we burn our fingers, is very different from the really behind this experience. Galilei assumed that the underlying reality in general consists of material units that can be described in terms of shapes, numbers, positions, and movements. As a consequence, the subjective experience of heat does not demonstrate any similarity with the underlying reality of heat which, however, could be described in terms of velocity of particles.

The distinction between appearance and reality applies to any natural phenomenon and is basic to the scientific outlook as proposed through the scientific revolution. We cannot expect to be able to grasp reality through its appearance. Sense experiences do not have any resemblance with the reality which stands behind the phenomena. But how to get beyond sense experiences, the emblematic form of subjectivism, and grasp what can be considered objective? If the human senses are mischievous, we need a different faculty that can ensure that we reach beyond mere appearance.

[2] See, for instance, Descartes (1998) and Galilei (1957). See also Skovsmose (2009) for a discussion of the appearance-reality distinction.

The basic claim of the scientific revolution was the underlying reality of any natural phenomena can be depicted through mathematics. Mathematics is crucial for describing shapes, numbers, positions, and movements. What our senses are missing, being trapped by the appearance of natural phenomena, mathematics can illuminate. As a consequence, mathematics can be nominated as the language of science. It provides a primary-secondary distinction by depicting what is primary (the real) and ignoring what is secondary (the appearance).

From this general understanding of mathematics, it is straightforward to move on to a general celebration of mathematical modeling. This ensures objectivity, as it will tend to eliminate subjectivity from the processes of description and present things as they really are. This interpretation provides an overall metaphysics for celebrating the canonical portrayal of mathematical modeling.

I find, however, this celebration to be highly problematic. I do not see mathematics as describing any preexiting reality as assumed by the scientific revolution and repeated many times since then. Let me explain:

One set of questions concerning appearance and reality emerge when one tries to describe a natural phenomenon like the movements of the planets. Thus the sun appears to be moving across the heavens, and by the end of the day, it appears to sink below the horizon. In such a case it seems clear that mathematics help to reveal a reality different from the appearance of the sunrise-sunset phenomena. However, mathematical modeling has been applied to all kind of phenomena, and the accompanying appearance-reality distinction might in many cases be highly problematic. Thus, mathematical modeling is applied within, for instance, business, marketing, promotion, managements, decision making, and production. In such cases, mathematics also establishes a distinction between what to consider primary and what to call secondary. Not that there needs to be any such differentiation *in advance* within the domain in question. Instead a primary-secondary differentiation becomes induced through the modeling process itself. Many features become eliminated from the domain that becomes re-presented through the modeling. For instance, in modeling a production process, the workers' social context becomes eliminated. Their family life becomes relegated as secondary. What becomes nominated as primary is how they can form part of an otherwise semi-automatized production process. It is the workers as potential components in a production process that is featured as primary.

In such a case, mathematical modeling does not simply *depict* what is primary and what is secondary: instead it *nominates* something as primary and ignores other things as secondary. One cannot assume the existence of such priorities in advance, but they become created during the mathematical modeling. Prioritization forms part of the construction of the model. This means that things become re-presented, and maybe more than that. The reality that is modeled may in fact become formed through the modeling process. The worker as a component of an automatic process is not really there in advance; the workers become configured as mechanical components. They turn into objects due to the modeling process.[3]

A mechanical worldview becomes imposed on reality, due to the mathematical modeling. It becomes established through the primary-secondary priorities that accompany mathematics. A mathematical model is not at all objective, as claimed by the classical portrayal, but rather constitutes the objects of the modeling process in a particular format. The underlying reality for a modeling process becomes an invented reality. It was not there waiting in advance; instead, it emerged as part of the mathematical modeling.

Modeling Is Not Descriptive, but Performative

According to the classic portrayal, modeling has been presented as a descriptive process. However, emphasizing that modeling means re-presenting, and that it produces its objects, bring us to the idea that modeling means acting. Modeling is not descriptive but performative. This idea resonates with a general move in the philosophy of language away from the analytic duality-interpretation.

However, let us start by taking one more look at this duality-interpretation, which was particularly elaborated within analytic philosophy. In his *Begriffschrift*, Frege (1967) provided the philosophy of language not only with a range of new ideas but also with an extreme degree of analytic exactness. The overall idea is condensed in the title, *Begriffschrift*. Directly translated, *Begriffschrift* means "concept writing": *Begriff* means "concept," and *schrifft* means "writing." What Frege had in mind was a language that provided an exact basis for addressing concepts. This language took the form of formal logic, in particularly apt for formulating set theory; and sets were seen by Frege as concepts. This language was further developed by Whitehead and Russell (1910/1913) in *Principia Mathematica* into a proper foundation for mathematics, while Wittgenstein (1992) in *Tractatus* gave it a unique role to play in his picture theory of language. According to this theory, language is picturing reality. Like any other map, a language could provide a more or less accurate picture. Language was creating a map, where one can see the 1 to 1 mapping as an extreme ideal. An assumption, also formulated by Wittgenstein, was added to the picture theory, namely, that some languages are doing better mapping than other languages and that the ultimate mapping, from a scientific perspective, was ensured by formal languages. I see Wittgenstein's picture theory from *Tractatus* as a principal formulation of the analytic duality-interpretation of language. Languages, in general, provide descriptions, and formal languages, in particular, provide descriptions that are appropriate for scientific enterprises.

However, the analytic duality-interpretation has been confronted by other interpretations of language. A provisory speech act theory was formulated by Austin. According to him, a statement has a locutionary content, an illucutionary

[3] See Chapter 15 for a further discussion of robotting and surveilling.

force, and a perlocutionary effect. As an example, we can consider the statement, "The worker could do his job faster." The locutionary content refers to the content of the statement: the speed of the particular working process. The illucutionary force refers to the power of the statements; for instance, when made as part of a management investigation, the statements can be turned into new demands to the workers. Finally, the statement has a perlocutionary effect, thus many workers might feel threatened by the statement. Austin's point is that any statement includes these three features. This way Austin confronts any analytic duality-interpretation of language.[4]

The speech-act theory became further developed by Searle, whose 1969 book is in fact entitled *Speech Acts*. From then on, speech-act theory proliferated. Discourse theory is one important feature of this proliferation, elaborating the performative aspects of languages. However, one should talk about discourse theories, as the social, political, and cultural ways of contextualizing speech acting have been elaborated in many different ways.[5] Discourse theories have been developed by Laclau, Mouffe, Žižek, and many others. Žižek's book *Violence*, from 2008, shows the profoundness of symbolic violence, which is still an example of linguistic performances. It also needs to be remembered that in a range of studies, Foucault pointed out the network of power structures that are integrated into scientific discourses. Much inspiration for the development of discourse theory has emerged from his work.[6]

Supported by Austin, Searle, Laclau, Mouffe, Foucault, Žižek, and many others, most analytic duality-interpretation of language has been left behind. We have moved towards a *discursive interpretation of language*, which we can try to summarize in the following way: (1) Reality and language are not independent entities; they are dialectically connected. Thus, reality is formed through discourses (although not necessarily only through discourses); and as multiple discourses can be operating, it is more appropriate to talk about realities in plural. (2) It is not possible to establish any affinity between reality and language, and any talk about language representing a reality is an illusion. Any so-called representation includes priorities and assumptions; it includes a reformulation according to certain assumed worldviews. These two assumptions express the idea that language is not just descriptive but also performative. Actions can be accomplished through language.

This shift from a duality interpretation to a discursive interpretation of language also concerns mathematical modeling. Modeling can be interpreted in terms of performance, as a form of speech acting. One could initiate a study of this acting by considering what could be the locutionary content, the illocutionary force, and

---

[4] See Austin (1962, 1970). During the same period, Wittgenstein (1958), in *Philosophical Investigations*, elaborated his ideas about language games, and such games also include actions. Thus, Wittgenstein himself provided a radical negation of the duality-interpretation of language.

[5] See, for instance, Torfing (1999).

[6] See, for instance, Foucault (1989, 1994, 2000).

elocutionary effect of mathematical modeling. Such a study could be conducted with reference to any example of mathematical modeling. There are, however, other ways of leaving behind any analytic duality-interpretation of mathematical modeling. Previously, I have explored the formatting power of mathematics,[7] but now I see these studies as generalized into studies of mathematics in action. However, there is a huge task ahead of us in developing a proper discourse theory with respect to mathematics. Such a theory would not only be a significant contribution to the philosophy of mathematics, it would also ensure a crucial development of discourse theory itself, which up to now has ignored formal languages.[8]

Mathematics-Based Discursive Acts

We have highlighted that a mathematical model does not represent, but rather re-presents, that it is not objective, but instead invents its objects, and that it is performative.[9] This brings us forward to the point that *mathematical modeling can be more adequately addressed through a discursive interpretation of language.*

After this observation, we are ready to characterize mathematics-based discursive acts. They represent the crucial elements in an interpretation of mathematics as discourse. As already indicated, such discourses operate in, for instance, engineering projects, economic strategies, organizational schemes, management principles, administrative procedures, business plans, and tax administrations. However, the discussion of mathematics-based discursive acts can be pursued in other directions as well. One can consider discursive acts associated with any form of ethnomathematics, street mathematics, everyday mathematics, and so on.

Let me just summarize the mathematics-based discursive acts, as I have presented them previously. Thus, I highlight the performative aspects of mathematics (at least some forms of mathematics) by commenting on (1) technological imagination, (2) hypothetical reasoning, (3) legitimation or justification, (4) realization, and (5) dissolution of responsibility.

Technological Imagination

An important feature of technological development is the conceptualization of new possibilities. Such a conceptualization is a discursive act. The space of possibilities is an expression of what could be called technological imagination, which is crucial for any form of technological innovation. For any technological construction, it is important to provide a blueprint of what is to be constructed. Mathematical models operate as such blueprints; not by reproducing something that exists, but by modeling something that one might be able to produce. In all

[7] See, for instance, Skovsmose (1994).

[8] In Skovsmose and Ravn (2011), we are exploring such new developments in the philosophy of mathematics.

[9] Several studies of mathematical modeling show a movement toward this deeper understanding. See, for instance, Barbosa (2006); and Jablonka (2010).

forms of technology, mathematics-based imagination opens spaces of possibilities.

As paradigmatic examples of such mathematical blueprint imaginations, one can refer to the conceptualization of the computer. The mathematical construct, in terms of the Turing machine, was investigated in every detail.[10] Even the computational limits of the computer were worked out before the construction of the first computer had taken place. One can also think of construction of the Internet, the proliferation of new forms of communication technologies, the implementation of technologies of surveillance, and so on. Within such domains, there is no commonsense-based imagination equivalent to a mathematics-based imagination. In general, technological development starts from a space of possibilities that is endlessly recast through mathematics-based imagination.

Mathematics-based technological imagination is crucial, not only for the construction of new technological artifacts, but also for the identification of new schemes for, say, production, management, decision-making, and such. It is an imagination, however, that exists within a certain space. It is an imagination that assumes the mechanical worldview, and it is an imagination that assumes rationality to be of a certain format.

Hypothetical Reasoning

Hypothetical reasoning is of the form, "if p then q, although p is not the case." Such counterfactual reasoning is essential in any kind of technological enterprises. If we do p, what would be the consequence? It is important to address this question before we in fact do p. Is a certain technological design feasible?

Hypothetical reasoning normally presupposes the use of elaborate mathematical models. Through an investigation of a mathematical blueprint of a possibility, one tries to identify the implications of actually carrying it out. However, the implications that are identified by investigating the mathematical blueprint are not real-life implications; they are just calculated implications. It is far from obvious what the relationship may be between calculated implications and real-life consequences of carrying out a technological enterprise. This observation applies to the implementation of any form of technological enterprise. The huge gap between calculated implications and real-life consequences provides a titanic entrance into the risk society. This is one example of where mathematics-based discursive acts could take us.

Legitimation or Justification

Discourses are brought in operation when we try to justify or to legitimate something, being a statement, an action, a decision. A discourse represents a worldview, which provides resources for both legitimation and justification.

[10] See, Turing (1965) as well as Skovsmose (2009) for a discussion for this example.

According to a classic philosophical perspective, justification refers to a genuine logical support of, say, an action, while legitimation might include dubious elements, for instance, by serving particular interests. The point of providing legitimation for an action is to make it appear as if it is justified. However, it might only be within an idealized philosophical framework that it is possible to distinguish between legitimation and justification. Discourses can be of various forms, and they might liberally support both proper and imaginary arguments.

Mathematical models provide discourses for both justification and legitimation. Furthermore, they might blur any such distinction. It has been pointed out that in much decision-making, for instance, concerning the implementation of a new healthcare program, mathematical modeling plays an active part in identifying potential implications. A broad spectrum of mathematics-based hypothetical reasoning is established in order to point out pros and cons with respect to the program. Mathematics is crucial for any such cost-benefit analysis, which informs the decision making.

Mathematics has been assumed to be objective by providing solid arguments for a decision. But a mathematical model is not a straightforward descriptive device. Thus cost-benefit analysis concerns a population, which is shaped through the modeling. It is not a real-life population but a systematized one. What to call costs, what to call benefits, and how to measure them, are all model-created measures. In general, modeling includes an invention of the object that becomes addressed. Thus, the pro and con argument, related to, say, the healthcare program, becomes situated in a model-invented reality.

Realization

One point of discourse theory is that the distinction between the discursive and the nondiscursive evaporates. What is considered reality becomes recast through discourses about reality. Philosophically speaking, one can claim that Descartes' distinction between *res cogitans* and *res extensa* becomes abolished. Reality becomes reworked through language, which in turn is formed through social practices. Reality includes materialized discourses. Thus, Gellert and Jablonka (2009, p. 20) emphasize that mathematics-based decisions operate on many levels:

> On the level of national policy, decisions about the distributions of state salaries, pensions, and social benefits rely on mathematical extrapolations of demographical and economic data. . . . On the level of interpersonal relations, mathematics-based communication technologies have already changed the habits and styles of private conversations.

And one could continue and refer to management principles, administrative procedures, organizational schemes, business plans, healthcare programs, military operations, and so on. In all such domains, mathematical procedures that might have been introduced as analytic devices now take the form of real-life operating decisions.

Our life-worlds are formed through categories and discourses, many of which emerge through mathematical modeling. These categories and discourses are put into operation. They are working. Thus, our reality includes materialized mathematical discourses.

The whole domain of relevant knowledge for decision making in the stock market—buying or selling—is mathematised and made available through figures and diagrams. In this way, mathematics can provide a highly relevant descriptive tool. One could also imagine that algorithms make proposals as to which decisions to make. However, there is a step more that can be taken. One might imagine that the very decision about selling and buying is in fact made by a mathematical algorithm. The Danish newspaper, *Politiken,* in its edition of February 24, 2010, contains an article, *Maskinen Overtager den Globale Børshandel* (*The Machine Takes Over the Global Stock Market*) by Jeremy Grant and Michael Mackenzie, whose point is exactly that the very decision-making is placed in the hands of algorithms. Furthermore, the newspaper contains an article by Per Thiemann stating that 20% of the selling and buying at the Copenhagen Stock Market is conducted by the computer. This is an example of mathematics coming to be a direct part of the economic reality.

Dissolution of Responsibility

Any action raises the question, Who is responsible for what is done? This question makes clear sense when an acting subject can be located. It becomes more difficult to address when the acting subject is an institution, an organization, a program. But it is still possible to find ways of defining responsibility in such cases. More complicated, yet, is defining responsibility with respect to speech acting. Naturally, there are procedures for locating responsibility when speech acts take the form of, say, a particular insult. But difficulties arise when we consider the general action-dimension of language. When one consider acts, as for instance emerging from a certain scientific discourse, the location of responsibility becomes difficult.

Let us again consider the situation of a rationalization of a production process. Such a rationalization includes a reconfiguration of the production process. But who is responsible for the implementation of the mechanical worldview with respect to this reconfiguration? It is difficult to identify a particular acting subject. Somehow responsibility seems to evaporate. Could the people who constructed the model be responsible? Or could it be claimed that the very mathematical discourse, as the carrier of the mechanical worldview, is responsible? Somehow discursive acts seem to be missing an acting subject. Actions based on mathematics appear to be conducted in an ethical vacuum. In this sense, a dissolution of responsibility is taking place.

An example of such a possible dissolution of responsibility is presented by Mario Sánchez (2009, 2010) in his discussion of a "marginalization index." This index has been applied in a Mexican context in trying to invent measures for the degree of marginalization which certain communities might suffer. Naturally, there can be many different ways of measuring marginalization, but whatever modeling is applied, some parameters have to be introduced and related, and some standards have to be introduced so that the entire social, political, and economic processes of marginalization emerge in a modeled format. Here may occur an extreme form of primary-secondary ranking, where the experienced characteristics of marginalization are "abstracted away" in favor of only concentrating on quantitative and "mechanical" features of marginalization. On this basis, political action might be taken, or not taken. Such an approach has many implications, one of which might be that new criteria are formulated and claimed to be "objective."

Additional Comments

There are a couple of remarks I want to add with respect to mathematics-based discursive acts in general. What I first of all had in mind was mathematical modeling as applied in a vast area of technology. One can, however, also try to investigate mathematics-based discursive acts with reference to pure mathematics. My point would be that the notion "pure" is mischievous: There is nothing pure in the sense that mathematics composes part of a discourse without any acting dimension. Also, pure mathematics is performative.

It is important to clarify the performative dimension of different forms of ethnomathematics. Thus ethnomathematics-based actions form part of a range of different cultural practices. Mathematics also forms part of everyday-actions, such as going shopping, controlling tax payments, making a budget, planning a holiday, and so on. When such different forms of mathematics are addressed, we gain a better understanding of mathematics as discourse.

Symbolic Power

Discourses can be powerful, mathematics can be powerful, knowledge can be powerful. The intimate connection between knowledge and power also applies to mathematics. In discourse theory, one refers to symbolic power, and the symbolic power of mathematics becomes realized through mathematics-based discursive acts.[11]

The discussion of knowledge and power has been stimulated by Foucault's work, which I find extremely inspiring. However, Foucault concentrated on certain domains of knowledge: psychology, psychiatry, and related discourses. Nowhere did he address knowledge of the exact sciences or mathematics. However, I find that a very important domain for exploring the relationship between knowledge and power is precisely the exact sciences and mathematics.

The development of new technologies has been decorated by an optimistic vocabulary: this development has been broadcast as both necessary and positive. Technology has been nominated as the motor of progress. This is, however, not

[11] See Skovsmose (2010b).

how I see things. Technological development can take any direction. It can be promising, risky, or disastrous. In whatever direction the technological development may happen to take us, it is propelled by mathematics in action. Technology is mathematics-based. It epitomizes mathematics-based symbolic power.

There are many ways the notion of ideology can be interpreted. It can refer to cultural, religious, and political ideas. Somehow an ideology represents a "worldview," and symbolic power operates through such views. This also applies to mathematics-based symbolic power. A classic example can be related to the formulation of the heliocentric worldview—a worldview in the most direct sense of the word. The formulation of this view had a huge ideological impact, confronting religious dogmas. The heliocentric worldview was constituted through mathematics. The mechanical worldview that accompanied the scientific revolution is also mathematics driven, as reality is defined in terms of numbers, positions, shapes, and movements of material units. Such a view has formed our conception of nature. A mechanical worldview can, however, also be imposed on social structures—on a production process, for instance, where the workers become conceptualized as more or less well-functioning units in an assembly line. Also, in this case, that mechanical worldview becomes facilitated through mathematics. In fact, whatever domain we consider, a mechanical worldview tends to be imposed when mathematics is brought in action. This is a most important example of symbolic power routed in mathematical discourses.

It makes good sense to talk about symbolic power with respect to mathematics, but does it also make sense to talk about symbolic violence? Such violence has been addressed with respect to language in general, by thinkers such as Žižek. Discursive acts can be violent, for instance, when one imposes a label with respect to certain groups of people. A racist vocabulary can serve as a paradigmatic example of such violence. Other forms of symbolic violence become acted out through labels like working class, lower class, deficient, primitive. Such labels compose part of different forms of discriminating discourses. Labeling often means a violent stigmatization formed through dubious metaphysical categories.

Mathematics-based discursive acts also include possibilities for symbolic violence. Let us consider again a mathematical model for productivity. Such a model includes a range of parameters, some of which refer to people and their performances. In order to be modeled, performances have to be expressed in quantitative terms. They have to be measured. This is the paradigmatic step for turning a qualitative phenomenon into a quantitative one. As soon as it has been measured, a phenomenon can be grasped by a model. It can be compared with some stipulated standards; it can be subjected to cost-benefit analysis; it can be manipulated within the universe of calculations. The human dimensions of the workers and what they experience due to the new productive schemes have become eliminated from all calculations. Their personal experiences have been disregarded as being subjective. These experiences do not belong to the mathematics-created reality. This is an illustration of symbolic violence rooted in mathematics.

This brings us to an important point with respect to social theorizing, as for instance formulated by Habermas, Giddens, and Bauman. They try to address some of the principal features of social development, and they explore notions like modernity, late modernity, postmodernity, liquid modernity, and so on. What surprises me in their formulations is that they do not see any particular need for addressing the role of mathematics. It is not possible to come across the word mathematics in the index in any of their numerous books. In fact, many other authors trying to formulate a broad social theorizing share the same ignorance of mathematics.[12] For me, it is in important to make a new turn in social theorizing by addressing the symbolic power that emerges though formal languages and mathematics in particular. Social theorizing will be incomplete without addressing this component of the knowledge-power dialectics.

NOTE

This paper was first published in 2012 in *Bolema* (Boletim em Educação Matemática—Mathematics Education Bulletin), *26*(43), 1–18. The paper is, however, partly reworked to include sections from "Symbolic Power and Mathematics," published in R. Bhatia (Ed.), *Proceedings of the International Congress of Mathematicians* (Vol. 1, pp. 690–705). Hyderabad, India.

REFERENCES

Austin, J. L. (1962). *How to do things with words*. Oxford, UK: Oxford University Press.

Austin, J. L. (1970). *Philosophical papers* (2nd ed., J. O. Urmson & G. J. Warnock, Eds.). Oxford, UK: Oxford University Press.

Barbosa, J. (2006). Mathematical modeling in classrooms: A socio-critical and discursive perspective. *Zentralblatt für Didaktik der Mathematic*, *38*, 293–301.

Christensen, O. R., Skovsmose, O., & Yasukawa, K. (2009). The mathematical state of the world: Explorations into the characteristics of mathematical descriptions. In B. Sriraman & S. Goodchild (Eds.), *Relatively and philosophically earnest: Festschrift in honor of Paul Ernest's 65th birthday* (pp. 83–96). Charlotte, NC: Information Age.

Descartes, R. (1998). *The world and other writings* (S. Gaukroger, Trans. & Ed.). Cambridge, UK: Cambridge University Press. (Original work published 1664)

Foucault, M. (1989). *The archaeology of knowledge*. London, UK: Routledge (Original work published 1969)

Foucault, M. (1994). *The order of things: An archaeology of the human sciences*. New York, NY: Vintage. (Original work published 1966)

Foucault, M. (2000). *Power* (R.Hurley et al. Trans.; J. D. Faubion, Ed.). New York, NY: New Press. (Original work published 1994)

Frege, G. (1967). *Begriffsschrift*, a formal language, modeled upon that of arithmetic, for pure thought. In J. van Heijenoort (Ed.), *From Frege to Gödel: A source book in mathematical logic, 1879–1931* (pp. 1–82). Cambridge, MA: Harvard University Press. (Original work published 1879)

[12] As an exception let me refer to Lyotard (1984).

Galilei, G. (1957). *Discoveries and opinions of Galileo* (S. Drake, Trans.). New York, NY: Anchor.

Gellert, U., & Jablonka, E. (2009). The demathematising effect of technology. In P. Ernest, B. Greer & B. Sriraman (Eds.), *Critical issues in mathematics education* (pp. 19–24). Charlotte, NC: Information Age.

Grant, J., & Mackenzie, M. (2010, February 24). Makinen overtager den globale børshandel [The machines take over the global stock market]. *Politiken.*

Jablonka, E. (2010). Reflections on mathematical modeling. In H. Alrø, O. Ravn, & P. Valero (Eds.), *Critical mathematics education: Past, present and future* (pp. 89–100). Rotterdam, The Netherlands: Sense.

Lyotard, J. F. (1984). *The postmodern condition: A report on knowledge*. Manchester, UK: Manchester University Press. (Original work published 1979)

Sánchez, M. (2009). Uso crítico de los índices y modelos matemáticos gubernamentales en el desarrollo de profesores en servicio. *Educación Matemática, 21*(3), 163–172.

Sánchez, M. (2010). Gobierno y matemáticas. *La Jornada.* Retrieved from http://bit.ly/9Kgkrz

Searle, J. (1969). *Speech acts*. Cambridge, UK: Cambridge University Press.

Skovsmose, O. (1994). *Towards a philosophy of critical mathematics education.* Dordrecht, The Netherlands: Kluwer Academic.

Skovsmose, O. (2005). *Travelling through education: Uncertainty, mathematics, responsibility*. Rotterdam, The Netherlands: Sense.

Skovsmose, O. (2009). *In doubt: About language, mathematics, knowledge and life-world.* Rotterdam, The Netherlands: Sense.

Skovsmose, O. (2010a). A critical rationality? *Philosophy of Mathematics Education Journal* (Special issue on Critical Mathematics Education), *25*. Retrieved from http://people.exeter.ac.uk/PErnest/pome25/index.html

Skovsmose, O. (2010b). Symbolic power and mathematics. In R. Bhatia (Ed.), *Proceedings of the International Congress of Mathematicians* (Vol. 1, pp. 690–705). Hyderabad, India: Hindustan Book Agency.

Skovsmose, O., & Ravn, O. (2011). *Matematikfilosofi* [Philosophy of mathematics]. Århus, Denmark: Systime.

Skovsmose, O., & Yasukawa, K. (2009). Formatting power of "mathematics in a package": A challenge for social theorising? In P. Ernest, B. Greer, & B. Sriraman (Eds.), *Critical issues in mathematics education* (pp. 255–281). Charlotte, NC: Information Age.

Thiemann, P. (2010, February 24). Maskiner handler danske aktier [Machines deals in Danish bonds]. *Politiken.*

Torfing, J. (1999). *New theories of discourse: Laclau, Mouffe and* **Žižek**. Oxford, UK: Wiley-Blackwell.

Turing, A. (1965). On computable numbers: With an application to the Entscheidungsproblem. In M. Davis (Ed.), *The undecidable* (pp. 115–154). Hewlett, NY: Raven. (Original work published 1936)

Whitehead, A. N., & Russell, B. (1910/1913). *Principia mathematica I-III.* Cambridge, UK: Cambridge University Press.

Wittgenstein, L. (1958). *Philosophical investigations* (2nd ed., E. Anscombe, Trans.). Oxford, UK: Basil Blackwell. (Original work published 1953)

Wittgenstein, L. (1992). *Tractatus logico-philosophicus* (C. K. Ogden, Trans.). London, UK: Routledge. (Original work published 1921)

Žižek, S. (2008). *Violence*. New York, NY: Picador.

CHAPTER 15

SYMBOLIC POWER, ROBOTTING, AND SURVEILLING

Cars can be numbered, cows can be numbered, and so can human beings. While cars have number plates in front and behind, and cows may have numbers fixed to their ears, human beings carry their identity card in their pockets or bags.[1] Any of these numbers constitute entrances in a variety of registers that comprise information of all different kinds: The time for the next mechanical check; the amount of milk given; income and taxes paid, just to indicate a few possibilities. Can the numbering of cars, cows, and people be an expression of symbolic power, or maybe even of symbolic violence?

As pointed out by Slavoj Žižek (2008), there can be different forms of violence rooted in language:

> Language simplifies the designating thing, reducing it to a single feature. It dismembers the thing, destroying its organic unity, treating its parts and properties as

[1] In a conversation, Ole Ravn has pointed out to me the formatting power of mathematics that can be associated with numbering.

Critique as Uncertainty, pages 215–230.

> autonomous. It inserts the thing into a field of meaning which is ultimately external to it. (p. 61)

Let us consider Žižek's observations step by step. *Language simplifies the designated thing and reduces it to a single feature.* To illustrate how language might provide reductions, one can think of the many labels through which groups of people are singled out. One find discourses which refer to "immigrants" when trying to shed light on street violence, "Blacks" when one addresses poverty in Africa, "slow learners" when trying to explain certain forms of educational problems. Simplification is also part of discourses referring to movie stars, business leaders, and cultural profiles. In fact, there is a close connection between designating, simplifying, and reducing a complex phenomenon to a single feature. *Language dismembers the thing and destroys its organic unity.* There are many ways that language, as an instrument for dissection, cut things into pieces. One need only to think of the discourses developed around production efficiency. Such discourses may refer to workers, but stripped of their human relationships, instead, highlighting them as more or less efficient elements of a production process. *Language treats parts and properties as autonomous.* One can think of the many databases, constructed through formal languages, that come to make part of expert systems through which people are managed. Through such systems, information about particular aspects of people's life can be processed, quite isolated from whatever else might be relevant to consider with respect to the person. *Language inserts the thing into a field of meaning which is ultimately external to it.* This observation may refer to almost any discourse, as a discourse includes a worldview. In this last statement, Žižek condenses the idea of linguistic relativism, according to which our life-worlds are formed through the language into which we are submerged.

Žižek talks about language in general. This provides an economic way of expressing oneself, as the word "language" could refer to a variety of different linguistic phenomena as "language game" or "discourse." Through Žižek's formulations, we highlight "violent" aspects of certain discourses. But naturally we need to keep in mind that discourses could operate in many different ways and that symbolic power has many instantiations other that violence. Thus, discourses can be illuminating, fascinating, innovative, and so on. I return to this complexity later, but first we consider the more violent aspects of symbolic power. (And as a parenthesis, we cannot ignore that whatever Žižek says about language applies equally to his own statements. It applies as well to what I, in the following, am going to say. Whenever we want to say something general about language, and we use language to express this, we are whirred around in a vicious circle. Nevertheless, what else could we use?)

Mathematics and Symbolic Power

We can recast Žižek's statement by substituting the word "language" by "mathematics." Then it runs as follows:

Mathematics simplifies the designating thing, reducing it to a single feature. It dismembers the thing, destroying its organic unity, treating its parts and properties as autonomous. It inserts the thing into a field of meaning which is ultimately external to it.

The aspiration of this paper is to illustrate that this sentence in fact makes sense. Thus, I want to provide an analysis of the symbolic power of mathematics. This analysis will take place in two steps. First, in the section "What Is Primary and What Is Secondary," I will try from a philosophical perspective to identify two basic nominations of what to consider primary and secondary, which become acted out through mathematical discourses. These nominations characterize the symbolic power of mathematics. Second, in the section "Two Examples," I will exemplify how this power becomes exercised through robotting and surveilling. Thus, I am not trying to apply any Žižek-generated argumentation in order to present this illustration; instead I will use arguments related to mathematics.

While Žižek discusses symbolic violence, I prefer to talk about symbolic power. Certainly, I see symbolic violence as one feature of symbolic power, also when we talk about mathematics and formal languages. However, by talking more broadly about symbolic power I try to refrain from *a priori* assuming a particular interpretation of how this power might operate.

I do not assume that the notion of mathematics can be captured by any particular definition. Instead, I find that mathematics can take many different forms, such as constructing a budget, calculating a salary, making an investment, reading a map, completing a design, solving school mathematics exercises, solving an engineering problem, not forgetting doing mathematical research. One can see mathematics as a range of discourses with some family resemblances, but certainly with many differences. As a consequence, one should not associate any essence to the notion of mathematics.

Sometimes I talk about mathematics, sometimes about formal languages, but I deliberately do not try to establish a line of demarcation between them. I find that it is important to discuss symbolic power with reference to this extended family of languages. Naturally, the discussion has to be instantiated with reference to particular mathematics discourses. One can consider mathematical discourses from engineering, management, economy, any form of applied mathematics, not to forget the so-called pure mathematics. One can as well consider any instantiation of ethnomathematics, everyday mathematics, and street mathematics with respect to symbolic power. In the following, I will illustrate the symbolic power of mathematics through features of mathematical discourses developed through the so-called scientific revolution and what can be referred to as the formal revolution.

What Is Primary and What Is Secondary?

We can imagine that symbolic power is exercised by inventing something that is not the case. It appears that by applying mathematical discourses one can invent measures, norms, and averages that were not really there before such discourses designated the entities. One can think about how achievements have become mathematized, for instance with respect to both production and learning. Thus, the mathematization of achievements brings about notions of, say, high or low productivity, efficiency, fast or slow learners, and achievement gaps. One can also assume that symbolic power takes the form of systematically overlooking particular features of certain phenomena. Thus, we can examine symbolic power in terms of both "seeing" and "overlooking." Symbolic structuring is a way of designating and also of ignoring; it is a way of nominating something as primary and other things as secondary.

In order to clarify further the primary-secondary nomination, I will consider two distinctions that have been associated with mathematics and formal languages: one between appearance and reality, and one between sense and reference. The first is related to the formulation of the mechanical worldview, while the second is related to the formulation of what I refer to as the formal worldview.[2] While the mechanical worldview accompanied the interpretation of nature in mathematical terms, the formal worldview went together with the digitalization of our environment. By paying attention to these two distinctions, I try to point out two "grammatical" features of mathematics-based symbolic power.

Appearance and Reality

The so-called scientific revolution is associated with the names of Copernicus, Galilei, Newton, and many others. New insights into nature and a new role for mathematics to play are constituents of this revolution. It was pointed out that nature operated according to a few basic laws that God had imposed on nature as part of his creative act. After God created the world as it should be, nature could do nothing but follow these laws. Furthermore, it appeared that the laws were few and simple, and this made it possible for humankind to understand nature.

The complete insight into nature, then, could be obtained by determining the implications of these laws. Thus, Isaac Newton's (1999) *Principia* symbolized the tremendous unification between mathematics and insight into nature by presenting some laws of nature as axiomatic and deducing a whole theory of nature from these.

This unification, however, is only made possible through the distinction between appearance and reality. Nicolaus Copernicus (2002) presented the heliocentric worldview that illustrates clearly what this distinction is about. Looking at the sun in the morning, one sees how it is rising in the sky. During the course of the day, one can follow the movement of the sun. In the evening, we can see the beautiful colors of the western sky, and we can see how the sun finally sinks below the horizon. Literature is awash with variations of sunrise-sunset descriptions. Let us imagine that we collect all these descriptions and from them try to

[2] For a discussion of these two distinctions, see Skovsmose (2009). My work with these distinctions is also inspired by my cooperation with Ole Ravn, see Skovsmose and Ravn (2011).

extract an insight into the sunrise-sunset phenomenon. According to the perspective of the scientific revolution, we would never establish any insight at all, as we would remain trapped by the appearance of the phenomenon. At the heart of the formulation of the heliocentric worldview is the assumption that one needs to get around the appearances to grasp the structures of reality. Only in this way will we be able to understand how the daily rotation of the Earth produces the apparently completely stable sequence of sunrises and sunsets.

Empirical observations played an important role as part of the scientific revolution, providing a contrast to free speculation based on holy or other authoritative texts, particularly those presented by Aristotle. However, to the proponent of the scientific revolution, the point was to recognize empirical observations not as duplicates of a reality but as signs which might not demonstrate similarities. The structure of appearance might be radically different from the structure of reality, and if one wanted to establish insight into nature, one needed to grasp the structure of reality. In other words, one needed an epistemic tool beyond the senses; and mathematics was assumed to do the job.

The distinction between appearance and reality has been emphasized by many, including Descartes, who can be considered the philosopher of the scientific revolution. However, I will refer to a particular text by Galileo Galilei as paradigmatic for formulating the appearance-reality distinction. In *The Assayer*, first published in 1623, Galilei discusses the notion of heat.[3] We all have experiences of this phenomenon. One may be burned by the sun, touch a warm kettle, come too close to a fire, and gulp a spoonful of scalding soup. One could try to register a broad variety of experiences with heat and on this basis try to formulate an insight as to what heat really is. Galilei's point is that such an approach brings us nowhere: Our experiences of heat do not reveal anything about the real nature of heat. The "mechanics" of heat, according to Galilei, is very different from whatever might be extracted from of our sense experiences, just as the mechanism of sunrise and sunset is different from experiences of these phenomena. In such cases, appearance and reality are radically different. According to Galilei, in order to provoke experiences of heat "nothing is required in external bodies except shapes, numbers, and slow or rapid movements" (Galilei 1957, pp. 276–277).

The mechanical worldview presents reality as a tremendous mechanism composed of material units, characterized by their shapes, numbers, and movements, governed by certain laws. Reality is conceptualised as a perfect *perpetuum mobile* created by God. Sense experience is not adequate for grasping the operation of this *mobile*. But what could be more adequate for describing size, number, and movement than mathematics? So, the appearance-reality distinction facilitates the formulation of the mechanical worldview and brings mathematics into a prominent position: It becomes the principal tools for grasping reality.

[3] *The Assayer* is reprinted in the book by Galilei (1957, pp. 229–280). See Skovsmose (2009) for a discussion of *The Assayer*.

So far so good. If we assume the outlook of the scientific revolution and that reality is in fact a mechanical structure, then mathematics can be assigned a tremendous descriptive power: It turns out to be not only necessary but also sufficient for grasping reality. If we think of this descriptive power as a form of symbolic power, the whole outlook of the scientific revolution would bring us to celebrate the power of mathematical symbolism—a celebration which in fact took place during the whole period of Modernity.

However, I find it important to consider a different interpretation. If we do not regard the mechanical world to be a given, but rather invented, and moreover, invented by means of mathematics, then we reach a quite different interpretation of the symbolic power that might be exercised through mathematics. The mechanical reality is not described by means of mathematics but rather established through mathematics as a projection of the grammar of mathematics. What can be represented in mathematics, such as shapes, numbers, and movements becomes nominated as "the real." The grand *perpetuum mobile* it not created by God, but installed through a mathematical discourse. This way we might give a particular interpretation of how mathematics might reduce a phenomenon to a single feature, treat its parts as autonomous, and insert it into a new field of meaning. This is what the mathematics-based mechanical worldview does. And this view is based on the primary-secondary priorities exercised through mathematical symbolism.

Sense and Reference

While in *The Assayer*, Galilei formulated a distinction between appearance and reality, Gottlob Frege (1969), in Über Sinn und Bedeutung, formulated a distinction between sense and reference. While the distinction between appearance and reality is linked to the scientific revolution, the distinction between sense and reference can be related to a formal revolution, including the formalization of logic and the construction of the computer. The formal revolution preceded the technologies of digitalization and automatization.

Both distinctions—between appearance and reality and sense and reference—specify what to consider primary and what to consider secondary. While the distinction between appearance and reality concerns our perception of nature and physical environment, the distinction between sense and reference concerns our perception of logic and rationality. Frege sought to grasp the nature of logical reasoning, and he wanted to identify the relationship between logic and mathematics.

The German word *Sinn* can be translated as "sense," and the word *Bedeutung* as "reference." To illustrate the distinction between sense and reference, we can look at the notion of a triangle. In order to indicate the sense of the word triangle, one could try to explain that we are dealing with a geometric figure composed of straight lines, and that it has three sides and three angles. And one can try to add further explanations if necessary. If one were to indicate the reference of the notion of triangle, one would look to the set of all triangles. Remembering that Frege was a Platonist, he would see the reference as the collection of ideal objects. More generally, the reference of a notion is the set of objects that "fall under" the notion.

Frege also applies the distinction between sense and reference to statements. If we state that "the sum of the angles in a triangle is 180°," then one could try to clarify the sense of that sentence by showing some of the steps in the proof of the statement. The sense of the statement has to do with the content of what is stated. However, according to Frege, the reference of the statement "the sum of the angles in a triangle is 180°" is something quite different. He suggests that the reference of a statement is its truth value. Furthermore, he assumes that there are only two such values: "true" (or T) and "false" (or F). This means that the reference of the statement "the sum of the angles in a triangle is 180°" is "true." If we were to consider all possible statements, they would have lots of different senses, but their references would be either "true" or "false." The domain of references of sentences would be a very small universe, namely, consisting of only two objects, the two possible truth values, "true" and "false."

Such a claim may appear absurd. However, it makes it possible for Frege to formulate his main point: In order to clarify the nature of logical reasoning and to specify the relationship between logic and mathematics, one needs to concentrate on the domain of references. References of both concepts and of statements can be considered primary, while senses are secondary: They can be ignored. In fact, when it comes to logical investigation, the dimension of senses only confuses the analyses. According to Frege (1978), many have suffered such confusion: John Stuart Mill (1970), for instance, who found that in order to understand the nature of logical reasoning and the foundation of mathematics, one had to grasp their inductive origin.[4] This is, according to Frege, the great mistake of psychologism, which locates the origin and foundation of logic and mathematics in the psychological domain. Frege strongly criticized any forms of such psychologism; instead, he proposed a strict logicism.

Frege's idea that the references of concepts and statements are crucial while their senses can be ignored was nicely condensed by Wittgenstein in the *Tractatus* (1992), where he presented a truth-table logic. This is a direct way of claiming that in order to clarify the functions of logical connectives like "and," "or," "if then," and "if and only if," one needs to consider only the truth values (the references) of the sentences that are connected by these connectives, and nothing else. The entire investigation of logic can concentrate on the domain of references. Thus, the notion of tautology, which is considered crucial for the characterization of logical reasoning, is a "representational notion" as it can be defined through truth-tables.

Through Frege's formulation of what to consider primary with respect to logical reasoning, a formal worldview was established. This view is beautifully con-

[4] See Mill's (1970) presentation in *A System of Logic* and Frege's (1978) harsh critique of Mill in *The Foundation of Arithmetic*.

densed by Wittgenstein in the 6th statement of *Tractatus*, where the universal form of a statement is presented. This form only recognizes references of concepts and statements as being relevant. There is no space for "senses" in this universal layout of a sentence. Thus, this 6th statement of *Tractatus* includes a drastic primary-secondary nomination, defining what I refer to as the formal worldview. This 6th statement is a manifestation of symbolic power.

This view facilitated a tremendous development of formal systems, formalization of deduction, automation of reasoning, and a proliferation of formal languages, including all variations of computer languages. The computer can be seen as a materialization of the whole formal approach. The development of artificial intelligence and expert systems draws on the same resources. The formal revolution became real.

If we allow ourselves the optimism of this revolution, one could claim that symbolic power, expressed through the formal worldview, ensures new powerful technological possibilities. However, also in this case, I find it important to consider a different interpretation. The sense-reference distinction may not reflect any "real" distinction. The nomination of references as "the real" and senses as a scientifically irrelevant psychological residue might be imposed on the domain of investigation. The nomination may not describe what is primary and what is secondary with respect to logic and reasoning, but instead prescribe such priorities. We might have to do with a configuration of the domain of reasoning and computation established through a total projection of a semantics of mathematics. Such a projection includes a principal reduction by leaving aside senses of concepts and statements as being secondary. As a consequence, a phenomenon that is addressed through a formalization becomes reduced to a single feature and inserted into a new field of meaning.

Two Examples

Let me start with a short summary: I find that the symbolic power of mathematics can be interpreted in terms of a nomination of the physical reality as a mechanical system, leaving aside the appearance of a phenomenon as secondary, and a nomination of the logical reality as we world of references, leaving aside any sense as secondary.

It has been broadly assumed that the task of science is to provide descriptions. This idea was captured by Wittgenstein's (1992) picture theory as presented in *Tractatus* and elaborated further by logical positivism, which found that a universal science could provide a universal picturing. Assuming a mechanical worldview, the reliability of this picturing was ensured by mathematics and by formal languages in general.

However, as pointed out, mathematical does not simply describe a domain; it configures this domain. Mathematics only describes as it has prescribed. Mathematics is allied with primary-secondary priorities as expressed by appearance-reality and sense-reference distinctions; and such priorities become acted out through the mechanical and the formal worldviews.

I will now try to illustrate this power by discussing two examples of how a mechanical and a formal worldview may be acted out, namely, through robotting and surveilling.

Robotting

Descartes formulated a dualism with respect to human beings: They are composed of two elements, mind and body. In his excellent biography of Descartes, Clarke (2006) points out that the construction of machines with a decorative and entertaining function, representing advanced engineering at the time, could have inspired Descartes to formulate a mechanical conception of the human body. By means of sophisticated hydraulic systems, puppets could come to perform human-like movements, and it makes sense to consider a human body to be a similar system, although far more complex. In fact, this was how Descartes (1998) presented his studies of the human body in *The World*.[5]

Descartes suggested a way of looking at human beings that recognized the body as a machine, while the soul functioned as the captain onboard. According to Descartes, and to very many others, an *insight* into nature was obtained by providing a mechanical outlook. However, we could formulate things differently: The body-as-a-machine perspective can be interpreted as *imposed* on nature. This imposition structures our understanding of what we see; it structures both what we come to see and what we come to overlook.

One could claim that nature might not suffer by being subjected to the mechanical worldview. However, if we see this view as imposed through mathematics, we have to be aware that such imposition can be exerted far beyond the range of the natural sciences. It can be exercised anywhere mathematics is brought in action.

In order to illustrate this point, we will take a look at scientific management, which addresses processes of production and human beings as components of such processes. Also in this case, mathematics facilitates the imposition of a mechanical worldview. It induces a primary-secondary nomination, stripping the processes of production of their human (and secondary) elements, and focusing on the mechanical (and primary) features of the situation.

Scientific management was presented by Frederick W. Taylor (2006) in *The Principles of Scientific Management*. This book can be read as a cruel continuation of Descartes' *The World.* In *The World*, Descartes looked into the human body and showed how physiological phenomena could be seen as pure physical phenomena. In particular, he showed how the human eye operates according to laws concerning the reflection of light. In *The Principles of Scientific Manage-*

[5] Animals, according to Descartes, can be seen as machines. This interpretation has been broadly accepted as part of the rational and enlightened outlook.

ment, Taylor presents the worker Schmidt, who is going to be subjected to the new principles of scientific management. Schmidt is going to operate according to laws governing efficient management. In his own both clear and cynical way, Taylor shows what a mechanical worldview might result in, when we see processes of production and the workers who make part of these processes as mechanical systems that need to be optimized.

The particular problem described by Taylor concerns how, in the most efficient way, to load pig iron onto a railway wagon. It had been observed that an average worker can load 12½ tons per day. However, Taylor and his staff had gone through a series of calculations and found that the number ought to be 47–48 tons per day. By means of numbers and figures, Taylor had X-rayed the working process and reached the conclusion that it could be conducted in a much more efficient way. A new way of organizing the process needs, however, to be implemented, and Taylor describes how Schmidt came to be working in accordance with the new measures:

> The task before us, then, narrowed itself down to getting Schmidt to handle 47 tons of pig iron per day and making him glad to do it. This was done as follows. Schmidt was called out from among the gang of pig-iron handlers and talked to somewhat in this way. (Taylor, 2006, p. 20)

At this point, Taylor presents the conversation taking place, of which I quote only the beginning:

> Schmidt, are you a high-priced man?
> Vell, I don't know vat you mean.
> Oh, yes, you do. What I want to know is whether you are a high-priced man or not.
> Vell, I don't know vat you mean.
> Oh, come now, you answer my questions. What I want to find out is whether you are a high-priced man or one of these cheap fellows here. What I want to find out is whether you want to earn $1.85 a day or whether you are satisfied with $1.15, just the same as all those cheap fellows are getting.
> Did I vant $1.85 a day? Vas dot a high-priced man? Vell, yes, I vas a high-priced man. (Taylor, 2006, p. 20)

Taylor continues the dialogue and soon comes the description of the behavior of a high-priced man:

> Well, if you are a high-priced man, you will do exactly as this man tells you tomorrow, from morning to night. When he tells you to pick up a pig and walk, you pick it up and walk, and when he tells you to sit does and rest, you sit down. You do that straight through the day. (Taylor, 2006, p. 21)

In short, a high-priced man works as a robot. What could be called the human elements of Schmidt, such as having his own priorities and ways of doing things, is relegated as secondary qualities. All human aspects of the person are ignored.

Recognized as important are only the actual physical operations that could be performed by Schmidt, and they have to be optimized. Through the new discourse of scientific management, Schmidt became identified as a mechanical system that could come to fit into the complex clockwork of the production system.

There are only a few figures and numbers in Taylor's presentation of the situation, but they help to set out the primary-secondary priorities. In general, such priorities are supported by the application of mathematics, and one has to remember that Taylor just introduced the approach of completing a mathematical X-ray examination of work processes and of forms of production of any kind. However, subsequently, mathematical numbers and figures came to dominate all forms of investigations of processes of productions and any form of optimizing such systems. Such an X-raying might be assumed to state something about the situation that is X-rayed. However, also in this case, we can see the configuration of the work process as a mechanical system as being imposed on the process. The whole terminology about, for instance, efficiency, output measures, productivity, and cost-benefit analysis becomes forced on the situation. It is a terminology that is facilitated, if not created, through the mathematical discourse that is brought in operation. This terminology and the decisions it facilitates are expressions of the symbolic power that is exercised by depicting a situation through numbers and figures, and subsequently assuming that one has to do with an adequate description.

The symbolic power of mathematics is exercised in the domain of production by representing human beings as robots. Schmidt, the robot, was represented in this way. But there is more to it. Taking a look into any production hall where cars are assembled, beers are bottled, and cell phones produced, one sees human beings, machines, and proper robots united in complex processes of production. The distinction between what is completed by human beings and what is done by machines and robots is not easy to maintain. In fact, it becomes irrelevant to make such a distinction, as it is the combined machinery that is optimized. That machines, robots, and human beings can be efficiently combined in schemes of production is partly due to the imposition of a mechanical way of looking at the whole process.

I have been presenting robotting as something negative. It is a symbolic power exercised when a mathematical discourse grasp only the mechanisms of work processes. However, one needs also to keep in mind that efficiency with respect to production will have a range of implications that in general are much appreciated. Many insecure, dreary, and degrading work processes have been substituted by automatized procedures. And one should always remember that automatization brings about cheaper items accessible to many more people. Robotting need not be a straightforward expression of symbolic violence. Still, it is an expression of mathematics-based symbolic power.

Surveilling

The distinction between sense and reference with respect to logic can be connected directly to symbolic power, determining what can be expressed and what cannot. One can associate this power with George Orwell's (1987) presentation of Newspeak and Oldspeak in *Nineteen Eighty-Four*. While Oldspeak was roughly associated with everyday English, including all the possibilities for being, for instance, ironic, mistaken, pointed, neutral, political, and critical, Newspeak did not allow for any ambiguities. Furthermore, the idea of Newspeak was to allow only the formulation of "politically correct" statements, due to its very grammar, which should filter what could be said. While Oldspeak allowed for the formulation of controversial issues as well as for any kind on small talk and babbling, Newspeak made possible only the formulation of "correct" thoughts. There are simply no grammatical possibilities in Newspeak for expressing deviant opinions. The imposition of Newspeak would prevent people from committing any form of thought crime against the given political order. Thus, Newspeak exemplifies a tremendous form of symbolic power.

This idea of regulating thought through a regulation of grammar was not invented by Orwell, but he might not be aware that Rudolf Carnap (1937, 1959) had formulated the same idea, although in a quite different setting. Carnap was a socialist, and he might well have shared many of Orwell's political priorities. Simultaneously, Carnap expressed the most radical position of logical positivism, and he was fascinated by the possibility of establishing a language of science along the lines Frege had proposed and as Wittgenstein had condensed in the 6th statement in *Tractatus*—a hard-boiled condensation of a scientific Newspeak. Carnap was eager to explore how a grammar of a formal language could ensure affinities with the overall outlook of logical positivism.

According to logical positivism, the grammar of natural language, Oldspeak, is too loose. This grammar allows for the formulation of many types of sentences, including statements that can be considered scientifically significant, but also sentences with allegorical meaning, poetic value, humorous effect, not forgetting pure nonsense. Thus, sentences like "Good is almighty" and "blue is better than the eleventh prime number" are both, grammatically speaking, completely correct, but according to Carnap it does not make sense to embark on any scientific investigation in order to clarify their truth-value. It seems consequential to try to formulate a language in which grammar ensures that what can be formulated grammatically correctly is exactly the set of proper scientific sentences. It was Carnap's ambition to characterize such a language, and his suggestion was that this was exactly the language for which Frege and Wittgenstein had paved the way. It should be a formal language where the dimension of "sense" was eliminated, only considering "references." Such a language would do the job and take up the role of a scientific Newspeak that would articulate exactly what was assumed relevant to articulate in science. In the book *The Logical Syntax of Language* from

1937, Carnap carried out a careful investigation of this idea.[6] He saw it as a tremendous scientific improvement to establish formal languages as the languages of science. This would eliminate a range of irrelevant notions and preoccupations, as they were connected to the sense-dimension of language. Thus, Carnap found that a formal language could represent a worldview, the mechanical, which was assumed to represents reality.

Formal languages play important roles in natural sciences, but the implementation of formal languages is far from restricted to this scientific domain. The symbolic power of such languages can be observed in very many other domains. When the primary-secondary prioritizing claims references to be primary and senses to be secondary, we witness an implementation of the formal worldview. Let me consider a particular domain subjected to such a worldview, namely, surveillance.

An actual technology of surveilling was outlined by Orwell (1987) in *Nineteen Eighty-Four*, where Big Brother, through the TV screen, could see whatever was going on in each individual home. The idea is recast in the popular TV show, *Big Brother*, where the audience plays the role of Big Brother in surveillance of a small community populated by beautiful young people. The popularity of *Big Brother* might have surprised Orwell, as this program profits from precisely what he finds repugnant: peeping into people's private lives.

However, surveilling is much more than "seeing"; it is also "remembering." The efficiency of surveillance presupposes that one keeps track of what is seen. Surveilling includes registering. One can register names of people; one can register their actions, and in particular their criminal acts; one can register people's economic situation, their debt, fortune, and income, which are all important in order to determine and control tax payment, but which are also important in order to judge people's credibility as loan takers; one can register people's economic actions, the transactions they are involved in; one can register people's preferences as part of the management of advertising; one can also try to register their opinions and political priorities; one can register students' performances in school as well as their competences in general. One can register whatever one desires. Making a register is a principal element of the technology of surveilling.

In order to make a register work, naming is necessary since the units that are surveilled need a label. A number is a straightforward label. However, the discussion of how privacy might be compromised by the introduction of civil registration numbers is now almost forgotten. Such a registration number is now an almost universally accepted phenomenon, and this facilitates the digitalization of surveillance. As soon as this is completed, the processing of information can begin; and through this processing, the technology of surveillance establishes its profile. One can establish all kinds of co-sets: the set of old people, women, diabetics, unreliable tax payers, prison inmates, people holding a driver's license, and

[6] See also Carnap (1959).

so on. And one can introduce all sorts of techniques for producing lists of people with certain characteristics. Furthermore, one can make correlations between all these different subsets. And on top of this, political, economic, and administrative decisions can be acted out.

Surveillance can take any form. One can, for instance, think of image recognition, which is employed in techniques for recognizing fingerprints and faces. Surveillance can also deal with the physical position and movement of the person, getting through passport control; moving from one country to another; going through the gate to the factory building; entering the supermarket; entering the parking lot, and such. Transactions on the Internet can be screened, and all e-mails can be logged. The frantic proliferation of surveillance is due precisely to the formal languages that facilitate the accompanying computational techniques.

The purpose of natural sciences, according to Carnap (1937), is to ensure extensive description of natural phenomena, and formal languages provide a grammar for doing this. The idea of establishing a surveillance of a certain domain appears not very much different from such a purpose. However, surveillance is much more than providing a description. It includes a configuration by means of labeling of the domain to be surveilled; it includes the elaboration of registers; it includes processing of information from these registers; and it turns into schemes of control, which in turn can be managed by any type of institution or organization and come to serve a range of political and economic interests.

The present-day scheme of computer-based surveillance would not have been possible without the imposition of the formal worldview, which configures the domain to be surveilled as a domain of "references." Surveilling is based on the Newspeak that accompanies the formal worldview. This language reduces a phenomenon to a single feature and in this way, subjects it to a scheme of control.

Surveilling can be described in negative terms. But there are many more comments to be added. Surveillance is far from a simple harmful phenomenon. It is crucial for safety, for instance with respect to transport and travels. One also needs to consider all the different procedures of surveillance that make part of healthcare programs. Scanning of the human body is an example of surveillance that would not have been possible without computerized digitalization put in operation. The symbolic power acted out through surveillance is far from just a violent power. It can have many different qualities.

Symbolic Power: A Challenge for Critique

I have tried to show that mathematical discourses provide conditions for exercising symbolic power. In particular, I have indicated how mathematics facilitates primary-secondary nominations with respect to appearance-reality and sense-reference. Through these nominations, mathematics fabricates worldviews, being mechanical or formal. Such worldviews are manifestations of the symbolic power of mathematics. They are crucial for robotting and surveillance as well as for all kinds of ramifications of these phenomena.

Thus, I have tried to show that mathematics simplifies the designating thing, reducing it to a single feature. It dismembers the thing, destroying its organic unity, treating its parts and properties as autonomous. It inserts the thing into a field of meaning, which is ultimately external to it.

However, as already mentioned, the language of mathematics is not just violent. It can be powerful in many different ways. Thus, we find automatizations that are not established in order to dominate workers. We find forms of surveillance associated with generally respected aims. There are in fact very many examples of symbolic power that are not possible to evaluate on any two-dimensional good-bad scale. Thus, symbolic power associated with mathematics can be problematic, tricky, cynical, suspicious, cooperative, brutal, supportive, dubious, loyal, helpful, benevolent, and, not to forget, violent. In fact, mathematics-based symbolic power can operate with all kind of qualities. As a consequence, it is not possible to provide any *general* critique of such symbolic power. Instead, the critique needs to be specific and contextualized. It becomes complex and uncertain as well. Thus, the symbolic power of mathematics is a real challenge for any critical enterprise.

Also for critical mathematics education. As mentioned previously, addressing the symbolic power of mathematics is crucial for leaving behind the modern conception of mathematics that includes a celebration of mathematical rationality. It is crucial for formulating a critical conception of mathematics that addresses the many actions that can be exercised through mathematics, like, for instance, robotting and surveilling.

NOTE

This paper was first published in 2012 in *Educational Studies in Mathematics, 80*(1), 119–132.

REFERENCES

Carnap, R. (1937). *The logical syntax of language*. London, UK: Routledge and Kegan Paul. (Original work published 1934)

Carnap, R. (1959). The elimination of metaphysics through logical analysis of language. In A. Ayer (Ed.), *Logical positivism* (pp. 60–81). New York, NY: Free Press. (Original work published 1932)

Clarke, D. M. (2006). *Descartes: A biography*. Cambridge, UK: Cambridge University Press.

Copernicus, N. (2002). *On the revolutions of heavenly spheres* (S. Hawking, Ed.). Philadelphia, PA: Running. (Original work published 1573)

Descartes, R. (1998). *The world and other writings* (S. Gaukroger, Trans. & Ed.). Cambridge, UK: Cambridge University Press, (Original work 1664)

Frege, G. (1969). Über Sinn und Bedeutung [About sense and reference]. In G. Frege, *Funktion, Begriff, Bedeutung: Fünf logishe Studien* [*Function, concept, reference: Five studies in logic*] (pp. 40–65). Göttingen, Germany: Vandenhoech & Ruprecht. (Original work published 1892)

Frege, G. (1978). *The foundations of arithmetic/Die Grundlagen der Arithmetik* (J. L. Austin, Trans.). Oxford, UK: Blackwell (Original work published 1884)

Galilei, G. (1957). *Discoveries and opinions of Galileo* (S. Drake, Trans.). New York, NY: Anchor.

Mill, J. S. (1970). *A system of logic*. London, UK: Longman. (Original work published 1843)

Newton, I. (1999). *The principia: Mathematical principle of natural philosophy* (I. B. Cohen & A. Whitman, Trans.). Berkeley; London, UK: University of California Press. (Original work published 1687)

Orwell, G. (1987). *Nineteen eighty-four*. London, UK: Penguin. (Original work published 1949)

Skovsmose, O. (2009). *In doubt: About language, mathematics, knowledge and life-world.* Rotterdam, The Netherlands: Sense.

Skovsmose, O., & Ravn, O. (2011). *Matematikfilosofi* [Philosophy of mathematics]. Århus, Denmark: Systime.

Taylor, F. W. (2006). *The principles of scientific management.* New York, NY: Cosimo Classics. (Original work published 1911)

Wittgenstein, L. (1992). *Tractatus logico-philosophicus* (C. K. Ogden, Trans.). London, UK: Routledge. (Original work published 1921)

Žižek, S. (2008). *Violence*. New York, NY: Picador.

CHAPTER 16

CAN FACTS BE FABRICATED THROUGH MATHEMATICS?

Can truth be fabricated? Can facts be fabricated? Can facts be fabricated through mathematics? In order to address such questions, I take the following steps:

First, as an introduction, I look at Descartes' assumptions that there exists certain Laws of Nature and that a perfect harmony can be established between knowledge and the reality that knowledge is about; in other words, the assumption that epistemic structures can represent ontological structures.

Second, we move on to a set of formulations about the notion of truth made by Foucault with particular reference to Nietzsche. These formulations represent an extreme opposition to the position of Descartes, as Nietzsche suggests that knowledge is not *about* anything. There is no reality that provides knowledge or truth with any substantiation. Knowledge is only an expression of something, namely, of a will to power.[1] This is the observation which leads Foucault to pronounce that knowledge and power operate in complex dialectics.

Third, with this observation in mind, we turn towards mathematics. In this domain, there has been much discussion about the nature of truth. What could a mathematical truth be about? The discussion about the nature of truth in the

[1] See, for instance, Nietzsche (1968, 2010).

Critique as Uncertainty, pages 231–245.

philosophy of mathematics appears completely different from the preoccupation with truth expressed by Nietzsche and Foucault. But we will be considering if one might establish any connections.

Fourth, we will address an example in mathematics regarding the formulation of a particular mathematical construct. By clarifying what this construct is about and what could be done by means of this construct, we will address more broadly what it could mean to put mathematics in action.

Fifth, we will indicate some of what can be done through mathematical constructs within technologic, economic, and sociopolitical contexts. This leads us to identify certain connections between mathematics and fabrication in a variety of domains. In this way, we might come to acknowledge that the Nietzsche-Foucault opening of the notion of truth also might be relevant with respect to mathematics and that a knowledge-power dynamics is in operation round mathematics-based fabrications, too.

Affinity

Euclid (2003) presented geometry in an axiomatic format that became paradigmatic. A system of knowledge has to be initiated by axioms; they have to be few and simple enough that their truth can be grasped by intuition. Intuition is not a reliable human faculty, and in many cases intuition has lead us astray, but in case we are dealing with sufficiently simple axioms, intuition is assumed to be able to grasp their truth with certainty.

From axioms, theorems can be deduced, and deduction has the property that if p implies q and p is true, then q is true as well. So if, on the basis of the axioms, one does a proper deduction, the truths of the axioms will be conveyed through the channels opened by deduction and extend to all the deduced theorems. Euclid expounds how certain knowledge can be obtained within mathematics, and, when generalized, this approach establishes a paradigmatic format for establishing knowledge within any domain.

According to Descartes, God is the creator of the universe, and he has imposed some fundamental laws which nature must obey. In *The World*, Descartes presented these laws.[2] The first law states that "each particular part of matter always continues in the same state unless collision with others forces it to change its state" (1998, p. 25). The *second law* states that "when one of these bodies pushes another it cannot give the other any motion except by losing as much of its own motion at the same time; nor can it take away any of the other's motion unless its own is increased by the same amount" (p. 27). The *third law* states that "when a body is moving . . . each of its parts individually tends always to continue moving along a straight line" (p. 29). Furthermore, Descartes claims that no other laws have been imposed on nature. This means that from these three laws, taken as axioms, one will be able to deduce all true statements about nature. As the laws,

[2] See also Skovsmose (2009).

claimed to be imposed by God, only concerns material properties, Descartes had established a materialistic perspective on Nature.

Descartes was inspired by the Euclidian paradigm, and he outlined how a complete insight into nature could be founded on a few axioms. It soon turned out that mathematics would be of tremendous importance for completing this deduction, as the basic law of nature, through Newton's reformulations (1999), was of a mathematical form.

The whole idea includes some principal assumptions about the nature of knowledge. There exists a reality that knowledge is about. Furthermore, it is possible to establish a close affinity between the structures of knowledge and the structures of reality. First, one has to do with an affinity with respect to truth, which means that a certain state of affairs can be represented in a statement. This statement can be declared true if the statement expresses a state of affairs which is in fact the case. Second, one has to do with an affinity with respect to deduction. This means that if, from true statements one makes a valid deduction, one will arrive at another true statement. Thus, deduction preserves truth-affinity.

Running through Descartes' interpretation of knowledge, one finds assumptions about affinity with respect to truth and deduction. Similar affinity assumptions have be elaborated on in different epistemological settings. As an example, one can refer to Wittgenstein's picture theory, as presented in the *Tractatus* (1992), and we may refer to logical positivism which assigned a particular role to mathematics, providing scientific theories with reliable descriptions. Assumptions about affinity have also been criticized, and a direct attack was initiated by Nietzsche and followed up by Foucault.

Illusion of Affinity

In the following, I want to refer to a few remarks made by Foucault in "Truth as Juridical Forms" (see Foucault, 1994, pp. 1–89). Foucault wants to show how social practices may produce domains of knowledge that include "new objects, new concepts, and new techniques" (Foucault, 1994, p. 2). And he wants to show how "truth itself has a history" (Foucault, 1994, p. 2).

There are many such underpinnings upon which Foucault could build, and he had prosperous sources of inspiration allowing him to leave behind any conception of knowledge and truth as representing a relationship between a domain of knowledge and a reality with which that knowledge is concerned. Knowledge can be seen as invented. It does not signify any form of relationship with any form of reality. Knowledge is not "even closely connected to the world to be known" (Foucault, 1994, p. 8); it has "no affinity with the world to be known" (Foucault, 1994, p. 9). This is really a profound claim. The formulation negates the whole idea, so clearly formulated by Descartes and integrated in the very vocabulary of the scientific revolution, that knowledge represents such an affinity. In particular, the relationship between the assumed mechanical reality and our knowledge about this mechanism was claimed to be established most adequately through mathematics. Mathematics represented the affinity-relation, both with respect to truth and deduction. However, when the notion of affinity is discharged from the consideration of knowledge and truth, we are really opening up a new epistemic terrain.

Foucault observes, "Nietzsche means that there is not a nature of knowledge, an essence of knowledge, of the universal conditions of knowledge; rather, that knowledge is always the historical and circumstantial result of conditions outside the domain of knowledge" (1994, p. 13). There is no particular nature in knowledge which a profound epistemological investigation can be expected to discover. The theory of knowledge cannot be lifted out of a discussion of what has been claimed to be called knowledge within a sequence of historical periods and sociopolitical settings. From an epistemic perspective, there is no different between a "theory of knowledge" and, say, a "theory of politeness" and a "theory of obedience."

This brings us, along with Foucault, deep into perspectivism: "knowledge is always a perspective" (Foucault, 1994, p. 14), and Foucault points out

> When Nietzsche speaks of the perspectival character of knowledge, he is pointing to the fact that there is knowledge only in the form of a certain number of actions that are different from one another and multifarious in their essence—actions by which the human being violently takes hold of a certain number of things, reacts to a certain number of situations, and subjects them to relations of forces. This means that knowledge is always a certain strategic relation in which man is placed. (1994, p. 14)

As a consequence, "One can speak of the perspectival character of knowledge because there is a battle, and knowledge is the result of this battle" (Foucault, 1994, p. 14).

It is important to consider the kind of notions with which such formulations surround the idea of knowledge. It is notions like perspective, action, violence, reaction, force, strategic relation, and battle. Out of this battle emerges knowledge. Traditionally, knowledge has been surrounded by domesticated notions like justification, truth, belief—to start off from the classic definition of knowledge as justified true belief—but we can add notions like argument, logic, rationality, representation, correspondence, and so on. However, when Foucault changed this classic conceptual environment into the radically different Nietzsche-inspired environment, he made it possible to address the knowledge-power dynamics.

This seems to bring about an extreme form of relativism, as one could claim that the only reason that a certain knowledge-claim can be maintained is that it supports or is supported by a dominant perspective. In his lecture, Foucault makes an interesting distinction between "powerless truth" and "truthless power" (1994, p. 33). I am uncertain how Foucault tries to position himself between the two notions, or if he finds that Nietzsche has annihilated any possible distinction between them.

Truths in Mathematics

When considering truth with respect to mathematics, we move into a completely different landscape of concepts. My point, however, is that truth with reference to mathematics might include issues of a Nietzsche-Foucault brand. Well, this observation is for later. Here we just try to get a hold of some features of the landscape where the discussion of mathematical truth has traditionally been located.

It has widely been assumed that there exists an intimate connection between truth and proof. Thus, a proof can be considered a procedure for establishing the truth of a theorem. This observation is in accordance with the Euclidian paradigm, according to which the axioms provide the starting point, and as deduction is a reliable carrier of truths, the truth of the axioms is distributed to all theorems. This paradigm has been presented in different ways, and it is part of Descartes' epistemological approach.

If one adds a dose of Platonism to the characteristics of the axiom-deduction-theorem device, then axioms represent insight into some simple truths about the mathematical world of ideas, while the theorems represent more complex truths. The pattern of deduction represents a sublime affinity between reasoning and the reality of ideas. Within this framework the notion of truth becomes safely located among notions like axiom, proof, theorem, certainty, and (mathematical) reality.

Let us take a look at a particular mathematical theorem: There exist an infinite number of primes. A classic proof starts by assuming that there is only a finite number of prime numbers. As a consequence, a greatest prime number must exist, which we could call P. It is then possible to define a new number, namely, $P! + 1$. This number cannot be divided by any of the numbers, 2, 3, 4, 5, . . . , P (we do not consider 1 a proper divisor), as P! is divisible by all these numbers and no number can be divisor of two consecutive numbers as P! and P! + 1. Therefore the number $P! + 1$ must either be a prime number itself, or be divisible by some prime number P^*. However, both $P! + 1$ and P^* must be greater than P. This contradicts our original assumption that P is the greatest prime number. So, if we assume that P is the greatest prime, we have reached a contradiction. We cannot assume the existence of a greatest prime number P, and therefore, infinitely many prime numbers must exist. This is a classic formulation of the proof.

We could, however, open up a more profound discussion of this proof. We are dealing with an informal proof, mixing natural-language formulation and symbols. Furthermore, it is not clear which axioms the proof assumes. What truths about the natural numbers are taken for granted as part of the proving? The layout of the proof does not meet any Euclidian standards. In fact, it was only through the formulation of Peano's axioms that the theory of numbers got a proper foundation.[3] These axioms can be presented in the following way:

[3] See Peano (1967).

$$(1)\ 1 \in N$$

$$(2)\ n \in N \Rightarrow S(n) \in N$$

$$(3)\ \forall n \in N\colon 1 \neq S(n)$$

$$(4)\ \forall n, m \in N\colon S(n) = S(m) \Rightarrow n = m$$

$$(5)\ \forall P\colon [P(1) \wedge (\forall n \in N\colon P(n) \Rightarrow P(\mathrm{S}(n)))] \Rightarrow \forall n{\in}N\colon P(n)$$

The particular symbols are N to be considered a set where 1 is a member; n which refers to an element of the set N; and S which is a function with N as domain, where $S(n)$ designates the successor of n. The symbol P refers to a property which numbers may hold, while the rest of the symbols belong to logical formalism. Axiom (1) states that 1 is a natural number.[4] Axiom (2) states that if n is a natural number then the successor of n, $S(n)$, is also a natural number. Axiom (3) claims that 1 is not a successor of any number n. Axiom (4) claims that if the successors of two numbers are identical, then the numbers themselves are identical. The last axiom (5) concerns mathematical induction, and it claims that if the number 1 has the property P, and if it is shown that if the number n has the property, then $S(n)$ will have the property as well, then all natural numbers will have the property.

Considering this axiomatic system for the theory of natural numbers, the proof of the infinity of prime numbers needs a substantial elaboration. There seems to be quite a number of missing links that need to be identified. Just to make a simple observation, a prime number is defined with reference to the notion of multiplication. However, Peano's axioms do not include any notions of addition, subtraction, multiplication, or division. They need to be introduced, thus addition can be defined in terms of the function S, and when addition is defined, the other notions can be defined as well. Next we need to establish proofs of simple properties, like the commutative law for addition: $a + b = b + a$. This law is not part of the Peano axiomatic but needs to be established as a theorem. So it is only through a thorough elaboration of the Peano axiomatic that we would reach the theorem that there exists infinitely many prime numbers (a theorem that should naturally be formulated in the symbolism of the system).

However, even if we imagine that we have established a proof that complies with all the Peano standards, there is much more to consider. In particular, does a Peano proof *prove* the *truth* of the statement that there are infinitely many prime numbers? It appears that if the Peano axioms are true, then the theorems of the systems are true. But are Peano axioms in fact true? Well, the first four axioms

[4] Peano assumes the set of natural numbers to begin with 1; in many recent presentations, it is common to let it begin with 0.

seem sufficiently simple that we might try to rely on intuition to provide a truth-injection. But what about the fifth axiom? How does intuition cope with this axiom?

This question brings us into logicism, as presented by Frege and by Whitehead and Russell.[5] According to logicism, we could claim that Peano's axioms are true. However, this claim needs the most elaborate substantiation. The logicist program is to present logic in an axiomatic format where Peano's axioms appear as theorems. This approach will reveal the logical complexity of mathematics. Furthermore, it will demonstrate the truth of the Peano axioms, at least if one can assume the truth of the axiomatics of logic. Russell found this to be the case, as the axioms presented in *Principia Mathematica* in fact appeared extremely simple and manageable, even for a feeble intuition. As a consequence, all theorems proved in the system of logic are true; therefore the Peano axioms are true; therefore they provide truth to all theorems of the systems; and therefore it is true that there exists infinitely many prime numbers.

However, if we do not follow logicism but formalism, the situation looks different. According to formalism, the only thing we can claim with respect to a formal system is that *if* the axioms are true, *then* the theorems are true. According to formalism, one cannot make any truth-claim with respect to axioms.[6] As a consequence, there is no truth for deduction to bring to the theorems. It appears that we have to do with a postal system where all envelopes are empty. One cannot claim that the theorem postulating an infinite amount of prime numbers is true, even after one has completed a proof of it within a formal system. The only thing one can claim is that the theorem can be proved within a given formal system.

Formalism also makes clear that the same statement need not be provable within another formal system. Formalism has in fact suggested that the notion of truth simply be interpreted as provability; this means that mathematical truth becomes a relative property. Truth only refers so some deductive properties of a formal system, is does not say anything about a mathematic reality. Formalism drained mathematical truth of any ontological elements. Mathematics is not about anything. Such knowledge becomes a perspective; it becomes located *within* a formal system.

Let us now consider the intuitionist perspective on the story.[7] The first proof of the existence of infinitely many prime numbers was indirect in the sense that it was assumed that only a finite number of prime numbers exists. Then it was shown that this assumption brings about a contradiction, and as a consequence, one concludes that there must exist infinitely many primes. More generally, the structure of an indirect proof is (1) one assumes the truth of a statement non-p; (2) one proves that assuming the truth of non-p implies a contradiction; (3) one concludes that p is true. However, according to the intuitionist approach, this form

[5] See Frege (1967, 1978), and Whitehead and Russell (1910/1913).

[6] See Hilbert (1967a, 1967b) and Curry (1951).

[7] See Brouwer (1975a, 1975b) and Heyting (1956).

of argument is not valid in mathematics. The indirect proving assumes the validity of the logical statement "*p* or non-*p*." But this logical property is questioned by intuitionism. One can not demonstrate a statement *p*, by shoving that non-*p* brings about a contradiction. If one wants to prove that there exists infinitely many prime numbers, this infinity should be directly constructed. According to intuitionism, the original proof would show nothing about the truth of the statement that there exists infinitely many prime numbers. And, furthermore, a formal elaboration of all the missing links makes no difference in case the pattern of indirect proof is preserved. In fact, intuitionism does not assume that formalism represents any deeper features of mathematics. So, according to intuitionism, none of the argument so far presented reveals anything about the infinity of prime numbers.[8]

We could continue the discussion of mathematical truth along the lines of logicism, formalism, and intuitionism, respectively. But we may broaden the discussion by taking a brief look at Lakatos' investigations in *Proofs and Refutations* (1976). Lakatos tries to characterise the logic of mathematics discovery according to the overall conceptual framework provided by Karl Popper. While Popper (1965) presents the logic of scientific discovery as a dynamic between conjectures and refutations, Lakatos presents the logic of mathematical discovery as a dynamic between proofs and refutations. This indicates an interesting status of the notion of proof. It plays the role of "conjecture," which appears very different from playing the role of truth-deliverer. Furthermore one should notice that Lakatos in *Proofs and Refutations* does not come up with a clarification of the notion of mathematical truth. Instead it seems to be dissolved in the complex dynamics between proofs and refutations.

If we were to try to draw a preliminary conclusion, it could be that mathematical truth is not a simple notion. It invites a variety of apparently contradictory considerations, even when we try to situate the discussion in a conceptual landscape together with notions as axiom, proof, theorem, certainty, (mathematical) reality, conjecture, refutation, and so on. Apparently we are still keeping the discussion in a safe distance from the Nietzsche-Foucault truth-hullabaloo made up of notions like perspective, action, violence, reaction, force, strategic relation, and battle. But let us now take a step further with respect to mathematics.

Mathematical Constructs

We are now going to take a different route into the discussion of truth with respect to mathematics, as we are going to consider the notion of fabrication. A first step on this route is to consider the idea of *mathematical construct*. This notion is

[8] However, it is possible to make the proof of infinity of primes constructive: Given any prime P_n, we can construct a greater prime P_{n+1}, and a proof including such a direct construction will be acceptable to intuitionism. In fact, intuitionism has made a huge effort to reformulate many classic proofs in a constructive way.

to prime us for a discussion of how mathematics comes to be part of a variety of forms of actions.

As an example of a mathematical construct, let us consider what could be referred to as a one-way function. This is a function, f, where it is easy from a given x to calculate $y = f(x)$, but impossible in any feasible way to calculate $x = f^{1}(y)$. A one-way function is not an axiom, neither a proof nor a theorem. It is a mathematical entity. However, we are not trying to instigate a discussion of the ontological nature of such entities, trying to position ourselves between, say, realism and nominalism. We are instead trying to consider the significance of such a mathematical construct in terms of what could be done with it. What space of possible actions becomes available with such a construct at hand?

The significance of a one-way function is related to modern cryptography.[9] Cryptography includes processes of encryption and processes of decryption. If we associate the calculation of $f(x)$ from the value of x to the process of encryption, then technically speaking, it becomes uncomplicated to encrypt. However, if we associate the processes of calculating the value of x from the value of $f(x)$ to the process of decryption, we have ensured that decryption is impossible in practice for an outsider, who has no chance of calculating $x = f^{1}(y)$.

In the context of cryptography, a principal example of a one-way function of two variables is $f(p, q) = pq$, where p and q are prime numbers of considerable size. Thus, when p and q are prime numbers of about 100 digits, it would be impossible to complete a factorization of the 200 digit number pq, and identify p and q. A factorization of such a number cannot be completed in historic time, even using all the computational power now available in the world. And as we have proved (one way or another) that there are infinitely many prime numbers, the exemplification of one-way functions in terms of the product of two large prime numbers makes sense.

Let us assume that the message to be encrypted is a number P. Turning a message in the form of a sequence of symbols into a number is a process that can be completed according to straightforward algorithmic procedures. The transformation of a string of symbols into a number can be considered a public and transparent process. The first step, then, is the encryption of the number P. The function of the encryption, E, will then transform P into $E(P)$. The essential point in modern cryptography is that E could be constructed as a one-way function. This means that while it is straightforward to calculate $E(P)$ from P, in practice it becomes impossible to calculate P from $E(P)$. In other words, it is impossible to calculate $E^{-1}(E(P) = P$; or if we refer to the process of decryption as D, we have $D = E^{-1}$ and that $D(E(P)) = P$. In order to calculate $D(E(P))$, additional information is needed. However, if such extra information is available, in this case, the value of the large

9 The significance of one-way functions for cryptography was pointed out by Diffie and Hellman (1976).

prime numbers that pay a crucial role in establishing E as a one-way function, then E becomes a transparent two-way function.

The properties of one-way functions build on number theoretical insight; however, the significance of this insight is not included in the variety of theorems which make part of number theory, The significance of a mathematical construct, like a one-way function, cannot be revealed through the clarification of the number theoretical insight that surrounds the construct. Its meaning is not composed of the regular mathematical meanings of the notions that surround the construct. Instead, the significance of a mathematical construct can be explored in term of its technological potential.[10]

Mathematical constructs make up a crucial and unique part of technological imagination. There are no other constructs, say, formulated in everyday language, which could substantiate the role of a mathematical construct as part of technological imagination. That mathematical constructs are unique to technological imagination is illustrated by the role played by the notion of one-way function in opening up new schemes of cryptography.

The significance of a mathematical construct has to be examined in terms of its potential in relation to technological design and fabrication. Let me add a few comments about my notion of technology: I use the term in a broad sense as referring to any form of technological machinery, be it an airplane, a computer, a bridge, a robot, or so on. Furthermore, I let technology refer to a range of procedures of management and structures of organization, for instance with respect to procedures for production, automatization, decision making, communication, money processing, and such. My point is that within such a broad definition of technology, we find that mathematical constructs have been brought into action. In this sense, I suggest exploring the significance of mathematical construct in terms of fabrication.

Fabrications

Let us return to the notion of truth. We have discussed truth in terms of affinity, and it makes sense to try to discuss the truth of a statement in this way. But it appears problematic when one tries to discuss the possible truth of a fabrication. Does it make sense to talk about the truth of technological innovation or of a design in general? It appears that when we pay less attention to mathematics as a logical architecture and instead focus on mathematical constructs and how they make part of processes of fabrication, we enter into a different conceptual landscape for reflecting on mathematics. We will try to illustrate this.

A cryptographic scheme is part of reality: It can be difficult to break. It could also turn out that one of the assumed prime numbers is in fact a product of other prime numbers, which would make it easier for an outsider to break the code. In

[10] See Skovsmose and Yasukawa (2009), as well as more general presentations like Schroeder (1997) and Stallings (1999).

such a situation, mathematics is more that a descriptive device. Mathematics is not just a tool for describing cryptographic process; instead it is part of the cryptographic processes themselves. In this sense, mathematics is part of the fabrication of a piece of our reality. And as such, one can address the mathematics construct and the function it fulfills is the same way as one can address other forms of fabrication. It could have many qualities: It could be original, expensive to buy and implement; easy to manage; or it could be possible to manage only for people with proper qualifications, and so on.

Mathematical constructs are essential for any processing of information and knowledge. We can see such processing as established through algorithms with all kinds of formats. Again, the role of mathematics is not simply to describe how information and knowledge are processed. Mathematical algorithms constitute the very processing; they perform the processing. Algorithms can have many different qualities: They can be fast, and this quality is crucial, since faster algorithms mean that knowledge and information processing is experienced differently by the users of the phone, the Internet, the TV, the credit card, and such. The possibility of establishing more efficient processing is of high value, just as it is with forms of transport, and as a consequence, one finds much research carried out with respect to mathematical algorithms.

Schemes for surveillance and monitoring are based on mathematical constructs. One important element here is the processing of huge amounts of data. Very many different mathematical techniques are brought together, and this mathematical complexity establishes more than a *description* of surveillance and monitoring. It constitutes the very act of surveillance and monitoring itself. And one brings together very diverse mathematical elements in order to establish a technology of control. In particular, image recognition can become a crucial feature of control, because with it, a machine will be able to recognize faces and fingerprints, and such recognition can be applied when a person passes a gate, wherever the gate might be positioned: in the airport, in the company, at the shopping center, or at the football stadium.

As mathematical constructs come to make up parts of our reality, this reality is changed. Through all forms of fabrication, it becomes a fabricated reality. It becomes a reality which includes many produced facts: It is possible to transfer funds from one's bank account to another sitting at home, as the transfer can be made via the Internet, and the processes are secure enough. One can drive in the city and find one's way around due to the installed GPS system. This changes the condition of maneuvering in a complex traffic situation.

One can conduct instant decision-making on the basis of the continually updated numbers, figures, and indices available to the global financial market. These forms of information are descriptive, as they condense certain economic transactions; at the same time, they are prescriptive with respect to the decision to be taken. In consequence, the procedures of the financial market are changed, becoming speeded up and more hectic. Such procedures are shaped by the available technology. In this way, we find mathematical constructs constituting part of a fact-fabrication taking place within the financial market. Procedures for production are continuously changed, not least due to the very many schemes for automatizing that can be implemented. Reality for the workers is changed. Processes of rationalization, of outsourcing, of restructuration, of firing, become part of their lived reality. None of any such restructuration with respect to production and work place takes place without mathematics having been brought into action. In this sense, one can claim that mathematics is part of the fabrication of facts. Our sociopolitical reality includes mathematics-based action and fabrication.[11] To acknowledge this brings the discussion of mathematical rationality to the center of knowledge-power considerations.

There are many ways in which mathematics may come to be part of our social reality as an ingredient of technological fabrication. Mathematics makes up part of the "grammar" of many discourses established around issues like economy, production, surveillance, communication, management, decision-making, and the like. Mathematics represents a symbolic power. Studies of this form of symbolic power will add an important dimension to the previous discussion of mathematical constructs; nevertheless, what has been said so far might be sufficient for jumping to some sort of a conclusion, by looking at a remark made by Nietzsche (1974):

> Let us introduce the refinement and rigor of mathematics into all sciences as far as this is at all possible, not in the faith that this will lead us to know things but in order to *determine* our human relation to things. Mathematics is merely the means for general and ultimate knowledge of man. (pp. 215, §246)

Nietzsche points out that we should introduce the refinement and rigor of mathematics. This should be done in all sciences where at all possible. This is in fact a strong recommendation of the use of mathematics. It appears to be completely in line with Descartes' priorities, with the ideas of the scientific revolution and with the recommendation of logical positivism which celebrated mathematics as the language of science. Nietzsche's remark seems to be in accordance with all celebration of mathematical rationality. However, Nietzsche goes on to note that this eager application of mathematics should not be carried out "in the faith that this will lead us to know things." This is really the crucial addition. Here, Nietzsche emphasizes that we should forget all about possible affinities between what is expressed in mathematics and any reality about which we might imagine we are obtaining knowledge. Mathematics, for all its rigor, does not bring us knowledge *about* something.

---

[11] Many have talked about our socially constructed reality. I completely agree with the perspective expressed through this notion. However, we have made an important additional observation, namely, that mathematics plays an important role in fabricating our socially constructed reality.

What then, according to Nietzsche, is the point of using mathematics? To determine our human relation to things! With this remark, Nietzsche points out the knowledge-power dynamics with respect to mathematics. Through mathematics, one can establish relations. So, quite contrary to an epistemic position assuming affinity, Nietzsche makes the point that mathematics tells about humankind and not about nature or any form of external reality. This brings us beyond any conception of truth as expressing a form of affinity between mathematics and something else.

Nietzsche's remarks prime us for acknowledging that we need to consider mathematics in addressing the knowledge-power problem. Foucault was inspired by Nietzsche, but he did not pay much attention to Nietzsche's remark about mathematics. But let us take this remark to suggest that an extensive discussion of knowledge and power could be explored with reference to mathematics. We might even assume that a careful reading of Nietzsche would have brought about the claim that mathematics is the most important site for a knowledge-power archaeology, as mathematics can be considered the means to an ultimate knowledge of humankind.

How, then, to explore mathematics as a general and ultimate expression of power? My suggestion has been that we start paying attention to mathematical constructs. This notion brings us away from the notions of axiom, definition, proof, lemma, theorem, which, one way or another, circle around a conception of truth as an expression of some kind of affinity. The notion of the mathematical construct brings our attention in a different direction. This notion makes us consider how mathematics can constitute part of fabrications of all sorts.

NOTE

This paper was published in *Philosophy of Mathematics Education Journal*. Special issue on Critical Mathematics Education, No. 25, October 2010.

REFERENCES

Brouwer, L. E. J. (1975a). Intuitionism and formalism. In A. Heyting (Ed.), *L. E. J. Brouwer: collected works, Volume I, Philosophy and foundations of mahematics* (pp. 123–138). Amsterdam, The Netherlands; Oxford, UK; New York, NY: North-Holland/American Elsevier. (Original work published 1913)

Brouwer, L. E. J. (1975b). Consciousness, philosophy, and mathematics. In A. Heyting (Ed.), *L. E. J. Brouwer: Collected works, Volume I, Philosophy and foundations of mahematics* (pp. 480–494). Amsterdam, The Netherlands; Oxford, UK; New York, NY: North-Holland/American Elsevier. (Original work published 1948)

Curry, H. B. (1951). *Outlines of a formalist philosophy of mathematics*. Amsterdam, The Netherlands: North-Holland.

Descartes, R. (1998). *The world and other writings* (S. Gaukroger, Trans. & Ed.). Cambridge, UK: Cambridge University Press. (Original work published 1664)

Diffie, W., & Hellman, M. E. (1976). New direction in cryptography. *IEEE transactions on Information Theory, 22*, 644–654.

Euclid. (2003). *Euclid's elements* (T. L. Heath,Trans.; D. Densmore, Ed.). Santa Fe, NM: Green Lion. (Estimated to be written around 300 BC)

Foucault, M. (1994). *The order of things: An archaeology of the human sciences.* New York, NY: Vintage. (Original work published 1966)

Frege, G. (1967). *Begriffsschrift*, a formal language, modeled upon that of arithmetic, for pure thought. In J. van Heijenoort (Ed.), *From Frege to Gödel: A source book in mathematical logic, 1879–1931* (pp. 1–82). Cambridge, MA: Harvard University Press. (Original work published 1879)

Frege, G. (1978). The foundations of arithmetic/Die Grundlagen der Arithmetik (J. L. Austin Trans.). Oxford, UK: Blackwell. (Original work published 1884)

Heyting, A. (1956). *Intuitionism: An introduction.* Amsterdam, The Netherlands: North-Holland.

Hilbert, D. (1967a). On the infinite. In J. van Heijenoort (Ed.), *From Frege to Gödel: A source book in mathematical logic, 1879–1931* (pp. 367–392). Cambridge, MA: Harvard University Press. (Original work published 1925)

Hilbert, D. (1967b). The foundations of mathematics. In J. van Heijenoort (Ed.), *From Frege to Gödel: A source book in mathematical logic, 1879–1931* (pp. 464–479). Cambridge, MA: Harvard University Press. (Original work published 1927)

Lakatos, I. (1976). *Proofs and refutations: The logic of mathematical discovery.* Cambridge, UK: Cambridge University Press.

Newton, I. (1999). *The principia: Mathematical principle of natural philosophy* (I. B.Cohen & A. Whitman, Trans.). Berkeley; London, UK: University of California Press. (Original work published 1687)

Nietzsche, F. (1968). *The will to power* (W. Kaufmann & R. J. Hollingdale, Trans.). New York, NY: Vintage. (Original work published 1901)

Nietzsche, F. (1974). *The gay science* (W. Kaufmann, Trans.). New York, NY: Vintage. (Original work published 1882)

Nietzsche, F. (2010). On truth and lie in a nonmoral sense. In T. Carman (Ed. & Trans.), *On truth and untruth: Selected writings* (pp. 17–49). New York, NY: Harper Perennial/ Modernthought. (Original work published 1873)

Peano, G. (1967). The principles of arithmetic, presented by a new method. In J. van Heijenoort (Ed.), *From Frege to Gödel: A source book in mathematical logic, 1879–1931* (pp. 83–97). Cambridge, MA: Harvard University Press. (Original work published 1889)

Popper, K. R. (1965). *The logic of scientific discovery.* New York, NY: Harper and Row. (Original work published 1934)

Schroeder, M. R. (1997). *Number theory in science and communication with applications in cryptography, physics, digital information, computing and self-similarity* (3rd ed.). Berlin, Germany: Springer-Verlag.

Skovsmose, O. (2009). *In doubt: About language, mathematics, knowledge and life-world.* Rotterdam, The Netherlands: Sense.

Skovsmose, O., & Yasukawa, K. (2009). Formatting power of "mathematics in a package": A challenge for social theorising? In P. Ernest, B. Greer, & B. Sriraman (Eds.), *Critical issues in mathematics education* (pp. 255–281). Charlotte, NC: Information Age.

Stallings, W. (1999). *Cryptography and network security: Principles and practice*. Upper Saddle River, NJ: Prentice Hall.

Whitehead, A. N., & Russell, B. (1910/1913). *Principia mathematica I-III.* Cambridge, UK: Cambridge University Press.

Wittgenstein, L. (1992). *Tractatus logico-philosophicus* (C. K. Ogden, Trans.). London, UK: Routledge. (Original work published 1921)

CHAPTER 17

MATHEMATICS AS PART OF TECHNOLOGY

Two key terms in discussing mathematics education are "modeling" and "problem solving." They are brought naturally into focus when discussion concentrates upon applicability of mathematics. It becomes reasonable for teachers to allow modeling and problem solving to take up a more central position in the educational process, especially when the teacher finds it necessary to argue in favor of the importance of the subject matter on the basis of its applicability.

In the following, I shall discuss ideas concerning the application of mathematics. I choose to concentrate on the concept "model" and only incidentally to take up a discussion of problem solving. This means that questions about the relationship between mathematics and "the real world" are in focus. Further, I have in mind mathematics education in upper secondary school and at the university level. I do not try to relate my viewpoints to primary school teaching, although it is also possible to point out several implications for that level.

My educational perspective is that of critical pedagogy, which involves a specific set of attitudes towards teachers, students, subject matter, and the aims of education, as well as their relations to the complex of social and political con-

Critique as Uncertainty, pages 247–262.

flicts and problems existing outside the school.[1] Briefly, not only teachers but also students are attributed a *critical competence* which is considered a resource to be further developed through their participation in the educational process. The content of subject matter is not (and cannot be) taken for granted on the basis of established tradition, but must continuously be subject to critical revision—there must be a *critical distance* to the curriculum. The teaching-learning process should be oriented towards the goal of providing the students with opportunities to develop their critical competence in the form of qualifications necessary for their participation in furthering democratization processes in society, and this includes developing qualifications that give them competence and capacity to deal creatively with problems in everyday life in such a way that they can support democratization processes in society. And finally, insofar as both teachers and students are oriented critically to traditional content and subject matter of education in order to develop their critical competence in a focus on problems outside the educational universe, there should be support for their efforts to have a *critical engagement* in their common educational (and social) endeavors.

The fundamental question to be answered in the following is thus: which attitudes towards and which approaches to including applied mathematics and the process of mathematical modeling are relevant, insofar as the ideas of critical pedagogy are to be taken seriously?

Pragmatism in Mathematics Education

It is easy to find recommendations in the literature for a more applied oriented mathematics education. An important documentation of all the possibilities for applying mathematics is found in "The Interaction Between Mathematics and Other School Subjects," where H. O. Pollak could state,

> Perhaps the outstanding feature of applications of mathematics in recent years is that the areas to which mathematics is applied have been increasing in number so rapidly. It is fair to say that no area of human endeavour is currently immune from quantitative reasoning or mathematical modeling. Besides the traditional physical sciences and engineering, the biological sciences, the social sciences, the management sciences, the humanities and everyday life are all arenas for interaction with mathematics. This is not meant to imply that mathematics is taking over all these other fields, but there *are* many interesting applications. (Pollak, 1979, p. 235)

It is an immediate implication of this observation that it is important to let students receive an impression of this extensive range of modeling activities; and the best way of doing this seems to be "the natural way": to let students learn about models by modeling, by making their own model, perhaps starting with problems identified by themselves.

---

[1] See Skovsmose (1984) as well as Volk (1975, 1980).

I do not say that this approach is irrelevant. Quite the contrary, I conceive modeling as a most useful activity in mathematics education. However, my thesis is that if such modeling activities are the only way in which students are becoming familiar with the problems of modeling, then simultaneously they are deprived of numerous possibilities for evaluating and criticizing the application of mathematics. Fundamental problems concerning the application of mathematics are not visible from "inside" the modeling process. That is, it is not possible to develop a critical attitude towards the application of mathematics solely by improving the modeling capability of students. In other words, some of the fundamental problems concerning mathematical modeling cannot be formulated in the conceptual framework which students develop through practical experiences organized solely to increase their capacity to construct models.

In fact, we are concerned here with a discussion of the teaching-learning process related to three different forms of knowledge. First, in relation to *mathematical knowledge* itself; second, in relation to *technological knowledge*, which includes knowledge about how to build a mathematical model and how to apply mathematics; and third, in relation to what I will refer to as *reflective knowledge*, which refers to a more general conceptual framework or metaknowledge about the nature of mathematical models and about the criteria used in their construction, application, and evaluation.

It is my thesis that the conceptual framework for discussing and criticizing a mathematical model cannot be reduced to the conceptual framework gained from a modeling process. In short, reflective knowledge cannot be reduced to technological knowledge. (Neither can technological knowledge be reduced to mathematical knowledge, an idea which is expressed in an educational context as when you learn mathematics you also learn how to apply it.)

Due to the fact that reflective knowledge constitutes a central dimension of a critical competence, and that reflective knowledge is partly independent of technological knowledge, it is not possible to remain satisfied with the pragmatic trend in mathematics education, which states that the important thing is to apply and to learn how to apply mathematics.[2] As already mentioned, I do not find the modeling capacity irrelevant, but it is insufficient. In the following we shall concentrate on the question, What is the more specific nature of this reflective knowledge?

Mathematics and Technology

The pragmatic philosophy of mathematics education takes the position that mathematics is useful; but we have to put the concept "useful" into a larger framework. (By doing so I hope to make it more obvious why reflective knowledge and technological knowledge are of different natures.) In general, what is to be con-

[2] This pragmatic trend is for instance found in Andrews and McLone (1976), Bender (1978) and Haberman (1977).

ceived as man's useful tool is technology. So when we concentrate on the applicability of mathematics, it is natural to look at mathematics as part of technology.

The application of the technologies of society requires specific qualifications of the people who employ them. The qualifications needed become formulated as demands for manpower which has acquired certain skills and capabilities relevant for use of technologies, both present as well as those being developed. These demands influence the school system and the curriculum. The school system comes especially under pressure to

- Introduce students to knowledge which enables them (to learn) to handle the necessary technologies;
- Give the students a "convenient" attitude towards technology. By "convenient," I mean "functional" seen from the perspective of the dominating power structures of society. This "functionality" includes a positive appreciation for and passive acceptance of technology, which takes the present lines of development for granted;
- Differentiate among students and select those who have achieved qualifications relevant for implementing the various technologies.

These three sorts of pressure are easily identified when the subject matter is mathematics. Developments take place in technology, and sooner or later implications for mathematics education are explicated, and changes are made to provide what are interpreted to be the relevant qualifications—not because of epistemological or educational reasons, but mainly because mathematics is part of technology. We find the school system under pressure to develop adequate technological knowledge, to develop a "convenient" attitude towards technology (one that involves very little, if any, interest in reflective knowledge about technology), and to sort and select among students with regard to these qualifications.

Previously, the general attitude towards technology and technological development was very optimistic. This development was conceived as the fundamental condition for solution of problems of society. But now the optimistic attitude is mingled with a much more skeptical and pessimistic one. Technology not only provides solutions to social problems, it also creates new social problems. Consequently, it is important to go into a detailed reflection and evaluation of the consequences of technology. I am not only thinking about "technological assessment" but also about a more fundamental and philosophical evaluation of the implication of the emergence of "the new technological society."

Reflective knowledge becomes important and essential, and to instill a "convenient" attitude into the students becomes not only questionable but also quite dangerous, seen from the stance of a critical point of view of education. That is why the pragmatic trend in mathematics education has to be denied. So, although Pollak is right in pointing out the importance of the wide variety of examples of applied mathematics, the conclusion to be drawn is not only that the students have to learn about all the different applications, but that they must learn, too, how to think about the different sorts of both use and misuse.

This conclusion is supported by a position which rejects the thesis of neutrality of technology. The conception of technology as neutral seems to be an implicit assumption—even at times made explicit—in many of the approaches to dealing with applied mathematics. The thesis of neutrality states that technology as such is value neutral, and that it is the intentions and/or the consequences of the application of technology which bear the responsibility for good or bad use. This thesis is incorporated not only in the (implicit) philosophy of technology of logical positivism, but also in the philosophy of orthodox Marxism.[3] Further, it is to be found precisely in the pragmatic approach to mathematics education.

Types of Technologies

One of the first to outline a philosophy of technology is Ernst Kapp (1978) in *Grundlinien einer Philosophie der Technik* from 1877. It is Kapp's idea that technologies could be considered as "projections" of our organs, the hammer as an extension of our arm, and so on. From that perspective, technologies could be identified with tools, and it is the fundamental thesis put forward by Kapp that these tools are used by man in his struggle against nature; in this sense, technology is an expression of a specific state of the man-nature relationship. This idea is quite common, expressed by Karl Marx and other philosophers with different points of view concerning technology. As an example, I quote from *Einführung in die Philosophie der Technik*, where Heinrich Stork as a preliminary definition states that "Technik" is an action

> by which mankind intelligently converts natural material and energies that they come to serve his needs and uses (technical actions); this action leads to an ever-growing collection of things and processes (technical articles). (Stork, 1977, p. 1, my translation)

However, a closer examination seems to disclose the fact that technology cannot be analyzed from one general point of view. There exist different types of technologies. I shall mention

- Manual tools (hammer, etc.)
- Technology of natural processes (steam engine, etc.)
- Social technology (scientific management, etc.)
- Information technology (computer science, etc.)

In the following, I do not take any further notice of manual tools, because this type of technology does not play any important social role in the application of mathematics today. We have to talk about different technologies because the

[3] For a discussion of the neutrality of technology see for instance Rapp (1981, pp. 53ff.).

metaconcepts to be used when analyzing the variety of problems connected to the different technologies cannot be unified into one conceptual framework.[4] The object of a natural technology is nature, but trying to understand social technology from that perspective will cause misunderstandings. The object of social technology is man and human relationships, and the object of information technology is communication. Technology is a way of handling power, but problems concerning the amount and distribution of power are different depending on the object of technology in question.

The Structural Context for Mathematical Models

It would in fact be wrong to conceive mathematics (to be understood as something in use, as a model-building capacity) as an independent technology. Mathematics could, however, be integrated into different types of technologies, and depending on the technological context, the mathematical model has to be analyzed, criticized, and evaluated in different ways. Even if two models seem to be nearly identical, and even if they could be handled mathematically in the same way, they could require quite different forms of evaluation insofar as they belong to different types of technologies. As a consequence, we do not find it possible to reduce reflective knowledge to technological knowledge.[5]

Two fundamental perspectives of an analysis of a mathematical model concern (1) its structural relationships to other elements in the situation, and (2) its developmental relationships. That is, its synchronic and its diachronic relationships.

The most general structural relationships for a mathematical model are shown in Figure 1. The model is constructed to represent an object, but (normally) not directly; the relation between a model and its object is mediated by a conceptual system (to be explained later). As mentioned, the object could be of quite different types: nature, human relationships or communication.

Next, the model is related to a theoretical scenery, so we have further to look at the relationship *model-theory*. The model could be related to a specific and unique theory, being the case if we have to do with a model of natural phenomena. In the natural sciences, it is possible to find a unique theory concerning the object in question: classical mechanics, thermodynamics, and so on. Alternative theories concerning certain natural phenomena could exist, but as a rule, we will be able to point out a single theory, which could constitute the basis for construction of a model to serve a technological need.[6]

---

[4] For a detailed discussion of that point of view see Skovsmose and Jensen (1986).

[5] The Lotka-Volterra differential equations are used both in biological and economic contexs. See Smith (1974) and Goodwin (1972) as well as Chapter 18. But the evaluations of the different models will be most different.

[6] This point has been stressed as a central question in Jensen (1980), a paper which reflects a very important and fruitful discussion of mathematical models taking place at the Roskilde University Centre.

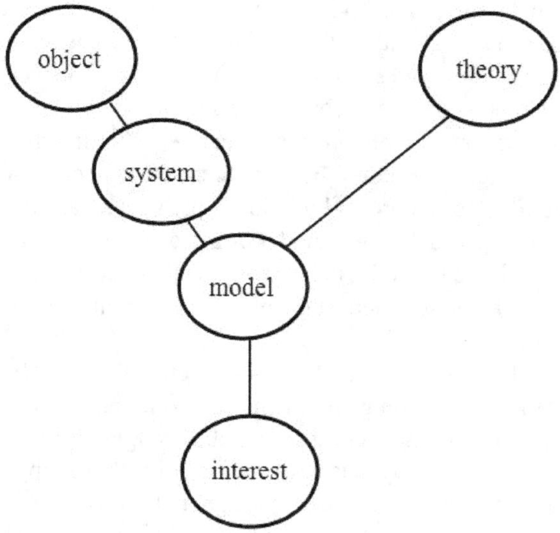

FIGURE 1. Structural relationships for a mathematical model.

If we have to do with social technologies, for instance, in form of macroeconomic models used in handling and discussing political and economic problems, the situation is different. The construction of such a model cannot be based on a homogeneous theoretical framework. Different alternative (and perhaps mutually inconsistent) theories are used to explain the economic phenomena, and so we have to choose a theoretical viewpoint which we want to use when creating a conceptual system to be modeled mathematically. That the model itself is the first "theory" concerning the object in question is another possibility, being the case if the object has not previously been investigated.

Summarizing: A construction of a model could be based (1) on a unique theory, in which case the model could be "deduced" from theory, and the foundation of the model will be of the same "quality" as the foundation of the theory; (2) on a specific theory selected among different alternatives; and on no theory at all, in which case the model therefore creates its own way of looking at its object.

Next we will look at the relationship *model-object*. Only in quite simple, and in this context uninteresting cases, do we find a direct and obvious one-to-one correspondence between model and object. In all other cases (and some will argue, also in the most simple cases), there is not a direct correspondence. Variations in this relation may be affected by factors such as the kind of technology involved (nature, human relationships or communication), the abovementioned three different forms of relationship between model and theory, and so on. The building of a mathematical model of an object is based on a specific way of looking at the object, that is, at "reality." The conception of reality must be so structured in our thinking that patterns can be identified. We have (1) to select elements from reality which are to be conceived as important, and (2) to decide which relationships among these elements are to be considered important. These two fundamental selections constitute an interpretation of "reality." This interpretation could be implicit or made explicit, it could be well-founded or accidental, reflecting the different possibilities for theoretical relationships. That means that a model is not a model of "reality" as such, but a model of a (conceptual) system, created by a specific interpretation based on a more or less elaborated theoretical framework. And these constructions of systems take place in different ways depending on the technology in question.

In the "Simulation Model of the Economic Council" (SMEC) used by Danish economists when advising the government and politicians concerning economic policy and its possible consequences, the Cobb-Douglas function of production plays a crucial role.[7] The model has been developed with that application in mind, and of course a specific choice must be made about how to conceive the economic reality. It is impossible to start model building without assumptions; we have to identify some relationships as fundamental. According to the Cobb-Douglas function, we have $Y = f(L, I)$; thus, the national product Y is a function of the labor force L and capital investment I.

If we are going to evaluate a model, we therefore must look at the relation *object-system*, asking questions like, Which presuppositions have made it possible to create the system? Could these presuppositions be verified in any independent way? Which theoretical frameworks are in accordance with the presuppositions? Which alternative theoretical frameworks do exist?

Next, we have to look at the relationship *system-model*. At this point, specific mathematical ideas are introduced. For instance, in SMEC, we specify the contents of the Cobb-Douglas function, and it is stipulated that $cf(L, I) = f(cL, cI)$ where c is a constant, and demands concerning the partial derivatives of f are stated. Partly these specifications are determined by our knowledge of economics and partly by our mathematical knowledge about how to obtain "reasonable" calculations.

If the model is based on a well-verified theory, which is often the case in natural technologies, we are in possession of two descriptions of the same object: the description via the theory, and the description via the model. So the precision of the model could be tested both via empirical investigations, if it is possible to carry them out, and via theoretical calculations. Due to the fact that theory itself must have an empirical foundation, we have to do with an indirect but very broad empirical test of the model. In this case, some of the key terms in the evaluation will be the model's approximation to the (conceptual) system, its idealization of

[7] SMEC is discussed in detail in Niss and Hermann (1982), especially with upper secondary school teaching in mind.

the object, and the like. If the system is constructed on the basis of a larger number of assumptions or stipulations which are dubious or controversial or obviously value-loaded (often the case in social technologies), a discussion about approximations could become rather misleading if it is conceived as the crucial point in the evaluation. In this case, we first of all have to learn about the differences between the different conceptual systems which could be identified depending on the point of view we want to adopt.

Consequently we have to look at the relationship *model-interest*. A model could be built with different, implicit or explicit, intentions in mind, and further, the use of a model could serve a variety of interests. Both intentions and interests could affect the building of the model not only directly but also indirectly through their influence on the theory chosen as well as on the construction of the system. The concept "interest" is broad and imprecise. In its most universal interpretation, we may consider the way Habermas (1968) relates interests to epistemological structures by talking about knowledge-guiding and knowledge-motivating interests. It is his main thesis that knowledge could not be constructed without interest as a structuring tool. He identifies three types of such interests: an instrumental interest in technical control of the environment; a hermeneutic (interpretative) interest in understanding, especially of the human environment; and an emancipatory interest in developing a constructive awareness of the possibilities for ethical action to transform the given situation.

I do not intend to elaborate Habermas' ideas in any more detail; however, the main point is that the development of mathematical models also involves a specific intention concerning their application, which must be taken into account when we evaluate the use of a model.

The influence of interests on model building also affects the role of models in both "the context of discovery" and "the context of justification" as they are referred to in the philosophy of science. A distinction between these two contexts is emphasized by logical positivism in the formulation of the thesis that the philosophy of science solely has responsibility for dealing with and analyzing the context of justification, in that the context of discovery is outside the scope of rational analysis. To quote one of the promoters of the positivist philosophy,

> The act of discovery escapes logical analysis; there are no logical rules in terms of which a "discovery machine" could be constructed that would take over the creative function of the genius. But it is not the logician's task to account for scientific discoveries; all he can do is to analyze the relation between given facts and a theory presented to him with the claim that it explains these facts. In other words, logic is concerned only with the context of justification. (Reichenbach, 1966, p. 231)

In the present context, I find this distinction important, not because I support the positivistic thesis but because a mathematical model in a technological context could play a role in both the discovery and justification processes. For instance, in macroeconomic theory, the Goodwin model has been elaborated in an attempt to explain the cyclic movements of a capitalistic economy. This model has no significant descriptive value, but its use is justified on the basis of its relevance for promoting the creation of economic ideas and hypothesis. So the evaluation of the Goodwin model is dependent on how it is conceived.[8]

In general, if the interest built into a model is disguised, not only the status but also the evaluation of the model can easily be misinterpreted and misused in both the discovery and the justification processes. SMEC could be used by politicians to give an idea of the "rationale" of a certain economics policy. However, it would be wrong to expect the predictions of a model of this type also to take place in real life, because the model is based on a system, created in the light of certain views, which need not be verified scientific theories. The predictions of the model cannot in themselves justify a political decision, but the prediction of the model could be useful when we discuss different political strategies. A mathematical model is a tool to handle and to investigate hypothetical situations, and it is very useful if it is used as if it belongs to a discovery context.

We are now able to identify a "fallacy" concerning the application of a mathematical model. Very often mathematics could be used to explicate (semi) hidden connections between certain elements in a system, and because the system is man-made this means that is explicates (semi) hidden connections between more or less explicitly stated suppositions about an object (nature, man, or communication). An investigation via a model could serve as a sort of test of consistency between fragment of theory, hypotheses, and assumptions, as illustrated in the case of the Goodwin model.

However, often when a mathematical model is applied successfully in a discovery context, it is supposed to have a similar importance in justifying statements about reality. I call this *the fallacy of supposed justification*. This fallacy appears in quite different ways depending on the technology in question. The fallacy is most often seen when we are dealing with social technologies.

Models in an E-M Process

If we want to identify the consequences of application of a mathematical model and to evaluate and to criticize it, it is not sufficient to investigate the structural relationships of the model. In addition, we have to analyze the model building as an integrated part of (but only a part of) a process of investigation and instrumental practice. During that process, the model interacts dynamically with the abovementioned four elements: object, system, theory, and interest. The process is partly an epistemological process and partly a process of manipulation constituting a foundation of a technological action; in short, an epistemological-manipulative process (e-m process).

The core element of such an e-m process is a *problem* (in this context a technological problem). It would be an oversimplification to conceive a problem as

[8] Groth (1981) gives a detailed elaboration of the Goodwin model. See also Goodwin (1972).

belonging to reality. It makes more sense to interpret a problem as a defect or a lack in a conceptual construction—a problem could be something theoretical. However, it makes even more sense to identify a problem as an inconsistency between a mental/theoretical construction and a real situation—an inconsistency conceived by man as important because of a particular interest. In other words, the concept "problem" does not belong to a single of the elements essential for a description and analysis of the structural or synchronic relationships of a mathematical model, but it constitutes a crucial element as the starting point in an e-m process in which a mathematical model plays a role.

If we apply a mathematical model in an e-m process, we are at the same time making use of a tool which is not neutral in relation to problem identification. An e-m process starts because of a (technological) problem in which we are interested and which becomes redefined within a conceptual framework of a system, the link between reality and model. When we apply mathematics as a tool for handling a problem, we again have to reformulate the problem as it has been defined in the terms of the conceptual system. However, it can be noted that the conceptualization of the system often is constructed in such a way that a mathematization is facilitated. A mathematical model is not a neutral tool; its use normally involves changing or reformulating the problem. So, when using a mathematical model, we should not expect to solve the initial problem. A similar observation is made by Imre Lakatos in *Proofs and Refutations*, and his comment is "After Columbus one should not be surprised if one does not solve the problem one has set out to solve" (Lakatos, 1976, p. 90).

Is it a good or bad thing that a problem changes because of the use of mathematics? This question cannot be answered in general; the answer depends on the specific situation. Application of mathematics could imply that important aspects are eliminated. Or the opposite: The application of mathematics could draw our attention to essential, not previously noticed, consequences of specific assumptions. Examples of both kinds of influences are readily found in connection with the model developed by the Club of Rome and described in *Dynamics of Growth in a Finite World*, by D. L. Meadows et al. (1974). In the discussion following the publication of this "World model," we find references to much more precise pictures of possible future states of the world, and at the same time critics have pointed out that the "World model" tends to eliminate consideration of certain problems, for instance those concerning North-South relations.

The analysis of the problem (the problems) in an e-m process presupposes a certain *structure of argumentation*. Insofar as it has to do with a scientific problem, the structure of argumentation (normally) must be in accordance with the paradigm of scientific methodology (although this methodology is not so easily delineated as previously assumed in the philosophy of science). In connection with technological questions, this structure of argumentation is not well defined. However, the structure of argumentation is highly sensitive to the use of mathematics. Where we try to handle a technological question by means of mathematics we are at the same time going to reshape the line of argumentation involved in problem solving. We find examples of this in connection with the development of SMEC. This model creates new possibilities for economic management; new arguments emerge about how to solve specific economic problems (for instance, concerning unemployment) because the model shows connections between economic elements not previously articulated in the same precise way. (Naturally it could be doubted if these connections in fact exist objectively, but they exist in the man-made conceptual system, which the model tries to copy.) In short, mathematics is not only a new language, but it also brings along new structures of argumentation.

According to Karl Popper, a scientific theory must be connected to a (not very well-defined) set of "falsifiers," not empty if the theory is to have scientific interest. It should be possible to falsify a scientific theory; it must be open to critique. This could be restated in sociological terms as follows: Every theory must be available to a group (of persons) for which it is possible to articulate a critique. That is, if the falsification of a scientific theory is to be a possibility, the critique to which the theory is subjected not only must have a form and content that is logical and relevant to the task of falsification (which is what Popper investigates) but also must in fact be socially manifest—actually embodied or incorporated in the actions of an identifiable group of persons who are capable of carrying out such a criticism.

The group of persons for whom it is possible to influence an e-m process we call the *basis for critique*. It consists of "the set of critics." However, the basis for critique is sensitive to the use of mathematics. If a mathematical model is introduced into an e-m process, the basis for critique connected to this process is changed; perhaps reduced because only a more limited number of persons is knowledgeable about the particular (mathematical) model introduced.

This is in accordance with remarks of Robert Lilienfeld about system theory, which he conceives as a construction supporting a new ruling class, not primarily as a tool to investigate and solve (technological) problems. Lilienfeld (1978) stresses that although system theory often has failed in solving practical problems, it has still been developed; he concludes that the reason must be that it also serves a political function: It gives more power to the class of technologists. To me, it is not important to verify whether Lilienfeld is right or not; but his idea here underlines the point that questions about the distribution of power are of interest when looking at the application of formal methods. That is why the basis for critique must be taken into account when we try to enumerate the functions of the use of mathematical models.

Besides the group of persons for whom it is possible to influence the e-m process, we are faced with a much larger group of persons, which is, more or less, influenced by the decisions and the instrumental actions carried out as a result of the e-m process. We have now to do with the fourth element in an e-m process: the *possible consequences for action*. Basically, a process of technological investigation results in an action, in contrast to a process of scientific investigation resulting in some sort of verification/confirmation/falsification.

We find that the space of possible actions is sensitive to the use of a mathematical model. It is significant in the field of administration and management that quite new possibilities for actions are identified because of the mathematical way of handling problems.

Summarizing, we find that the four elements of an e-m process (problem identification, structure of argumentation, basis for critique, and space of possible consequences for action) are all sensitive to and influenced by application of mathematics. In that sense, mathematics is not a neutral tool in a technological investigation.

Conclusions

A mathematical model is connected to its "surroundings" in a very complex manner. It is related to an object (via a system), to a complex of theories and to different interests, referred to as the synchronic relationships of the model. In addition, we found that these relationships interact: The system is designed in the light of both theory and the interests of the model builders and of the model users.

Only in simple cases do we find an unchanged, stable formulation of a problem for which a solution is found via a model. In general, mathematics plays an integrated part in an e-m process, in relation to the core elements: the identification of the problem, the structure of argumentation, the basis of critique, and the possible consequences for action, referred to as the diachronic relationships of the model. It is our thesis that all these relationships are sensitive to influence from the application of mathematics. Further, we maintain that an analysis of synchronic relationships is a necessary condition for an understanding of a model, but not a sufficient condition. No critique or evaluation could be carried out unless we examine the consequences of the model and then we have to include the diachronic relationships. They constitute an unavoidable element to be considered in a critique of the use of a mathematical model. (In the literature, the discussion of a mathematical model is usually limited to the relation *model-object*, which in addition makes up only a part of the synchronic relationships.)

The contents of both the synchronic and the diachronic relationships depend upon the type of technology in question. In the case of a technology of nature, we are faced with a rather straightforward specification of a system, and the process of investigation is less problematic (seen from a philosophical and epistemological point of view), partly because the mathematical terminology is already intimately integrated into the natural sciences, which constitute the theoretical basis of the identification of the system (to be modeled).

It has not been the intention to specify in detail which particular problems may emerge when mathematics is involved in social or information technology. Such specifications must be carried out as an investigation of particular model-building processes. In conclusion, I only want to stress that application of mathematics involves a whole range of questions, and let me emphasize:

1. These questions do not have much similarity to those normally discussed in relation to the application of mathematics. For example, questions concerning approximation and accuracy constitute only a minor part of the great variety of modeling questions.
2. These questions involve a wide range of issues because both the synchronic and the diachronic relationships are of great complexity. Furthermore, the questions concerning the applications depend on the type of technology; this becomes obvious alone by the fact that the construction of the system to be modeled cannot be done in any uniform way; on the contrary, the construction depends on the object of the technology.

In the diversity of the creation of the system, we find a main reason for the thesis that the application of mathematics in different technologies gives occasion for different questions.

3. Partly because of (2), we conclude that the knowledge, necessary for a discussion of the applicability of mathematics, is in no way reducible to knowledge about how to build a model. In other words, reflective knowledge is of a different nature and cannot be reduced to technological knowledge.
4. Because the pragmatic trend in mathematics education—primarily focusing on the development of a model-building competence—ignores the fact that reflective knowledge cannot be reduced to technological knowledge, I find the pragmatic trend unable to cope with important aspects of the analysis and critique of mathematical model building. Instead, pragmatism too easily turns out to become a propaganda for the subject matter: mathematics.

What are the conditions of providing opportunities for developing not only technological but also reflective knowledge in an educational context? Such conditions will be difficult to fulfill because

5. If we want to provide opportunities for development of reflective knowledge, we have to look for a different organization of the educational process. The traditional organization of the educational process is so structured that it focuses primarily, if not exclusively, on transmitting fact, skills, and such.

This is just an alternative formulation of the thesis that the fundamental structuring principles of the curriculum are derived from or are in accordance with the dominant power relations of society. Especially, mathematics education serves as the most significant introduction to the technological society. It is an introduction which both provides (a part of) the students with relevant technical skills and at the same time provides (all) the students with a "convenient" or "functional" attitude towards the technological society. ("Functional" is seen from the perspective of the dominating power structure; which aims at an understanding for an acceptance of the necessity and the givenness of the present trends in technological development.) The main reason for these statements is that in mathematics education, the development of reflective knowledge is partly ignored and partly impeded.

NOTE

This paper was published in 1988 in *Educational Studies in Mathematics*, *19*(1), 23–41.

REFERENCES

Andrews, J. G., & McLone, R. R. (Eds.). (1976). *Mathematical modeling.* London, UK: Butterworths.

Bender, E. A. (1978). *An introduction to mathematical modeling.* New York, NY: Wiley & Sons.

Goodwin, R. M. (1972). A growth cycle. In E. K. Hunt & J. G. Schwartz (Eds.), *A critique of economic theory: Selected readings* (pp. 442–449). Harmondsworth, UK: Penguin.

Groth, C. (1981). *Cyklisk Kapitalakkumulation I - II* [Cyclic accumulation of Capital, I – II]. Økonomisk Institut, University of Copenhagen.

Habermas, J. (1968). *Erkenntnis und Interesse* [Knowledge and human interests]. Frankfurt-am-Main, Germany: Suhrkamp Verlag.

Haberman, R. (1977). *Mathematical models: Mechanical vibrations, population dynamics, and traffic flow*. Englewood Cliffs, NJ: Prentice-Hall.

Jensen, J. H. (1980). *Mathematiske modeller: Vejledning eller vildledning* [Mathematical models: Guidance or misguidance]. *Naturkampen, 18*, 14–22.

Kapp, E. (1978). *Grundlinien einer Philosophie der Technik* [Principles of a philosophy of technology]. Düsseldorf, Germany: Publisher. (Original work published 1877)

Lakatos, I. (1976). *Proofs and refutations: The logic of mathematical discovery.* Cambridge, UK: Cambridge University Press.

Lilienfeld, R. (1978). *The rise of systems theory: An ideological analysis.* New York, NY: John Wiley & Sons.

Meadows, D. L. et al. (1974). *Dynamics of growth in a finite world.* Cambridge, MA: Wright-Allen.

Niss, M., & Hermann, K. (1982). *Beskæftigelsesmodellen i SMEC III* [The employment model in SMEC III]. Copenhagen, Denmark: Nyt Nordisk Forlag Arnold Busck.

Pollak, H. O. (1979). The interaction between mathematics and other school subjects. In *New trends in mathematics teaching IV* (pp. 232–248). Paris, France: UNESCO.

Rapp, F. (1981). *Analytical philosophy of technology.* Dordrecht, The Netherlands: D. Reidel.

Reichenbach, H. (1966). *The rise of scientific philosophy*. Berkeley and Los Angeles: University of California Press. (Original work published 1951)

Skovsmose, O. (1984). *Kritik, undervisning og matematik* [Critique, education, and mathematics]. Copenhagen, Denmark: Lærerforeningernes Materialeudvalg.

Skovsmose, O., & Jensen, J. H. (1986). *Teknologikritik* [Critique of technology]. Herning, Denmark: Systime.

Smith, J. M. (1974). *Models in ecology*. Cambridge, UK: Cambridge University Press.

Stork, H. (1977). *Einführung in die Philosophie der Technik* [Introduction to the philosophy of technology]. Darmstadt, Germany: Wissenschaftliche Buchgesellschaft.

Volk, D. (1975). Plädoyer für einen problemorientierten Mathematikunterricht im emanzipatorischer Absicht [Plea for a problem-oriented mathematics education with an emancipatory interest]. In M. Ewers (Ed.), *Naturwissenschaftliche Didaktik zwischen Kritik und Konstruction* [Natural science education between critique and construction] (pp. 203–234). Weinheim, Germany: Belz Verlag.

Volk, D. (1980). *Zur Wissenschaftstheorie der Mathematik: Handlungsorientierende Unterrichtslehre* (Band B) [To the philosophy of mathematics: Action-oriented education (volume B)]. Bensheim, Germany: Päd extra Buchverlag.

CHAPTER 18

REFLECTIVE KNOWLEDGE

Its Relation to the Mathematical Modeling Process

The interest in an applied oriented mathematics education has greatly increased within the last 15 years, partly as a reaction to the structuralist movement in mathematics education, partly as a consequence of new educational insight, and partly as a consequence of new demands of manpower.

Very close relationships exist between the structure of society, its technological development and mathematics education. Technology is a dominant feature of civilization, and man is completely immersed in technology. The structuring principles of society are determined by technology, and technology is influenced by mathematics.

The concept "technology" is general, and we can make a distinction between different types of technologies: manual tools (hammer), technology of natural processes (steam engine), social technologies (scientific management), information technology (computer science). The object of manual tools is nature, and much philosophy of technology is based on reflections on this type of technology, for instance, technology seen as a tool in man's struggle against nature. The object of a natural technology is nature too, but now the source of energy is not

Critique as Uncertainty, pages 263–280.

manpower but machine power. The object of social technology is man and human relationships. The object of information technology is communication. Technology is a way of handling power, but problems concerning the amount of and distribution of power depend on the object of technology in question.

The relationships between types of technologies and natural sciences, including mathematics, vary greatly. Manual tools are developed quite independently of science. The development of steam engineering takes place parallel to the theoretical understanding of thermodynamical questions. Information technology is totally based on developments in mathematics, while social technologies, for instance, economical management, are often closely related to mathematics. Mathematics is of absolute importance for the development of today's technology.

The fundamental structuring principles of the curriculum are derived from or are in accordance with the position in society of the different types of technology. These structuring principles determine the position of the subjects in the educational system, and to some degree the main features of the curriculum, but normally details in the curriculum are determined by other factors as well. In particular, mathematics is given a special role to play in today's education.

Mathematics education serves as the most significant introduction to the technological society. It is an introduction which both endows (a number of) students with relevant technical skills, and at the same time endows (all) students with a "functional" attitude towards the technological society. "Functional" is defined from the perspective of the dominant power structures.

However, these theses need not be interpreted as "necessary truths." The theses state what I find to be the case, if not the practice, of mathematics education is shown a special and progressive concern. That is the reason why I am concerned with critical education and trying to identify possibilities for a critical mathematics education.

It is most important to discuss the new trend towards an applied oriented mathematics education. Does this trend imply that the students will be able to understand better the social practice of mathematics? Understand the potentials of mathematical model-building activity in society? Understand the nature of problems connected to the use of formal methods such as a problem-solving technique? Is it at all possible to develop such an understanding in a mathematics education?

In a great number of books and papers about mathematical modeling, we find a presentation of four steps involved in a modeling process.

- We start with a "real world problem."
- However, it is impossible in any simple way to identify a solution, and consequently we have to make simplifications leading into a formal mathematical model.
- The model makes a deduction possible, and we reach a mathematical solution.

- By an interpretation, we transform the mathematical solution into a "real solution."

Of course we often have to repeat the process a great number of times until a satisfactory approximation and a real, acceptable solution are obtained. This is the standard description of the main steps in a modeling process, and it is suggested that students must learn to move around in this 4-step process.

However, I find this conclusion, as well as the description of the four steps, misleading when we are trying to develop a critical attitude towards mathematical model building. The description could become useful if we want only to focus on a pragmatic aspect; by pragmatism in mathematics education, I summarize the idea that it is most important for the students to learn about model building and that the best way of learning this is to build models.

If we want to keep the critical dimension in mind, and if we do not wish to let mathematics education degenerate into a "blind" introduction to the highly technological society, we have to introduce a perspective other than pragmatism into the application orientation.

Reflective Knowledge

That we have to pay special attention to mathematics education in a highly technological society is underlined in Keitel (1989):

> If mathematics is a necessary and essential—although not the only—precondition of technology, then mathematics teaching and learning is a necessary prerequisite for everyone who wants to understand and reconstruct or develop technology—and to judge its use or abuse. (p. 8)

That it is important to differentiate between different sorts of knowledge in relation to mathematics is stated when Keitel continues,

> On the other hand one can only partly be introduced to an understanding of mathematical technology by referring to mathematics itself, since the means-and-ends-relation stringently requires knowledge about both the objective and subjective contexts of the interference as well. Hence an introduction to the understanding and evaluation of technology within mathematics education cannot be restricted to mathematical techniques or theorems but must constantly refer to a broad understanding of the subject of the context. (p. 8)

It is precisely the nature and scope of this "broad understanding of the subject" that we are going to investigate.

A similar intention is stressed in Niss (1989), underlining that one of the arguments for including application and modeling in the curriculum must be to "generate, develop and qualify a critical potential in students towards the use (and misuse) of mathematics in extra-mathematical contexts" (p. 23). Similar opinions are expressed by Blum and Niss (1989), Booss-Bavnbek and Pate (1989), and

Hanna (1989). To specify these ideas, we shall make a distinction between three different types of knowledge related to a process of mathematical modeling.

- Mathematical knowledge itself.
- Technological knowledge, which in this context is knowledge about how to build and how to use a mathematical model. We may also call it pragmatic knowledge.
- Reflective knowledge, to be interpreted as a more general conceptual framework, or metaknowledge, for discussing the nature of models and the criteria used in their constructions, applications, and evaluations.

An important insight of the pragmatic trend in mathematics education is that mathematical knowledge does not automatically develop a technological competence in using mathematics. The assumption that mathematical knowledge is a necessary and sufficient condition for using mathematics is the implicit assumption in structuralism in mathematics education, stressing the importance of raising the architecture of mathematics. However, out of the numerous unsuccessful attempts to make pure logic structures useful grew the understanding that technological knowledge is different from mathematical knowledge. If students wish to be able to handle modeling processes, they particularly need to learn about modeling; and if the best way of learning is by doing, the solution must be to engage students in modeling processes.

Further, it is my thesis that reflective knowledge is different from technological knowledge. If you learn how to build up mathematics, you do not necessarily learn how to conceive your model. Reflective knowledge cannot be reduced to technological knowledge (as technological knowledge cannot be reduced to mathematical knowledge). This idea is equivalent to the core problem of the technological society: The knowledge by which we create technologies is insufficient as a foundation for an evaluation of the technologies created. An additional interpretation of our constructions is necessary.

In what follows, we shall investigate the thesis of the "independence" of reflective knowledge: A foundation of a critical attitude towards the application of mathematics is not developed in mathematics education even if it is application oriented. My intention is to investigate the thesis from an epistemological point of view.

Reflective Knowledge: An Example

In Fischer and Malle (1985), we find a discussion of a suggestion for project work in mathematics, addressed to mathematics teachers in primary and secondary schools. The authors indicate the possibility for developing reflective knowledge (without using that term) in mathematics education.

The example concerns the regulation of traffic at a crossroads by means of a traffic light, and it is shown how this theme could be handled in a project-organized education, characterized by open problems, which can be investigated and handled in different ways; more intensive studies of a specific subject; cooperation; and partial self-organization by the students.

At the beginning of the project example, a reference is made to an (imaginary) discussion in a "town council" concerning a dangerous crossroads. For instance, too many cars are crossing when the traffic light is yellow; what is the best thing to do? The discussion ends with the suggestion, let us ask some students of mathematics. Based on calculations, what would they recommend as a solution?

Fischer and Malle make proposals for problems and questions for the students to deal with, and some principal results are made explicit. For instance, the connections between speed, braking, and such. Naturally "everything" cannot be expounded (Fischer & Malle, 1985), but things are elaborated so carefully that it is possible for a teacher to imagine what could happen in the classroom if he wanted to realize the suggested project work.

Finally, which recommendations are to be given to the town council? It is a most crucial point, because Fischer and Malle stress both what is to be said and also what should not be concluded from the mathematical modeling. It is not adequate in mathematical teaching just to concentrate on showing how useful mathematics can be; to do so would have been in accordance with the pragmatic trend. Instead, a discussion concerning both possibilities and the limitation of the applicability of mathematics should be started.

By their example of project work, Fischer and Malle invite such reflections and draw attention to the development of reflective knowledge in mathematics education. The problem is that it is not pointed out what to look for if we want to reflect on models.

As mentioned before, in the description of the project work, Fischer and Malle put forward different characteristics, all concerning the organization of the educational process. These demands, which I call secondary demands, are important, but it is more important to try to specify primary demands concerning the content of the education. This could characterize project work aiming directly at developing reflective knowledge.

However, my aim is much more modest. I hope to indicate some sort of problem relating to the modeling process which is important to analyze in an attempt to develop reflective knowledge. The problems will be summarized under four headings:

- Concealment of preunderstanding
- Confusion of guiding interests
- Creating a nonexisting object
- Reflective knowledge is much more . . .

Concealment of Preunderstanding

Every mathematical model must be based on a specific interpretation of reality. Other possibilities do not exist. We cannot come into contact with any "reality" without structuring it. This statement has been stressed very much in the philosophy of science as a reaction to the doctrine of logical positivism, that objectivity and neutrality is a possible aim, although sometimes difficult to obtain. In Hanson (1958), it is explained why observations in general, and also scientific observations, will become prestructured in accordance with our conceptual patterns; it is not possible or even important to avoid these patterns. Instead, we may use investigations of our prestructured world as a basis for restructuring activities, perhaps leading to an improved understanding. This epistemological enterprise has much in common with the hermeneutic tradition, as for instance developed in Gadamer (1960), hermeneutics explains that the existence of preknowledge is an absolute condition for obtaining additional knowledge. It is a variant of this idea that Habermas elaborates into the theory of knowledge constituting interests.

This underlines that a model is not a model of reality as such. We have to make an interpretation of reality; that means making a structuring of reality which creates patterns. We have to identify elements from reality which are to be conceived as important; we also have to decide which relationships among these elements are essential. These selections constitute an interpretation of reality. We may talk about a process of system development.

The created system is not a real construction; it is a conceptual system, created by means of a certain interpretation of reality, that is, by means of a certain theoretical framework for looking at reality and by having a certain knowledge guiding interest in mind. The preconditions for the process of system construction could be implicit or made explicit. They could be well founded or accidental, but they are inevitable. A conceptual systematization of reality is a prerequisite for a modeling process. Only a highly developed system could facilitate a modeling process.

A mathematical model is not a model of reality; it is a model of a conceptual system created by a specific interpretation, based on a more or less elaborated theoretical framework, and based on some specific interests. However, this fact is often forgotten and eliminated during the modeling process. Instead, a strange metaphysic develops, ascribing objectivity to the model itself. This phenomenon has to do with the language of mathematics, eliminating its relationship to the created conceptual system.

The language of mathematics makes the very process of system construction invisible, and by doing this it is made difficult to identify the type of system development that had preceded the model in question. Quite different types of system development exist, depending on which relationships the process has to theoretical structures and complexes of interests. An evaluation of a model must be based on a critical investigation of the preceding system construction.

The first main problem related to mathematical modeling we shall summarize as concealment of preunderstanding, to be understood as the phenomenon accompanying a mathematization of disguising the complexity of the construction of the conceptual system which constitutes the very foundation of the model itself. To tackle this problem, the identification of preunderstanding is a main objective for reflective research.

This could be illustrated by an example. In economic analyses, at the micro- as well as the macrolevel, the Cobb-Douglas function of production plays a crucial role. It determines the (national) product as a function of two variables.

$$Y = f(L, I)$$

That is, the product Y is a function of the labor force L and the capital investment I. Economical models are developed with this function of production as the kernel. The first step in the process of mathematization is to specify the function f, and it is stipulated that where c is a constant. This equation is based upon a rather simple interpretation: If we shall increase the production by a factor c, it is necessary to increase both the number of workers and the means of production with the same factor.

The next step in specifying f covers the ideas, if we increase the numbers of workers, the production will increase; if we increase the means of production, the production will increase also. In other words

$$\frac{\partial f(L,I)}{\partial L} > 0 \qquad \frac{\partial f(L,I)}{\partial I} > 0$$

However, the increase will slow down. Thus, if we increase the number of workers by 1,000, the production will increase, but if we again increase the number by 1,000, the new increase in production will be less than the first one. The same will be the result with relation to investment. Stated as additional conditions concerning the partial derivation, we have

$$\frac{\partial^2 f(L,I)}{\partial L^2} > 0 \qquad \frac{\partial^2 f(L,I)}{\partial I^2} > 0$$

A function fulfilling these conditions is $f(L,I) = \alpha L^{\alpha} \cdot I^{1-\alpha}$ where a is a constant and $\alpha \in]0;1[$.

From this, a specific model building could proceed. In the "Simulation Model of the Economic Council" (SMEC) used by Danish economists when advising the government and politicians concerning economic policy and its possible consequences, the Cobb-Douglas function of production plays a crucial role (Niss & Hermann, 1982). With this function as point of reference, the model is specified into a great number of equations, making the model useful in an attempt to perform economic forecasting.

At the same time, the mathematical language hides the fact that a specific system development has taken place. This development starts by paying special attention to a preinterpretation of the economic reality, and according to the Cobb-Douglas function, the national production is determined by the labor force and the capital investment, and nothing else. A whole economical metaphysic is built into the economic model via this kernel function, reflecting, for instance, the assumption that ecological costs are out of the question when discussing the economy. Some sort of preinterpretation is always necessary in developing a system, but not the interpretation in question. Economical models could be developed in relation to other basic interpretations of the economic reality. This situation has to be outlined by a reflective investigation.

Of course, the situation could have been different. Sometimes we find that the model-building process could rely on well-established theories, for instance, the case if the model concerns phenomena in the physical world already described in classical mechanics. In this situation, the preinterpretation will be of quite another nature (Jensen, 1980).

The aim of a reflective investigation is not to try to eliminate any preinterpretation but to identify the nature of the preinterpretation which has proceeded the mathematical modeling. This preinterpretation is hidden behind the mathematical language, making the different models look identical if from a mathematical point of view they are identical.

Confusion of Guiding Interests

A system development, and hence a mathematical model, could be built with different intentions and interests in mind. This observation is stressed in Davis and Hersh (1988) where the distinction is made between descriptive, predictive, and prescriptive functions of applied mathematics.

A mathematical model could be used to summarize some aspects of reality, and a key question of course is what aspects of reality are possible to describe in the specific language of mathematics.

The model could be used to make a prediction, and naturally it makes a big difference which part of reality such a model describes, the reliability of a prediction depends on the theoretical basis for the interpretation of reality; it depends on the type of preceding system development. The relation to theory differs greatly if we have a model from classical mechanics expressing connections between distance, gravity, and time, or a Cobb-Douglas function of production expressing a relationship between the national product and work and investment. The quality of the prediction is very different in the two cases.

Finally, a mathematical model could be used as a prescription exemplified by cases where the use of the model leads directly to human actions or technological actions. Concerning the prescriptive use of mathematics,

> We are born into a world with so many instances of prescriptive mathematics in place that we are hardly aware of them, and, once they are pointed out, we can hardly imagine the world working without them. Our measurements of space and mass, our clocks and calendars, our plans for buildings and machines, our monetary system, are prescriptive mathematisations of great antiquity. To focus on more recent instances, and in this way to get a feeling of what living without them might imply, think of income tax. This is an enormous mathematical structure superposed on an enormous pre-existing mathematical financial structure. (Davis & Hersh, 1988, p. 120)

It is important to have an understanding of the dimension of the extension of implicit mathematics in society. Further, it is important to understand the difference between descriptive, predictive, and prescriptive uses of mathematical models. This exemplifies a reflective understanding.

A perplexity arises if we are unable to distinguish between the different ways of use, in other words, if we are unable to identify the different complex of interests related to a modeling process. It creates a serious misunderstanding if we identify a prescriptive use and a descriptive use. If we try to draw a descriptive conclusion from a model developed for prescriptive purposes, we will make the descriptive fallacy committed too often, owing to conflicting interests built into the (invisible) process of system development.

This constitutes the second main problem related to mathematical modeling: the confusion of guiding interests for mathematical modeling. An aim for reflective knowledge is to identify precisely the specific interest incorporated into the system development foregoing the mathematization.

We are able to exemplify this confusion by looking at the classical distinction between "the context of discovery" and "the context of justification." This distinction was emphasized by logical positivism by the thesis that the act of discovery escapes logical analysis; the philosophy of science in the shape of a logical analysis of the language of science has to concentrate upon the context of justification.

Using this distinction, it could be claimed that a mathematical model should be carried out in connection to the question: How does the model facilitate the justification of some assumptions? However, I find that mathematical models could serve important functions, not only in a context of justification, but also in a context of discovery. They could be related to quite different domains of interests, also in a scientific context. The problem is, as mentioned before, that this fact is hidden behind the mathematical appearance of the model. A confusion of interest takes place if results from a mathematical model belonging to the context of discovery are used as if they may serve a verification. However, we often meet this mistake exemplifying the descriptive fallacy.

The importance of distinguishing between the context of discovery and the context of justification could be made obvious when looking at a mathematical model elaborated in an attempt to explain some cyclic movements in a capitalistic economy (Goodwin, 1972). These movements have been described by Karl Marx, but in colloquial form. The main question concerns the relationship between the level of wages and the level of employment. The more intuitive ideas of Marx are the following: If the level of employment increases, the power of workers increases too, and so it is possible to increase the salary. However, this has implications in a capitalistic economy. If the salary increases too much, the rate of employment tends to decrease, the workers lose power in relation to the employers, and a cyclic movement is started, only to be neutralized if the capitalistic economy breaks down.

Goodwin makes some assumptions too simple to represent reality, but he finds that they represent the most essential dynamic aspects of capitalism. Goodwin puts this ideal and nonreal system into mathematical garb, and he obtains the following equations:

$$\frac{dx}{dt} = ax(t) - bx(t)y(t)$$
$$\frac{dy}{dt} = -cy(t) + dx(t)y(t)$$

where $x(t)$ and $y(t)$ are functions of time t, and a, b, c and d are parameters, all positive.

> In this form we recognize the Volterra case of prey and predator. To some extent this similarity is purely formal, but not entirely so. It has long seemed to me that Volterra's problem of the symbiosis of two populations—partly complementary, partly hostile—is helpful in the understanding of the dynamical contradictions of capitalism, especially when stated in a more or less Marxian form. (Goodwin, 1972)

First, let us look at the biological interpretation of the differential equations. Let $x(t)$ be the prey population at time t and let $y(t)$ be the predator population at time t. In the absence of any predator, the increase of the prey population will be proportional to its size, and we have

$$\frac{dx(t)}{dt} = ax(t)$$

where a is a positive constant. The number of contacts between prey and predators will be proportional to $x(t)y(t)$, so we have

$$\frac{dx(t)}{dt} = ax(t) - bx(t)y(t)$$

for some positive constant b. The number of predators will decrease with a rate proportional to their number, if no prey population exists, but the number of predators will increase at a rate proportional to the number of contacts between prey and predator; therefore we find

$$\frac{dy(t)}{dt} = -cy(t) + dx(t)y(t)$$

where c and d are positive constants.

Goodwin's economic interpretation of the parameters a, b, c, and d is rather complicated, but not important in this context, $x(t)$ is interpreted as the level of employment and $y(t)$ as the level of wage. All solutions of the differential equations, with $x(0) > 0$, are positive and periodic functions of time; thus, for some $T > 0$ we have $x(t + T) = x(t)$ and $y(t + T) = y(t)$.

The Goodwin model has no significant descriptive value. It has been further developed and modified (Groth, 1981), but this is not important in this context. The relevance of the Goodwin model is found in its ability to create, provoke, and develop economic ideas. That is, in its ability to develop a discussion initiated by Marx.

By putting up a formal setting, the model could explicate some of the logical relationships between economic concepts and assumptions, and by doing so outline some consequences of economic assumptions. The model has much more to do with the conceptual system created by economic theoretical assumptions than with any economic reality as such, a point also stressed (Goodwin, 1972). The model draws attention to features of the conceptual system which are an attempt to interpret the economical reality. In this way, the model belongs to the context of discovery. The mathematization constitutes a test of the consistency of the conceptual system used in discussing economical phenomena.

If we are going to evaluate the model, we will presumably find that its descriptive value is very poor. However, to focus upon that is to put the model into a wrong perspective. Its value must be discussed in a context of discovery. To what degree could economic ideas grow out of the model construction?

In general, mathematics could be used to explicate (semi) hidden connections between certain elements in a system. Because the system is man-made, this means that it explicates (semi) hidden connections between more or less explicitly stated assumptions about an object. An investigation via a model could serve as a sort of test of consistency between fragments of theory and different assumptions. That is what Goodwin is trying to do to Marx, and that is the reason why we, in discussions of models, must take the domain of interests into consideration. If we interpret the Goodwin model as belonging to the context of justification, and if we look only at its descriptive values, our discussion of the model will become distorted.

Creating a Nonexisting Object

Often we talk about an object of a model: The model describes an object up to a certain degree of approximation. We talk about the model as a picture of reality, and in numerous introductions to the modeling processes it is explained that modeling must start in reality. We have to model something.

However, this way of interpreting models easily repeats the "fallacy of *Tractatus*," which refer to the assumption that language (always) provides a picture of reality. Wittgenstein (1953) himself criticizes the descriptive theory of language developed in Wittgenstein (1992). Instead of a theory of language as a picture, he develops a theory of language as a collection of a variety of games, and by doing this he introduces a theory of language as action.

The descriptive interpretation of language causes a multitude of philosophical difficulties and turns out to be a too simple way of looking at "the jobs of language." Language is not a picture and hence not a picture of "something." In the same way, a mathematical model is not a picture, it is not a picture of some part of "reality." To conceive a model as (always) a model of something real will be to repeat the fallacy of *Tractatus*.

The concept reality is very diffuse. It is impossible, as pointed out in the discussion of the concealment of preunderstanding, to build any mathematical model without looking at reality by means of a complex and constructed system. A model is a mathematization of a conceptual structure, and to nominate any real object as the object of the model will often become a fallacy.

By this, we have identified the third main problem related to mathematical modeling: The nature of mathematical language makes it tempting to accept a picture theory and by doing this to invent an object to be pictured by the mathematical model. That is the phenomenon of objectivation connected to mathematics (Bishop, 1988).

Of course, this phenomenon is not always a problem. To identify the shape of the Earth with a sphere will become a reasonable description with lots of purposes in mind (although in some situations an ellipsoid will give a better description). To use a model as a picture of "something" seems not to be a failure in this connection. However, the problem is that sometimes a model is not a model of "something," although the mathematical language is easily given a descriptive interpretation. The "fallacy of *Tractatus*" is ways a possibility; to prevent that is an aim of a reflective analysis.

Say we have a phenomenon P, which we will try to explain by relating it to some causes C. We are unable to observe directly how P is produced by C, but our hypothesis is that some real mechanism M_t does exist. A possibility is to invent some hypothetical mechanisms Mh and try to find out if M_h will produce P when C. The hypothetical mechanism M_h is expressed in a (mathematical) model.

Now two different possibilities exist. The relation between M_r and M_h may become impossible to investigate in ways other than looking at C and P. Will M_h always produce the same result as M_r, given a specific cause? If that is the case, we have a model of the black-box type. That means that the structure of the mechanism M_r can in no way be compared with the structure of the mechanism M_h expressed in the model.

The other possibility is that the structures of M_r and M_b could be compared in some ways. Perhaps it turns out that the way M_h explains how C produces P is similar to the way M_t in fact produces P from C. In this case, we have a white-box model.

Again, the problem is not that a model sometimes is a black box and sometimes a white box. The problem is that the mathematical language makes it tempting always to interpret a model in a descriptive way; that means always to interpret M_h as a description of something. That is the phenomenon of "objectivism," and a reflective analysis has to make a warning if we have black boxes.

Reflective Knowledge Is Much More . . .

We have identified three main problems related to the mathematical modeling process: the problem of the concealment of preunderstanding, the problem of the confusion of guiding interests, and the problem of objectivism. To develop reflective knowledge, we have to investigate these problems, but this does not in any way include all that is important for developing reflective knowledge.

The modeling process could create a specific way of interpreting the problem; it could cause a limitation on the group of people who have the possibility of participating in the discussion of problem solving, and as a consequence of a rationalistic and calculatoric way of arguing, a specific type of solution could become created. However, also by looking "inside" a mathematical model, it is possible to obtain reflective knowledge.

This could be illustrated by the North Sea model, which was constructed by the Danish Institute for Fishery and Marine Research to give an explanation of some of the dynamical relationships between different species of fish in the North Sea. Previously, assessment of the quantity of fish had been based on one-species models, which describe the quantity as determined by the growth and the mortality totally determined by the quantity of the species of fish in question. An important step forward would be to consider the growth and mortality also in relation to the size of the stocks of other species in the same area, and that is what is done in the North Sea model. The model consists of a great number of differential equations, not to be taken into consideration in this context.

The problem initiating the development of this model has primarily been theoretical, concerning the limited descriptive value of the different one-species models. Further, it is obvious that a dynamical interaction between the different species of fish could exist, strongly indicated by the original use of the Lotka-Volterra differential equations.

From a theoretical point of view, it is important to integrate biological assumptions about the interactions of the species into a single model; but when elaborated, the North Sea model could also tackle the technological problem: how to increase the fishing (in Denmark). The model could be used in an attempt to estimate the highest possible yield of North Sea fisheries. The increase of North Sea fisheries from 1960 to 1970 has in fact been from 1.5 to 3.0 million tons, and it is reasonable to ask if this increase could continue any longer.

Andersen and Ursin (1977) state that it is impossible to give any single answer to that question

> because different combinations of fishing mortality and size of first capture for individual species might give the same result and there is no way to know if the highest estimate found is near the maximum. A procedure investigating by small steps all feasible combinations of fishing mortality and the size of first capture for eleven species at the same time is impossible at the present stage of computer technology. However, a few simple considerations can indicate a type of fishing strategy which would lead to large catches. (p. 428)

Later they indicate that an increase of the fisheries to more than 5 million tons per year will be possible.

By such conclusions, the model becomes involved in a political discussion between England and Denmark concerning the right policy of fishing. We see a twist in the interpretation of the model concerning the confusion of interests. Even if the model could become very useful in identifying important biological connections, and in that way raise important theoretical question, it is not certain that it could be used with success with technological aims. As stated previously, although the model is useful in the context of discovery, it may not be useful in the context of verification.

Looking into the details in the model, it is possible to identify phenomena important for developing reflective knowledge. In a complicated model such as the North Sea model, we find a great number of parameters, the value of which has to be estimated. Now, it is possible in relation to a great variety of mathematical models to identify different types of parameters. We could talk about empirical parameters, theoretical parameters, pseudoempirical parameters.

We call a parameter empirical if its value could be estimated by a well-defined empirical investigation. The parameter is theoretical if its value could be estimated by relating it to theoretical investigations from which its value could be calculated. Naturally, these theories must have an empirical basis, but that is a different question. The pseudoempirical parameters have a different status. Here we are concerned with parameters whose values are difficult to estimate in an independent way; instead their values must be estimated as an integrated part of the modeling process. Andersen and Ursin state about the North Sea model,

> Even if all parameters were independently estimated outside the model it would have been necessary to adjust the parameter values to produce output similar to observed values of yield, body size, etc. . . . Many parameters, however, are estimated within the model . . . Adjustments of parameter values are made by hand and the program is run again until the agreement with observations is satisfactory. (Andersen & Ursin, 1977, pp. 402–403)

We find that the value of different parameters could become connected to "reality" in quite different ways. If the parameter is empirical, we have a specific empirical test determining its value. If the parameter is theoretical, its value is based on a theoretical construction, and if there only exists one theoretical construction of relevance for interpreting reality, it is possible to relate very much to such values. If the parameter is pseudoempirical, the situation is quite different. Then it is impossible to determine the value of a specific parameter by means of a specific test or calculation. Instead, the value of the parameter is estimated, together with the values of other parameters, and the total "valuation" constitutes the model, which now turns out to be a black box without any structural similarities with real structures. The interpretation of pseudoempirical parameters is problematic if we tend towards an objectivism. On the other hand, we have to realize that we are dealing with a model of the black-box type.

Instead of looking at the foundations of the parameters, it is possible to look at the sensitivity of the parameters. The question is, if we change the value of a parameter, how much would it influence the values, functions, results, and such determined by the model? Normally, we find that some parameters are sensible and that some parameters are not. Perhaps it will be possible to differentiate between different situations; for example, if we change the parameter a, the model will change but keep its original structure; if we change the parameter b, the model will change and also change its original structure. Some models possess a "structural stability," others do not; when this is pointed out, we know more about what is important to be aware of when discussing results from the model.

An important question is now, in which way is the sensitive parameters related to their basis? In particular, how large is the sensitivity of the pseudoempirical parameters? If the pseudoempirical parameters are the sensitive ones, the status of the model would be strongly influenced. It would be difficult to falsify the model by its consequences, because the sensitivity of the pseudoempirical parameters assures that it is easy to correct the outcome of the model, and the pseudoempirical status of the parameters assure that the values of the parameters could be changed without getting into a conflict with specific empirical tests or specific theoretical interpretations. The model receives a new status by being not possible to falsify.

Does this imply that the model in question is bad? No, that is not the conclusion to be drawn, but it illustrates the importance of looking into details in the model to obtain a critical understanding of its status, and the status of the conclusion drawn because of the model. It is important to notice if the model tends to tackle scientific or technological questions. To use a model with sensitive and pseudoempirical parameters in relation to a set of scientific problems could imply that new hypotheses are created about certain relationships, and especially about what is important to investigate. The application of the same model in a technological context could turn out to be problematic because it is difficult to criticize, it cannot be falsified by its consequences. The problem is now that the inner structure of the model loses its connection with reality. The model turns out to be a black box.

Being aware of such phenomena is an important part of a model investigation. In particular, observations concerning the status of its parameters could become essential when developing reflective knowledge.

Conclusions

The observations concerning the different types of parameters could be related to the three main problems mentioned concerning mathematical modeling. If all the "essential" parameters could be identified and estimated in relation to a specific theory, in other words, if the model makes up a specification of a well-based system, we may talk about a theoretical model. On the other hand, if an essential parameter in the model is pseudoempirical, we have a model of the black-box type, which means a model that does not relate to reality in any way. The only way to evaluate a black-box model is by its output. It does not specify connections which mirror any real functional relationship. This implies that any objectivism will become a catastrophe.

The understanding of these phenomena exemplifies reflective knowledge; and such knowledge does not develop along the same lines as pragmatic knowledge. We have to make a new focus. Of course, it is an open question if it is possible to establish reflective knowledge without a pragmatic competence. In my opinion, reflective knowledge is related to pragmatic knowledge, but not in any direct way: First, the pragmatic knowledge does not automatically lead up to reflective knowledge; second, reflective knowledge does not presuppose a well-developed pragmatic competence; if that were the case, only "experts" would be able to criticize "experts."

The scope of reflective knowledge is much broader than the scope of pragmatic knowledge. In a reflection about modeling, we have to look at the model in relation to a construction of a conceptual system, itself related to a complex of theories and a complex of interests, needed in a specification of the system related to "reality"; while a pragmatic understanding concentrates on the relationship of models to "reality." In short, the object of reflective knowledge is much more complex that the object of pragmatic knowledge.

Another way of making a distinction between different types of knowledge is to look at the process of the development of knowledge. By doing this, we can make a distinction between a monological and a dialogical interpretation. A classical monological theory of epistemology is empiricism. To obtain knowledge is seen as an individual process; it is not necessary to interact with others. The source of knowledge is the senses. Likewise, the rationalistic epistemology is monological. Knowledge is produced by "individual thinkers"; the thinking must be clear and (mono) logical.

In a dialogical epistemology, the discussion comes to play a crucial role. The main idea is simple. Any specific individual understanding is inadequate, but in any attempt to improve our understanding and to move in the direction of more knowledge, we depend on each other. Nobody can state anything true, but if we interact in dialogical relationships, we will be able to move in the direction of more knowledge. The condition for obtaining knowledge is not that we obtain additional true information but that we interact in a unique way, characterized as a dialogical relationship. These considerations have to be elaborated to outline the dialogical nature of reflective knowledge. This is the ambitious epistemological program for describing reflective knowledge.

NOTE

This paper was published in 1990 in *International Journal of Mathematical Education in Science and Technology*, *21*, 765–779.

REFERENCES

Andersen, K. P., & Ursin, E. (1977). A multispieces extentision to the Beverton and Holt theory of fishing, with accounts of phosphorus circulation and primary production. *Meddelelser fra Danmarks Fiskeri-og Havundersøgelser*. Danish Institute for Fishery and Marine Research. N.S., 7, pp. 319–435.

Bishop, A. (1988). *Mathematical enculturation*. Dordrecht, The Netherlands: Kluwer Academic.

Blum, W., & Niss, M. (1989). Mathematical problem solving, modeling, applications, and links to other subjects: State, trends and issues in mathematics instruction. In W. Blum, M. Niss, & I. Huntley (Eds.), *Modeling, applications, and applied problem-solving: Teaching mathematics in a real context* (pp. 1–21). Chichester, UK: Ellis Horwood.

Booss-Bavnbek, B., & Pate, G. (1989). Information technology and mathematical modeling, the software crisis, risk and educational consequences. *Zentralblatt für Didaktik der Mathematik, 89*(5), 167–175.

Davis, P. J., & Hersh, R. (1988). *Descartes' dream: The world according to mathematics*. London, UK: Penguin.

Fischer, R., & Malle, G. (with Burger, H.) (1985). *Mensch und Mathematik* [Man and mathematics]. Mannheim, Germany: Bibliographisches Institut.

Gadamer, H.-G. (1960). *Wahrheit und Methode* [Truth and method]. Tübingen, Germany: Mohr.

Goodwin, R. M. (1972). A growth cycle. In E. K. Hunt & J. G. Schwartz (Eds.), *A critique of economic theory: Selected readings* (pp. 442–449). Harmondsworth, UK: Penguin.

Groth, C. (1981). *Cyklisk Kapitalakkumulation I—II* [*Cyclic accumulation of Capital, I–II*]. Økonomisk Institut, University of Copenhagen.

Hanna, G. (1989). Teaching the appropriate use of a mathematical model. In W. Blum, J. S. Berry, R. Biehler, I. D. Huntley, G. Kaiser-Messmer, & L. Profke (Eds.), *Applications and modeling in learning and teaching mathematics* (pp. 312–316). Chichester, UK: Ellis Horwood.

Hanson, N. R. (1958). *Patterns of discovery*. Cambridge, UK: Cambridge University Press.

Jensen, J. H. (1980). *Mathematiske modeller: Vejledning eller vildledning* [Mathematical models: Guidance or misguidance]. *Naturkampen, 18*, 14–22.

Keitel, C. (1989). Mathematics education and technology. *For the Learning of Mathematics, 9*(1), 7–13.

Niss, M. (1989). Aims and scope of applications and modeling in mathematics curricula. In W. Blum, J. S. Berry, R. Biehler, I. D. Huntley, G. Kaiser-Messmer, & L. Profke

(Eds.), *Applications and modeling in learning and teaching mathematics* (pp. 22–31). Chichester, UK: Ellis Horwood.

Niss, M., & Hermann, K. (1982). *Beskæftigelsesmodellen i SMEC III* [The employment model in SMEC III]. Copenhagen, Denmark: Nyt Nordisk Forlag Arnold Busck.

Wittgenstein, L. (1953). *Philosophische Untersuchungen* [Philosophical investigations]. Oxford, UK: Basil Blackwell.

Wittgenstein, L. (1992). *Tractatus logico-philosophicus* (C. K. Ogden, Trans.). London, UK: Routledge. (Original work published 1921)

PART 5

CRITICAL MATHEMATICS EDUCATION

CHAPTER 19

EXPLOSIVE PROBLEMS IN MATHEMATICS EDUCATION

Dear Richard, Thank you for drawing my attention to David Wheeler's collection of problems. It has been fascinating to reconsider them.[1]

Wheeler did ask for problems "whose solution would be likely to advance substantially our knowledge of mathematics education." Very many formulated such problems, and as examples, I consider the following three formulated by Geoffrey Howson, John Mason, and Alan Bishop:

> Howson: "It would help me to have a better theoretical framework within which to consider/study/investigate mathematics education."[2]

> Mason: "Is mathematics really essential?"[3]

[1] This paper was published as a letter to Richard Barwell, the editor of *For the Learning of Mathematics.*

[2] Research Problems in Mathematics Education – I, *For the Learning of Mathematics*, *4*(1), 1984, p. 41.

[3] Research Problems in Mathematics Education – III, *For the Learning of Mathematics*, *4*(3), 1984, p. 24.

Critique as Uncertainty, pages 283–287.

Bishop: "How is mathematical meaning shared?"[4]

These formulations are short and clear. Nevertheless, I have the feeling that it is impossible to reach any solution to be generally agreed upon. Howson might have had the same feeling, as he modified the task by adding, "Here I am for the moment not asking for a theory, just a framework which might help to develop one." But even with such modification, his problem appears gigantic.

Naturally, I came to think of David Hilbert's 23 problems presented in 1900, which, since then, have provided an ongoing challenge to mathematics research. I looked at a web page[5] where the problems are nicely classified: Some were solved in a particular year, some are partially solved, one is too vague in its formulation to be solved, while three remain unsolved. Progress in mathematics has, at least partially, been related to the resolution of these problems.

Hilbert's famous second problem was "Prove that the axioms of arithmetic are consistent." Gödel's second incompleteness theorem, however, showed that no proof of this consistency can be carried out within arithmetic itself. This was a most alarming result considering Hilbert's metamathematical aspirations. When I was young, I spent half a year studying Gödel's proof. I was fascinated by the perspectives it provided on mathematics.

While Hilbert's problems concern mathematics, Wheeler's problems concern mathematics education. Linguistically speaking, there is not much difference between "mathematics" and "mathematics education"; epistemically speaking, there is. The difference becomes clear when we consider the difference in nature between Hilbert and Wheeler's problems. The formulation of Hilbert's problems seems be based on some shared conceptions of what a solution could mean, while what to consider a solution to a Wheeler problem appears to be ambiguous.

This brings me to consider the notion of explosive concepts. By this I understand a concept that can be defined only through concepts just as open and vague as itself. As an example, one can think of "democracy." In order to clarify what democracy means, one may consider the meaning of, for instance, equity, justice, and inclusion. However, these notions are no easier to define than the notion of democracy. Thus, a definition of an explosive concept does not "narrow down" its possible meanings. Instead, it opens into new landscapes of possible meanings.

Hilbert's problems are formulated through concepts that one could call solid, like integer, real number, axiom, function, arithmetic, consistency, polynomial group, algorithm, prime number, on so on. Naturally, the definitions of such concepts are not fixed; they develop during history. What might be considered a proper definition in one period might appear inadequate later. Think of the notion of function, for example, which has been part of a fascinating conceptual development. But, although "function" demonstrates a historicity, it is not explosive. Its

[4] Reseach Problems in Mathematics Education – II, *For the Learning of Mathematics*, *4*(2), 1984, p. 41.

[5] (http://en.wikipedia.org/wiki/Hilbert's_problems)

definitions and redefinitions are based on notions generally accepted in the mathematical community at a given time. In this sense, we can consider "function" a solid concept. (Let me just add that the notion of a proof being "finite," as used by Hilbert in formulating what methods could be allowed in metamathematics, can hardly be called solid. Let this, then, indicate that explosions might also occur in mathematics.)

Contrary to Hilbert's problems, many of Wheeler's, including those presented by Howson, Mason, and Bishop, are formulated through explosive concepts. In order to define a "theory," one needs to address broader questions about knowledge, justification, and power. A discussion of mathematics being "really essential" leads us to consider the possible roles of mathematics in society, in technology, and in people's life-worlds. A discussion of "meaning" brings us into a turbulent trip through the history of philosophy, even before we come to address "shared meanings."

Let us call a problem that is formulated through explosive concepts an explosive problem. Thus, many of the problems collected and presented by Wheeler are explosive. Hilbert's problems, however, can be characterized as being solid. They are formulated through solid concepts, and there exists a general agreement, within the mathematical community, of whether a particular problem is solved or not. No such solved/not-solved duality seems applicable to the Wheeler problems.

What to think of this situation? Should mathematics education try to formulate its fundamental problems in solid concepts? My answer is "no." I find it healthy for mathematics education to grapple with explosive problems.

I do not find it promising to follow the trend that tries to introduce operationalized definitions of educational phenomena, say, students' performance and learning gains, and, in this way, try to deactivate explosive concepts. Instead, it is important to acknowledge the explosive nature of crucial problems in mathematics education.

Consider again Howson's problem, which includes the notion of theory. Sure, there have been many attempts to define what a theory is, as well as attempts to try to close the concept, for instance, by defining a theory as an instrument for making predictions. However, I find it much more productive to acknowledge the explosive nature of "theory" and explore its relationships to other open concepts. Thus, we can consider relationships between knowledge and power as well as relationships between theory and power. We can ask, What interests might a theory in mathematics education serve? What functions might the school mathematics tradition, with all its rituals and exercises, serve? In what way does this tradition accommodate the economic order? Does mathematics education have a disciplining function, as implied by Foucault? All such questions can be related to Howson's problem of establishing some adequate categories for investigating mathematics education.

Mason's problem brings us further: Is mathematics really essential? This question opens up a range of perspectives. We could start with the socioeconomic order of today and ask if mathematical knowledge is essential for increasing productivity. We could consider if mathematics education serves a disciplining function, for instance, by cultivating a prescription readiness. Naturally, one could also ask to what extent mathematics is essential for human dignity and for human life in general. One could ask if mathematical rationality is essential for developing and maintaining democratic institutions in society. One could consider in what sense mathematics education contributes to the development of citizenship, and also of critical citizenship. All such issues whirl us into different open conceptual landscapes.

The same can be said about Bishop's problem. The very notion of meaning has a long history in philosophy. Many different interpretations of "meaning" have been formulated. And, certainly, no general agreement is in sight. Meaning has been interpreted in terms of reference, while Wittgenstein opened a quite different line of thinking by associating the meaning of a concept with its use. Discourse theories are going further by addressing meaning in terms of actions and power relationships. With respect to mathematics education, one can consider meaning a relational concept, thus, the meaning of a classroom activity has to do with its relationships to other practices. I have suggested that students' experiences of meaning have to do with relationships to their foregrounds. The notion of meaning is explosive, and all such considerations are only introductory for addressing Bishop's concern about "shared meaning."

It is important to address the issues raised by Howson, Mason, and Bishop. In general, it is important for mathematics education to grapple with explosive problems.

Let me just say a couple of words about theoretical progress in mathematics education. Naturally, we immediately have to recognize that the very notion of progress is explosive. So when we talk about progress in mathematics education, we cannot be supposed to know what we are talking about. Let me, however, acknowledge this fact by suggesting that progress in mathematics education is related to a readiness to grapple with explosive problems. So, while progress in mathematics might be related to the solution of (solid) problems, progress in mathematics education, seen as a theoretical discipline, can be related to grappling with explosive problems. This could bring about, for instance, dramatic conceptual changes, new discourses, reconstructions of perspectives, and provocative changes of concerns. This means that mathematics education should not consider solid problems, as formulated by Hilbert, to represent any epistemic ideal. I see instead the explosive nature of problems in mathematics education as being a remarkable strength.

Ah, I almost forgot to say anything about what kind of fundamental problems might be relevant to address in mathematics education, now, almost 30 years after Wheeler collected his problems. I would need more time to think about this, but let me just mention some concepts that I, for the moment, think could be relevant for formulating such problems: globalization, ghettoizing, inclusion, exclusion, democracy, citizenship, social justice, equity, democracy, mathemacy, students' foreground, intentionality, imagination, hope, dialogue, empowerment, despair, meaning, action, rationality, discipline, power, capitalism, suppression, exploitation, colonization, racism, sexism, interest, critique.

Sure, this list needs to be continued. My point, however, is not to try to provide an exhaustive list but to point out the relevance of explosive concepts in the formulation of important problems in mathematics education. (Naturally, there are many other nonexplosive notions that could be relevant to use as well like, say, statistics, probability, algebra, landscapes of investigation, algorithm. But this is another issue.)

I have written this letter to you enjoying an excellent view from my flat, 4th floor, here in Rio Claro. It is warm, more than 30 degrees. I will stop now. Then you can read this letter with a cooler and clearer mind.

NOTE

This paper was published in 2012 in *For the Learning of Mathematics*, *32*(2).

CHAPTER 20

CRITIQUE, GENERATIVITY, AND IMAGINATION

One of the grand narratives of Modernity tells us about the objectivity and neutrality of science; in particular, it celebrates natural science as demonstrating the ideal scientific format. During Modernity, mathematics developed a special status; it was considered a sublime, if not a divine, rationality. Another grand narrative, liberalism, conveys the general celebration of liberty, equality, and fraternity; and it cultivates ideas about social justice. Liberalism has, however, also turned into a general celebration of the powers of the market, claiming that they need to be set free in order to secure the general welfare of society. Simultaneously, Marxism tells us about the increasing exploitation of workers and points out political actions necessary to change society. There are grand narratives related to colonialism and to racism, narrating, even with enthusiasm, about the "white man's burden."

The notion of *critique* has grown up and multiplied along with such grand narratives. It has become an integral part of some while simultaneously challenging others. Whatever it has done, critique is not independent of such narratives. The grand narratives of Modernity have been challenged, and this means that the concept of critique must also be challenged. Critique needs critique.

Critique as Uncertainty, pages 289–299.

In this article, I want to point out that some important changes in the conception of critique are taking place. These changes also apply to the notion of critique used in critical education, as well as in critical mathematics education. Thus, I will address conceptual issues that I find are crucial for the further development of critical mathematics education. Let me also emphasize that I consider critical mathematics education and mathematics education for social justice as being, in practice, synonymous, although there are differences in the conceptual histories of critique and of social justice.[1] In the following, however, I concentrate on "critique."

The Emergence of Modern Conceptions of Critique

Modernity represented a step away from dogmatism. By dogmatism I refer to the claim that some authoritative texts contain truths about certain issues. The Bible may serve as an example, but many texts other than the Bible have been read in a dogmatic way. In general, dogmatism assumes that some truths have been encoded in certain texts and that a proper interpretation of these texts will bring their truths to light.

The emergence of what can be referred to as modern conceptions of critique can be related to the principle of universal doubt, as proposed by René Descartes: If it is possible to doubt a statement, one cannot consider it certain and therefore it cannot count as knowledge.[2] This doubt confronted any form of dogmatism, since dogmas can be doubted. As a consequence, any dogmatic assertion had to be removed from the assumed stock of knowledge. The principle of universal doubt caused an epistemic tsunami so forceful that, according to Descartes, nothing was left except a single stony statement: *cogito, ergo sum*. This statement represented the solid ground upon which knowledge could be built.

Once this ground had been identified, the construction of the whole edifice of knowledge could begin. As an instantiation of knowledge, *cogito, ergo sum* demonstrated, according to Descartes, the criteria for knowledge: Only a statement as evident as *cogito, ergo sum* could be counted as knowledge. In other words, the justification of knowledge had taken a radical antidogmatic format: Justification came from the perception of certainty. Knowledge should be evident and this evidence was a human experience. No justification of knowledge need be looked for outside of human rationality.

What I refer to as modern conceptions of critique can be summarized in the following way: A critique should prepare the ground for obtaining *certainty* and *truth*; it should be an expression of *rationality*; and it should include a search for the proper *foundation* of knowledge. I choose to talk about "modern conceptions

[1] A genealogy of "social justice" will immediately bring us to Rawls' (1999) important study, *A Theory of Social Justice*. Rawls represents a radical trend within the peaceful analytical approach. Further back in this genealogy we get to the works by Mill, Bentham, Locke, and many others.

[2] See Descartes (1993) for his principal discussion of epistemic issues.

of critique" in the plural, as we are dealing with a multitude of critical approaches woven together with different aspects of Modernity. To talk about modern conceptions of critique is just a convenient way of emphasizing some (sometimes rather feeble) resemblances among a range of language games, including "critique."

Descartes prepared the layout of modern conceptions of critique, although he did not explicitly talk about critique; that was to come. In particular, Immanuel Kant and Karl Marx added two heavy, almost nonoverlapping chapters to the grand history of critique. One can consider Kant and Marx to be the critical hardliners of Modernity.

Kant's main idea was that some fundamental notions were not derived from empirical observations but instead preceded such observations. Thus, the notions of space and time were not distilled from experience. Instead, they established universal categories for organizing experiences. And according to Kant, the principal task for a critique was to identify any such *a priori* categories. Kant turned critique into a principally epistemic project through which one could identify the universal human conditions for obtaining knowledge.

Inspired by Hegel, who in turn drew heavily on Kant, Marx took critique to a quite new location. On the one hand, Marx found it important to criticize all established economic theories, as they neglected the principal features of economic reality. The formation of capital, for example, had not been grasped correctly, and Marx pointed out that this formation was a direct result of the exploration of labor. One goal of Marx's critical enterprise was to reveal the hidden and brutal mechanism of capitalism. On the other hand, Marx wanted to criticize not only economic theories but also economic reality itself. He wanted to address the very exploration of labor. In this way, critique became not just an epistemic project but a political one as well.

Different as they are, there are important similarities between the critical approaches of Kant and Marx. They each wanted to provide solid foundations and to establish certainty and truth. Thus, Kant wanted to provide a foundation of knowledge in general, while Marx wanted to provide a solid foundation for political activities. To both Kant and Marx, critique was a rational affair. Therefore, I consider Kant and Marx to be two hardliners in the development of modern conceptions of critique.

Such conceptions have proliferated in many ways within education, and some trends in critical education have assumed different features of them. Some approaches try to establish critique as a rational affair by showing how arguments and recognized ideological positions can be questioned. Some demonstrate the inspiration of Marxist terminology by assuming that the aims and tasks of critical education can be formulated with direct reference to the class struggle. Other trends in critical education have drawn on Jürgen Habermas's studies. In *Knowledge and Human Interest*, Habermas (1971) connects the social sciences with an emancipatory interest—for some the defining interest of critical education. Habermas tries to define the true nature of knowledge-constituting interests and to provide the notion of emancipation with genuine clarification. Thus, Habermas unites aspirations from both Kant and Marx, and one can consider *Knowledge and Human Interest* to be a closing chapter in the history of modern conceptions of critique.

Challenging Modern Conceptions of Critique

Modern conceptions of critique have been challenged, and I want to mention two examples: one provided by Michel Foucault, another by Walter Benjamin. While Foucault challenged critique as a pursuit of truth, certainty and foundations, Benjamin challenged critique as a directly rational affair. These challenges are important for opening up new features of critical approaches in general and to mathematics education in particular.

Foucault (1989, 1994) wanted to dig into the foundations of knowledge, although with important differences compared with Kant or with Habermas. First, Foucault addressed particular instantiations of knowledge. Second, he was excavating foundations, not with any aspiration of revealing their universal nature or to identify proper knowledge-constituting interests, but in order to reveal how fragile and mischievous they were. Through a genealogy, an archaeology of knowledge, Foucault tried to show how the foundation of knowledge was first of all made up of a collection of presumptions. In this way, he demonstrated how foundational assumptions include particular interests of the most dubious kind. Instead of seeing foundations as a noble epistemic ideal, they were revealed as serving particular interests. In this way, a dialectic between knowledge and power was laid bare. Foucault's genealogy represented a new form of critical enterprise.

What was established and maintained as knowledge included a powerful insistence on what to call true. Foucault talked about the "regime of truth," which is a metaphoric expression for knowledge-power dialectics. In the interview "Truth and Power," Foucault states that "each society has its regime of truth, its 'general politics' of truth—that is, the types of discourse it accepts and makes function as true" (Foucault, 2000, p. 131). In this statement, "society" can refer, for example, to a scientific community or to a period of time, and naturally, it can have all its normal meanings. Grand narratives, including the grand narratives of modernity, define regimes of truth.

Regime means domination, while truth has been assumed to signify genuine epistemic ideals. Thus, by talking about a regime of truth, Foucault highlights the intimate relationship between knowledge and power. Regimes are powerful; still they are temporary. Regimes try to dominate; they set standards; they make exclusions; they make alliances; they can be overcome anyway. And such observations apply to regimes of truth as well.

Foucault departed from certain instantiations of knowledge: psychology, psychiatry, social sciences, and humanities in general. He revealed how notions related to norms, standards, deviation, madness, and sexuality were all integrated into regimes of truth. In this way, he confronted directly any modern celebration of science, including its objectivity and neutrality. However, Foucault did not address the so-called exact sciences and mathematics. As I will discuss in the next section, science and mathematics also represent a massive domain of knowledge-power interactions. Thus, there is much more to be expected from an archaeology of knowledge than the excavations of Foucault.

Nevertheless, through his genealogy, Foucault made an important addition to the notion of critique. He did not let critique operate in any sanitised *a priori* landscape. Instead, he turned it into a highly contextualized activity. A critique does not provide any genuine foundation for knowledge; it serves instead to show the dubious character of assumed knowledge. To Foucault, a critique is no search for certainty. It is, indeed, almost the contrary: Through a critique, one may reveal the temporality and doubtfulness of established knowledge and of any truth claims. In this way, Foucault helped to cut the connections between critique and some of the grand narratives of Modernity.

Benjamin contributed to the Critical Theory movement. But a closer look at some of his writings suggests that he was far from submitting himself to any recognized analytic pattern. He took, instead, an important step, different from and somehow more radical than Foucault's, beyond modern conceptions of critique. Benjamin was engaged in a radical, even anarchic, transgression of established critical rationalities.

In *The Arcades Project*, Benjamin (1999) developed the idea that the arcades—the glass-covered streets constructed in Paris during the 1820s and 1830s where people could walk from shop to shop without being concerned about weather or transport—displayed much about the society in development. Benjamin saw the arcades as indicative of new features of society. They were an expression of a new luxury in consumption and, to Benjamin, they also provided a window on society. The arcades were the modern bazaars and the forerunners of the shopping centers now spread all over the world.

Benjamin presented his investigations in a surprising format. He did not try to develop a critique of society in any systematic way; there were no theoretical clarifications accompanying his walk through the arcades. His investigations did not result in any coherent text written by himself. Instead, he operated with an inventive form of noticing. He made a note of this and a note of that; he copied a piece of this text and a piece of that text. He ended up with a monstrous pile of notes, most of which were quotations. These notes were organized in "convolutes," through which he tried to provide some order to his observations. (Benjamin referred to what must have been huge folders as convolutes, as do the editors of *The Arcades Project.*) In *The Arcades Project*, there are convolutes about arcades, fashion, boredom, eternal return, iron construction, exhibitions, the collector, dream city, the flâneur, theory of knowledge, prostitution, gambling, mirrors, Marx, *and more.* And in each convolute, Benjamin's own notes became mixed up with all kinds of copied texts. Like a surrealist artist using bit and pieces from posters, journals, and magazines to make a collage, Benjamin tried to organize his pile of notes into a panoramic insight. This collage constitutes *The Arcades Project*.

Through this project, Benjamin opened up a new critical approach. Critique need not be systematic; it need not be placed on any platform. Horkheimer and Adorno had emphasized to Benjamin that a study of Marx was indispensable for understanding the time period Benjamin was addressing. Sure enough, there is a convolute labeled "Marx." However, leafing through this convolute, one does not find anything resembling a systematic reading. Instead, there are here-and-there notes. There is no picture of Marx in total. Instead, Marx is torn to pieces, some of which Benjamin found gaudy enough for his collage. His critique takes an anarchistic format, methodologically speaking.

Through *The Arcades Project*, Benjamin takes critique far beyond Kant's transcendental accuracy and Marx's profound systematics. But Benjamin's critique does not take the format of a genealogy, as proposed by Foucault. Benjamin's critical approach opens up a multiplicity of perspectives, associations, connections, trends, possibilities, and uncertainties. His critique is not regulated by well-defined procedures. It includes imagination and creative features.

What should we call the outcome of the critical turbulence to which Foucault and Benjamin contributed? Postmodern notions of critique? I prefer, however, not to use the label "postmodern," as it draws on a range of other assumptions which should also be questioned.

Critique and Generativity

Let us now consider what might happen to the notion of critique when its modern roots are challenged and we pursue further the ideas proposed by Foucault and Benjamin. To keep this article short, I restrict myself to two issues: In this section, I discuss critique and *generativity*, while in the next section, I discuss critique and *imagination*. In this way, I will try to articulate some new features of critique. I will also indicate what these ideas could mean for critical mathematics education.

It is a principle idea of critical mathematics education to question any celebration of mathematics. However, for centuries, the rationality of mathematics has been celebrated. Thus, during the so-called scientific revolution, mathematics was assumed to provide a genuine insight into nature. This claim took on a profound religious significance. Descartes, Galilei, Newton, and their contemporaries were all deeply religious and took God's existence as a given. According to them, nature was created by God, and the laws of nature were an expression of his creative power. As these laws had a mathematical form, God must have created nature according to a mathematical blueprint. This idea gave mathematics a sacred significance. Mathematics reflected not only the structure of nature, it also represented the rationality of God. And human beings shared this rationality with God! There were really good reasons for celebrating mathematics. In fact, one can see this celebration as one of the grand narratives of Modernity.

A genealogy may serve as a deconstruction of the authority of mathematics. Let us, however, take a closer look at the following remark made by Foucault: "If, concerning a science like theoretical physics or organic chemistry, one poses the problem of its relations with the political and economic structures of society, isn't one posing an excessive complicated question?" (Foucault, 2000, p. 111). Sure. However, it is important for me to emphasisze that the natural sciences, including mathematics, are just at forcefully linked to the political and economic structures of society as any of the sciences addressed by Foucault. Thus, with respect to mathematics, the economic requirements are extreme, both in terms of the need for funding, as well as for profitable outputs. Here, the knowledge-power interaction has turned into an extreme. We therefore have to take a step further than Foucault, who did not pay much attention to the so-called exact sciences in his archaeological studies.

A critique of mathematical rationality can take the form of a genealogy. And it is crucial that mathematical ideas, procedures, and perspectives are deconstructed by revealing the socioeconomic complexity of their origin. This is, in fact, the basic idea of social constructivism, which interprets the construction of mathematical ideas as a social process (see, for instance, Ernest, 1998). Such a genealogy may reveal powerful regimes of truth operating in the name of mathematics that have not been addressed by Foucault.

However, a critique of mathematics needs to address not only its genealogy but also its *generativity*. While a genealogy looks backwards, a study of generativity looks forwards, addressing questions like what could be the implications of operating with a mathematical rationality? What are the potentials of this rationality? What could be achieved by mathematics-based action? What could be achieved by whom? What kind on dominance could be exercised through mathematics? What kind of risks accompanies mathematics-based actions? And so on.

I see a genealogy and a study of generativity as supplementing each other. They are important approaches for addressing mathematics critically. Thus, I suggest that Foucault's critical approach be extended in two ways: First, it has to encompass the exact sciences, including mathematics. Second, it has to address not only the genealogy of notions and ideas but also their generative potentials. These two extensions are important for the further development of critical mathematics education.

The point of investigating *mathematics in action* is precisely to reveal to what extent mathematics-based actions forms our social reality. Mathematics is not pure and detached. Quite the contrary: It comprises huge generative potentials. In any technology, including construction, schemes of production, management, forms of communication, knowledge and information processing, techniques for surveillance and control, economic planning and transaction, risk estimation, and so on, we find mathematics in action. And such actions can have all kind of qualities. They can be inventive, brutal, cynical, money saving, risky, and so on. There are no qualities, good or bad, that, in general, can be attributed to mathematics in action. The qualities of such actions have to be addressed in each particular case.

Mathematics in action is part of the formation of new technologies. This formation depends on mathematical conceptions of possibilities. The explosive development of all forms of communication and surveillance technologies is a principal example of mathematics-based innovation. Through technological innovation, mathematics and power become intimately connected. There are many examples of the use of mathematics-based analysis designed to anticipate the implications of major technological initiatives. Thus, any risk assessment of a proposed new power plan in an area at risk of earthquakes is mathematics-based. By conceptualizing what is feasible and what is not and by providing a discourse of justification, mathematics and power are further integrated. Mathematical rationality is often materialized through technology; thus, all forms of automation embody mathematical algorithms and procedures in action. In this way, mathematics becomes part of our reality. Thus, mathematics can be seen as the iron constructions of the epoch we are entering. We see arcades not only covering streets and shops but in vast landscapes of production, consumption, communication, and the like.

It is important to criticize mathematical rationality whatever form this rationality might take: pure mathematics, applied mathematics, everyday mathematics, any form of ethnomathematics. No form of mathematics in action is immune to critique. Such a broad critique of mathematical generativity is crucial for any further development of critical mathematics education.

Critique and Imagination

Mathematics education may serve many different functions, and there may be a huge gap between the explicit formulated aims of mathematics education and its actual sociopolitical, economic, and cultural roles. According to Louis Althusser's structural Marxism, education is instrumental for the ideological structuring of society. Thus, the education system will prepare young people to enter the capitalist order of society, to operate as workers, and to do so in a submissive way. And mathematics education is part of this preparation for obedience. Furthermore, this structuralism claims that the very notion of educational improvement is, at best, meaningless; it is more likely, however, that any educational "improvement" simply facilitates a further submission of educational practices to the capitalist order of things. This observation has strong implications for the understanding of critical education. Critique at any local magnitude does not add up to any proper critique. The capitalist order of society has to be changed in order to make sense of small-scale educational changes. I am, however, not convinced of the irrelevance of small-scale educational changes.

Critical mathematics education can draw on conceptions of critique that are not derived from grand narratives like Marxism. It is important not simply to understand critique as a transparent analytical device or as a preguided political activity. A critique need not be regimented by any analytical order. In particular, an important step beyond modern conceptions of critique is taken when critique is related to *imagination*. Critique can mean an exploration of possibilities which need not be related to sociopolitical changes overall: One can consider possibilities in everyday situations, possibilities in the personal life of students, possibilities in the classroom, and so on (see Skovsmose & Penteado, 2011).

For an imaginative exploration of possibilities, one should not expect to draw on any well-defined methodology. Critique can include many creative features, and inspiration can be found in Benjamin's analytical anarchism. Metaphorically speaking, a critique can be composed of many convolutes, including sketchy notes, singular observations, and loose ideas. Thus, a critique of educational practices might take the form of an artistic patchwork, to which students, teachers, and researchers may contribute. Whatever imaginative elements a critique may include, they are important to a critical mathematics education.

I have tried to facilitate the imaginative dimension of critique through a discussion of *landscapes of investigation* (see Chapter 1). Such landscapes refer to different teaching-learning milieux. They represent different ways of engaging with mathematics, and they provide different framings of teacher-student interactions and interactions among students. The discussion of landscapes of investigation serves the purpose of inspiring pedagogical imagination within a specific context. It might be possible to open a particular exercise for further exploration; to consider how a mathematical theorem can be illustrated through different examples; to explore mathematical strategies for certain games; to explore what exercises might grow out from reading a newspaper; and to explore how project work may address real-life problems. Many teachers and students contribute to a pedagogical imagination by illustrating how classroom practices could be different. I have personally learned much from noticing how teachers and students, in collaboration, explore features of a landscape of investigation.

An *exploration of possibilities* can also be seen as an important feature of critical research. Much research, however, is merely descriptive, as is much research inspired by Foucault's genealogical approach. Descriptivism in education has been justified by the claim that proper research needs to be grounded in empirical investigation, that identifies, for instance, the atmosphere of the classroom; the students' activities as well as their conceptions of certain mathematical ideas; the teacher's communicative strategies; the evaluation of the students; the economic regulation of schools; the political and ideological domination of schools; and so on. I agree that it is important to research what is the case, the given, the actual, including all its constraints. However, it is important for critical research to move beyond descriptivism. Thus, it is important to imagine alternatives, which can be specific, general or illusory. Imagination can have many different qualities, but it always confronts the given.[3]

[3] For a general presentation of researching possibilities, see Chapter 9.

A feature of critical practice and research is to address not only the given situation but imagined situations as well. Furthermore, it is important to consider what could be done differently, acknowledging the constraints that condition the particular teaching-learning processes. Perhaps only some smaller changes are possible, but small does not mean irrelevant. Thus, my point is that critical practice and research address tensions between (1) what is actual; (2) what could be imagined; and (3) what could be tried out.

Explorations of possibilities, both in the classroom and in research, bring critique and imagination together. To be critical means to recognize that what appears as given might be contingent: It could be different. Daily educational practice includes many constrains set by the socioeconomic situation. But it also includes contingencies that mean that something could be done differently. However, contingencies are not revealed through strict descriptivism; imagination is necessary. There is no straightforward analytic procedure to identify new possibilities, and pedagogical imagination could be as undisciplined as Benjamin's approach. One need not assume that small educational changes add up to huge social changes, but educational changes could still make a difference for some students.

Critique: An Explosive Concept

One cannot expect critical education, including critical mathematics education, to provide solutions to socioeconomic and political problems on any larger scale. Education is hardly a world-changing institution. Still, this does not mean that educational fatalism is the consequence. Education might change something for some students. It might, in fact, change a range of life worlds. Thus it makes sense within education to work for social justice and, in particular, to do so in an unjust society, even knowing that the particular effort will not bring about a justice.

One element of modern conceptions of critique was a search for genuine foundations. However, Foucault demonstrated the critical force of a genealogy of knowledge which revealed foundational solidity as illusory. A genealogy, however, points back in time, while a discussion of the generativity of knowledge is points towards the future. For critical mathematics education, it is important to formulate a critique not only in terms of a genealogy but also in terms of generative investigations.

To include imagination as an important feature of critique is also important in moving beyond modern conceptions of critique. A pedagogical imagination is crucial for exploring educational possibilities. And one should not expect such explorations to be organized in any particular way or to be guided by particular methodological principles. They could, instead, take anarchic forms, as anticipated by Benjamin. Imagination is tentative; it is preliminary; it takes place in the open.

Critique must include generative as well as imaginative explorations. This means that the very notion of critique escapes being captured and nursed by well-intended definitions. Moving beyond the modern outlook means that critique becomes recognized as an open concept, a fluctuating concept, and a contested concept. In other words, critique becomes an explosive concept.

NOTE

This paper was published in *For the Learning of Mathematics*, *31*(3), 19-23, 2011.

REFERENCES

Benjamin, W. (1999). *The Arcades project*. Cambridge, MA: Belknap Press of Harvard University Press.

Descartes, R. (1993). *Meditations on first philosophy in focus* (S. Tweyman, Ed.). London, UK: Routledge. (Original work published 1641)

Ernest, P. (1998). *Social constructivism as a philosophy of mathematics*. Albany: State University of New York Press.

Foucault, M. (1989). *The archaeology of knowledge*. London, UK: Routledge. (Original work published 1969)

Foucault, M. (1994). *The order of things: An archaeology of the human sciences*. New York, NY: Vintage. Books (Original work published 1966)

Foucault, M. (2000). *Power*. New York, NY: New Press.

Habermas, J. (1971). *Knowledge and human interests*. Boston, MA: Beacon. (Original work published 1968)

Rawls, J. (1999). *A theory of justice*. Oxford, UK: Oxford University Press. (Original work published 1971)

Skovsmose, O., & Penteado, M. G. (2011). Ghettoes in the classroom and the construction of possibilities. In B. Atweh, G. Graven, W. Secada, & P. Valero (Eds.), *Mapping equity and quality in mathematics education* (pp. 77–90). New York, NY: Springer.

CHAPTER 21

BEYOND POSTMODERNITY IN MATHEMATICS EDUCATION?

While postmodernity has brought new profoundness to critical activities, it has also created some of its own limitations. I consider it important to address both aspects, and in so doing, try to move beyond postmodernity. I will discuss this possible move with reference to mathematics and mathematics education.

The label postmodernity has been used widely with reference to new trends in architecture, art, and literature which break with modernist principles; it has been used with reference to new conditions for knowledge production; and it has been used in social theorizing to refer to new social and cultural phenomena. I am going to use postmodernity as a reference to a critique of Modernity, and I will concentrate on the specifically philosophical aspects of this critique.

Modernity itself can be associated with some general conceptions of science: that the natural sciences are able to reveal to us the secrets of nature, and that science in general can be both objective and neutral. An integral feature of Modernity is the recognition of the importance of science, knowledge, and education, and the whole Enlightenment project integrates these aspects of the modern outlook. Furthermore, Modernity can be associated with the emergence of ideas about de-

Critique as Uncertainty, pages 301–316.

mocracy, freedom, and equality. Clusters of such different ideas are brought together by the notion of progress: Self-improvement is a human possibility.[1]

If we take a closer look at the modern period, we find that very different sociopolitical phenomena developed together alongside scientific discoveries, the formulation for democratic ideals, and the preoccupations about enlightenment and progress. Colonialism and exploitation took many forms, as did racism. During the age of Modernity, one finds laborious attempts to provide a scientific underpinning of racist classifications of human beings as being more or less developed. Similar classifications were applied to languages: Some were identified as being more developed and complex than others. Such ethnic and linguistic classifications were weaved together with notions of progress and development and theorized into grand discourses that could legitimate colonialism and suppression.[2] Zigmunt Bauman made a gloomy addition to such observations by emphasizing that even the Holocaust can be seen as being made possible through discourses established within modernity.[3] Through the very act of making such observations, we enter into a fierce critique of Modernity, and this is what postmodernity is principally about.

Let me try to summarize how we are going to proceed. I will present the most radical form of postmodernity by outlining some of Nietzsche's ideas about truth, knowledge, science, progress, democracy, and ethical values in general. Then I will present an example of the archaeology of knowledge by referring to studies by Foucault. This brings us forward to the notion of *genealogy*, which is a defining construct for the postmodern conception of critique. I will discuss genealogy with respect to mathematics education and emphasize the extreme importance of this line of critique. However, I will then emphasize that a critique addressing *fabrications* that can be related to mathematical rationality is crucial as well. Finally, I want to add *imagination* as crucial for any critical activity. By bringing together genealogy, fabrication, and imagination, I try to broaden and deepen critical enterprises and in this way, try to move beyond postmodernity.

Nietzsche's Radical Postmodernity

Friedrich Nietzsche confronted the official grand narratives of Modernity.[4] He found that Modernity, although it had taken steps away from orthodox Christianity, was still lingering in a religious worldview. Nietzsche acknowledged that since Descartes, all philosophers had tried "to assassinate the old concept of the soul," and in this way, philosophers have attempted to "assassinate the basic assumption

---

[1] For a discussion of the notion of progress, see Bury (1955), and Nisbet (1980).

[2] See, for instance, Said's study of orientalism (1979).

[3] See Bauman (1989).

[4] In *Ecce Homo,* Nietzsche (2009) makes comments on his previous publications and about *Beyond Good and Evil* he states that it is in essence a critique of modernity (*Ecce Homo, Beyond Good and Evil*). It could naturally be added that Nietzsche confronted much more than the Modern outlook: Christianity in general and the whole philosophical tradition since Plato.

of Christian doctrine." However, Nietzsche's point is that "overtly or covertly modern philosophy is anti-Christian, although it is by no means anti-religious" (1998a). And this is the problem.

Maybe modern philosophy has assassinated the soul and other religious notions, but it maintains a religious outlook by installing new idols. The modern idols have the form of philosophical ideas through which reality becomes judged. Nietzsche's postmodernity hammers into pieces any such idols. Let us look at them one by one.

The idea of truth. According to Nietzsche, there is no truth to be found anywhere.[5] We talk, and not least through modern philosophy, as if we are able to discover truths. The distinction between appearance and reality plays a fundamental role in the modern worldview, as presented by Descartes, Galilei, Locke, and many others; it was crucial as well for the formulation of Platonism which Nietzsche was also attacking. This distinction points out that behind appearance there is reality, and the claim is that true knowledge concerns this reality. But, according to Nietzsche (2003), there is no reality behind appearance: "The 'apparent' world is the only one: the 'real' world has only been *lyingly added*" (2003).[6] The reality behind appearance is thus a myth. There is no reality which truth refers to. Truth has no permanence; it is an illusion. Truth is not about anything; it is a way of talking; it is a "mobile army of metaphors" (Nietzsche, 2010, p. 29). Truth is just a modern idol that has substituted ideas about the existence of paradise and of an eternal life after death.

The idea of knowledge. Nietzsche introduces a radical perspectivism.[7] There are no points outside the stream of life from where one can look at things, judge things, and formulate true statements. One is instead submerged into this stream. Nietzsche's perspectivism destroys any hope of establishing knowledge, according to any Modern aspirations. What is called knowledge does not represent any "insight" about certain states of affairs. As a consequence, Nietzsche reaches an epistemic Darwinism:

> It is improbable that our "knowledge" should extend further than the strictly necessary for the preservation of life. Morphology shows us how the senses and the nerves, as well as the brain, develop in proportion to the difficulties in finding nourishment. (1968, p. 272)

Formulated differently, one can see the brain as an organ among other organs. It is an organ necessary for our survival. The secretions produced by the brain are no more unique than the secretions produced by any other of our organs. Sweat is necessary for our survival, and so is knowledge. That we happen to call the output from the brain "knowledge" does not make it any different from any other forms

---

[5] See also Clark (1990).

[6] In this and all other citations, the italics are made by Nietzsche in the original.

[7] See also Hales and Welshon (2000).

of biological extracts.[8] The problem, however, is that the notion of knowledge, through immense philosophical misinterpretation, has become reified as an idol.

The idea of science. If knowledge is not about anything, and certainly not about truth, what then to think of science, the celebrated institution of Modernity? Just a different idol. Science cannot represent any search for truth as there is no such thing to search for. What then to think of the phenomenon that is in fact referred to as science? Nietzsche simply statesz "A 'scientific' interpretation of the world, as you understand it, might therefore be the most stupid of all possible interpretations of the world, meaning that it would be one of the poorest in meaning" (1974).

The idea of progress. It has been claimed that progress is a defining element in Modernity. In the essay *A Discourse on Inequality* from 1755, Rousseau emphasizes that, contrary to animals, human beings have a "faculty of self-improvement."[9] This idea is axiomatic for the whole Enlightenment project, which claims that knowledge and education are crucial for the common welfare and for progress in general. The *Encyclopaedia*, to which Rousseau also contributed, is an expression of the idea that knowledge should be collected and divulgated broadly.[10] Nietzsche just shrugs his shoulders: "The last thing I promise is to improve humanity" (2009, my translation). And even more explicit, "'Progress' is merely a modern idea, that is to say a false idea." (2003).

The idea of democracy. As progress is an illusion, one should not expect Nietzsche to think any better of democracy, which has been celebrated as part of modern progress. Nietzsche has not anything positive to say about the French Revolution, and in general he finds nothing worthwhile captured by notions like liberty, equality and fraternity. Instead Nietzsche sees democracy as a miserable expression of a herd mentality:

> Indeed, with the help of a religion that played along with and flattered the most sublime desires of the herd animal, we have reached the point of finding an ever more visible expression of this morality even in the political and social structures: the *democratic* movement is Christianity's heir. (1998)[11]

Furthermore, he characterizes "modern democracy . . . as the decaying form of the state" (2003). Nietzsche would certainly not be found in any democratic camp.

---

[8] Nietzsche states that the "sense for truth" will have to legitimize itself as a "means for the preservation for man, as *will to power*" (Nietzsche, 1968). The sense for truths is useful like any other senses. It helps to ensure our survival.

[9] See Rousseau (1984, p. 88).

[10] It should be noted that Rousseau not only presented the notion of self-improvement; he also developed a counterstory to the general celebration of progress by commemorating the natural state of affairs.

[11] Nietzsche also observes: "The *overall degeneration of man*, right down to what social fools and flatheads call their 'man of the future' (their ideal!); this degeneration and diminution of man into a perfect heard animal; this bestialization of man into a dwarf animal with equal rights and claim is possible, no doubt about that!" (1998).

Ethical values. According to Nietzsche, the slave morality put values upside down by celebrating the poor, powerless, suffering, deprived, sick, ugly, and such, and by nominating the noble, powerful, and beautiful as evil, cruel, lustful insatiable, and godless (*On the Genealogy of Morals*, First treatise, §7). This slave morality is, according to Nietzsche, contrary to life as "life itself *in its essence* means appropriating, injuring, overpowering those who are foreign and weaker; oppression, harshness, forcing one's own form on others, incorporation, and at the very least, at the very mildest, exploration" (1998b). Thus Nietzsche emphasizes that

> "exploitation" is not part of the decadent and imperfect, primitive society: it is part of the *fundamental nature* of living things, as its fundamental organic function; it is a consequence of the true will to power, which is simply a will to life. (1998b)

After such claims, it is not surprising that Nietzsche sees democracy as an idol; just a modernized version of slave morality.

It would be a risky business to try to adopt the whole Nietzsche package including all its premodern postmodernity. And this is not what has been done, but Nietzsche's radical postmodernity has inspired many.[12] His radicalism invites a *deconstruction* of celebrated ideals. The notion of deconstruction was formulated by Jacques Derrida (1974) in *On Grammatology*, and I find that this notion describes precisely what Nietzsche is suggesting as well as what many postmodern investigations are completing.[13] I will use it as a general characteristic of postmodern critical endeavors, naturally acknowledging that Nietzsche talked about genealogy when identifying ideas as idols. The general claim is that behind apparently decent scientific standards, aspirations about progress, ethical values, and the like, one will find much less noble forces acting out. The aim of a genealogy (a deconstruction) is to reveal this reality, so different from any idealist reality that has just been "lyingly added.".

Foucault's Archaeology of Knowledge

According to modern standards, science should not be subjective, but objective; it should not include special priorities or particular perspectives, but distance itself from any interests. Furthermore, it was assumed that science, due to its own intrinsic dynamic, does in fact acknowledge such demands. So, if one just lets science develop according to its intrinsic standards, it will reach objectivity and neutrality. Thus, science is an expression of solid epistemic standards.

---

[12] See Deleuze (1986) for a careful interpretation of Nietzsche's ideas in general and in particular between the Apollonian and Dionysian ways of life.

[13] Derrida is inspired by Heidegger who uses the notion *Abbau*, for deconstruction ideas. This German word resonates with *Aufbau*, which means construction. Thus, *Aufbau* and *Abbau* can be seen as an inverse processes. *Abbau* can been read as destruction, but it is better understood as an unmaking, thus Derrida emphasizes, "The movements of a deconstruction do not destroy structures from the outside" (Derrida, 1974, p. 24).

Inspired by Nietzsche, one should consider the trust in science as being a variation of a religious trust, just worshipping different idols, as false as any other idols. In "Truth and Juridical Forms," Michel Foucault refers several times to Nietzsche as an important source of inspiration.[14] Foucault does not address natural science but concentrates on psychology, psychiatry, and related discourses. He pursues the idea that the polished surface of science covers a muddled situation. He presents his archaeology of knowledge in terms of a genealogy. Through detailed historical studies, he demonstrates how any claim about objectivity and neutrality evaporates. What constitutes a fact at a particular historical moment becomes questioned in the next moment. Facts appear time-dependent, as do truth and scientific categories.[15] It appears that science is developing not in any continuous process of accumulation but sometimes in surprising and apparently irregular ways. In the interview "Truth and Power," Foucault refers to a particular observation that illustrates his point:

> In a science like medicine, for instance, up to the end of the eighteenth century one has a certain type of discourse whose gradual transformation, within a period of twenty-five or thirty years, broke not only with the "true" propositions it had hitherto been possible to formulate but also, more profoundly, with the ways of speaking and seeing, the whole ensemble of practices, which served the purpose of medical knowledge. (Foucault, 2000, p. 114)

One has to do with some overall and radical changes of a scientific outlook. In some places, Foucault talks about an episteme as comprising a broad set of ideas, assumptions, conceptions, priorities that function as a scientific worldview for a particular science at a particular moment. Foucault also uses the notion of discourse for such a worldview. So what Foucault is referring to here is a change in a medical discourse taking place within a relatively short period. This change included changes of what were considered true propositions, and as emphasized by Foucault, this change included not only a change in ways of talking, it also included changes of medical practices. As science becomes constituted through discourses, and discourses are condensing complexities of ideas, priorities, interests, and such, sciences becomes deprived of all idealized standards. Discourses are an expression of power, and as a consequence knowledge and power becomes intimately connected. Thus, through his genealogy, Foucault deconstructed any idealized conception of science. In fact, he brings us close to Nietzsche's perspectivism, including the observation that behind any idealized surface there exists a chaos of power.

This dynamic becomes condensed by Foucault through the notion of "regime of truth."[16] What Foucault observed with respect to medicine was a change of a

[14] "Truth and Juridical Forms" is included in Foucault (2000, pp. 1–89).

[15] It is interesting to compare this observation with Grabiner (1986), who asks if mathematical truths are time-dependent.

[16] See Foucault (2000, p. 131) as well as Chapter 8 in this book.

regime of truths. And, as he pointed out, it is possible to observe such changes with respect to many different scientific discourses. Through the notion of "regime of truth," Foucault draws directly on Nietzsche's characteristic of truth as a metaphoric way of talking.

Through his investigations of discourses, Foucault made an important contribution to critical enterprises. He showed that what is assumed to be stable is far from being so. Truth is time-dependent; it is context-dependent; it is power-dependent; it is dependent on whatever. The notion of fact has simply become dissolved through Foucault's acidic deconstruction.

Genealogy

Foucault's studies of genealogy have inspired many critical investigations of education and of mathematics education. Let me, however, concentrate on the later.[17]

Many general formulations that accompany mathematics education are in need of being deconstructed. Thus, in official documents, one can read, most often in a highly elaborated vocabulary, that the aim of mathematics education is to provide mathematical knowledge that can be of general interest for society and of personal relevance for the students. Such formulations assume that knowledge and education provide an important source of welfare, both for the society and for the individual. With reference to educational practice and acknowledging the formulated aims, one finds many suggestions for improving classroom practice. This reflects another of the grand ideas of Modernity, namely, that improvements and progress always have to be worked for.

Genealogies have pulled aside such decorative formulations, and there is much inspiration to be found in Foucault's work for doing so. In *Discipline and Punish*, Foucault talks about discipline, disciplinary society, and disciplinary institutions, and among them the school. With reference to the military, Foucault states that "discipline increases the skill of each individual, coordinates these skills, accelerates movements, increases fire power" and so on (Foucault, 1991, p. 210). Discipline is a defining element in military efficiency. Similar observations, however, apply equally as well to workplaces, hospitals, prisons, and schools: "The disciplines functions increasingly as techniques for making useful individual" (Foucault, 1991, p. 211). Usefulness can be applied to any category of people and to any institutions in society. However, at the same time, "usefulness" is defined with reference to a specific social order. Thus, Foucault emphasizes that the "growth of a capitalist economy gave rise to the specific modality of disciplinary power" (Foucault, 1991, p. 221). This power imposes a "political anatomy" in all kinds of institutions, also in schools.

To this observation one just need to add that mathematics as a school discipline also functions as a discipline in Foucault's interpretation of the word. With refer-

---

[17] For Foucault-inspired discussions of education see, for instance, Popkewitz and Brennan (1998).

ence to Foucault, one could claim that mathematics education is observing the political anatomy necessary for establishing usefulness. This observation implies that the formulated aims and good intentions that accompany mathematics education are only serving a decorative purpose. They are lyingly added. Our society is organized according to the demands of the free market, and the educational institutions are docilely responding to the definition of "usefulness" as exercised by this market; mathematics education as well. Such observations bring Thomas Popkewitz to summarize and deconstruct the Modern self-understanding in the following way:

> Mathematics is one of the high priests of modernity. Mathematics education carries a salvation narrative of progress into the upbringing practices of schooling. The mathematics of the school is told as a story of progress about the cultured, modern individual whose reason is bounded by the rules ands standards of science and mathematics. The narrative is of the enlightened citizen who contributes to the global knowledge society. (Popkewitz, 2004, pp. 251–252)

This is a self-understanding that covers the disciplinary reality of mathematics education.

Let me now try to summarize some of the very many observations addressing mathematics education as a disciplinary institution. One characterizing feature of mathematics education is the many exercises that, lesson after lesson after lesson, are presented to the students. Considering general formulations about, say, learning mathematics and creativity, the regime of exercises seems paradoxical. However, this regime can be interpreted as preparing for what I have called a prescription readiness.[18] This readiness refers to a capacity of completing small tasks that are defined from outside, that should not be questioned, and that should be completed promptly and with accuracy. Such a prescription readiness makes part of the political anatomy, which defines what is useful and what is not. It can be identified as one of the disciplinary features of mathematics education.

Mathematics education makes use of an extended system of tests, examinations, and labeling, which can be considered crucial for establishing the anatomy of the coming labor force. This workforce needs to be useful; however, there are very many forms for usefulness. For the functioning of the whole capitalist machinery of production, administration, surveillance, transport, economic transactions, and such, one needs a differentiated labor force with very different qualifications. Thus, the capitalist machinery operates with a matrix of different forms of usefulness, and this makes a meticulous labeling of individuals crucial. The educational system is taking care of this. As a consequence, students leaving the education system can be located in the proper cell in this matrix of usefulness. A careful labeling of products put out at the market makes things easier for the

[18] See Skovsmose (2008).

consumers. This also applies to the labor market of the capitalist society. Among the different school disciplines, mathematics appears eager to provide labels.

Mathematics education forms part a network of power relations. Thus, to understand what is taking place in the mathematical classroom, one cannot simply be concerned with studies of the classroom. One needs to consider the whole context of schooling. On several occasions, Paola Valero has presented a diagram with a teacher-student-mathematics triangle in its center, representing the activities in the mathematics classroom. However, she emphasizes that this triangle only represents a small part of a grander network, including youth culture, labor market, family, community, teacher education, staff, school leadership, mathematics education research, international comparisons, policymaking, academic mathematics, technological and scientific development, and so on.[19] All such factors, and in particular the power relations between them, are relevant for interpreting the actual disciplinary functions of mathematics education.

Many postmodern analyses have tended to stop at this point. This would in fact be in accordance with Foucault's approach. A genealogy reveals the powers that operate behind the educational scenery. Naturally we could ask, And what then? Considering the result of the deconstruction, how are we to improve the educational reality? By formulating such questions, we might be sliding away from a postmodern approach. Thus, we have to remind ourselves of Nietzsche's remark: "The last thing I promise is to improve humanity." And this formulation can easily be translated into, The last thing a deconstruction could promise is to improve mathematics education. In fact, a proper deconstruction—not assuming any of the modern ideals (or idols), including a notion like progress—must leave things as deconstructed.

Many postmodern approaches have assumed this position. I acknowledge that any form of genealogy in terms of a deconstruction is important. However, I do not want to assume that a genealogy represents the whole range of critical approaches. I want to address two more elements by addressing fabrication and imagination, and in so doing, I want to bring critical enterprises beyond a postmodern outlook.

Fabrication

As the notion indicates, a genealogy is a form of analysis or deconstruction that works its way back in time. Through his archaeology, Foucault presented a history of "the given"; he demonstrated how a scientific discourse, a particular regime of truth, become established as a conglomerate of preconceptions. I fully acknowledge that such critical investigations are powerful. I want, however, to go further and point out that there are other forms of critical investigations also relevant for mathematics education. Let me illustrate what I have in mind by referring to mathematics. An archaeology with respect to mathematical ideas will

[19] See Paola Valero's figure as shown in Menghini, Furinghetti, Giacardi, and Arzarello (2008, p. 286).

dig into their historical origin and try to reveal how knowledge-power dialectics constitute a part of their genealogy.

One can, however, also address the rationality of mathematics by investigating what can be accomplished through this rationality. What does it mean to bring this rationality into operation? What can be generated through this rationality? What possibilities can be established? What dangers and risks can be established as well? This brings us to consider what can be *fabricated* by means of mathematics, and in what follows this notion has to be interpreted in a general philosophical way as referring to *homo faber*. I see questions about fabrications as being different from those addressed by a genealogy. While a genealogy concerns the past, fabrications concern the future. Thus, fabrications are addressed through studies of *generativity*, which refers to what can be generated or fabricated through mathematical rationality.[20] Such fabrications can have many forms, and I will provide just a brief summary.[21]

Fabricating possibilities. Mathematics makes part of a range of design processes within any technological domain, and here I use technology as referring not only to technological artefacts, like bridges, airplanes, computers, and such, but also to less tangible constructions like schemes for production, health programm, techniques for surveillance, economic strategies, management principles, and such. In all such areas, mathematics provides powerful means for identifying new possibilities. Furthermore, no other natural-language formulation of possibilities operates in an equivalent way. For fabricating new possibilities within our already highly mathematized environment, mathematics is unique.

Fabricating facts. It is common knowledge that mathematics makes it possible to describe facts. To claim that mathematics fabricates facts might sounds strange, but let us consider an information system through which knowledge and information become processed. This system operates as if it were a computer language, including several layers of formal languages. It can be considered a conglomerate of mathematical algorithms. Such a system might, for instance, describe a tax system. This system not only describes taxes, it also prescribes taxes. It condenses decisions about what people have to pay in tax. Furthermore, there are information systems that not only prescribe taxes but also medications, treatments, loan conditions, working standards, and so on. Such prescriptions become life conditions for a range of people. They turn into facts.

Fabricating risks. Development of technology depends on simulations. This applies to the construction of cars, airplanes, transport systems, houses, shopping centers, robots for production, tax systems, and so on. Blueprints for such constructions have the form of mathematical models. One could assume that a mathematical simulation model provides a description of the construction when completed. This is, however, only partly true. The completed construction contains a range of

[20] For a discussion of the notion of generativity, see Chapter 20.

[21] The following summary follows Skovsmose (2009).

properties which are not anticipated by the simulation model. Many of the risks we are now facing—for instance, with respect to atomic energy, economy, transport, health care—are connected to the unanticipated implications of implemented technologies. A mathematical model however only anticipates and predicts outcomes that fall within a particular range of its algorithmic rationality.

Fabricating objectivity. Turning an entity into an object of study means that we strip it for a range of properties. This process is supported by mathematics, which highlights precisely the mechanical elements of what we are studying: being a new architecture, a medical treatments, a new configuration of a production line, new security matters, and so on. Mathematics helps to fabricate what becomes addressed as an object. In many cases, this "objectification" makes one assume that we are addressing things in an "objective" way; but objectivity can be seen as a mathematics-based fabrication.

Fabricating life-worlds. Our life-worlds are submerged in mathematics-based fabrications.[22] If necessary, our body becomes treated according to medical standards about blood pressure, cholesterol, glucose, and so on. They are all calculated standards, with reference to which diagnoses are formulated, medical treatments are prescribed, and health is monitored. In addition, standards for food quality are calculated: What should be considered acceptable levels of pollution is determined through statistical investigations, cost-benefit analyses, market interests, modelings of consumers' behavior, all condensed into well-defined numbers nominating what to call nonproblematic. Very many work processes take place in highly elaborated technological environments, where automatic processes and human activities are combined into an efficient production system. The market, including the labor market, is structured according to a huge set of mathematically formulated laws and principles. Our transport systems and our means of communication are all expressions of mathematics-based fabrications. In this sense, our life-worlds are themselves fabricated.

Naturally, one can see some overlapping between a genealogy and studies of generativity, but still these studies are different. Through a genealogy, one sees what has been incorporated, or *acted-into* a phenomenon, while a study of fabrications addresses what might be *acted-out* from the phenomenon. Both forms of investigation play important roles in critical investigations. We have illustrated generativity in terms mathematics-based fabrications. This way we have addressed mathematics rationality critically. Naturally, we can also discuss generativity with respect to mathematics education. This is a crucial feature of addressing mathematics education critically. However, we now move on to a third feature of critique: imagination.

---

[22] The notion of life-world is used by Husserl, but I am using the notion rather liberally (see Skovsmose, 2009).

Imagination

Beyond genealogy and generative studies, there is one more element that I want to consider part of a critical approach, namely, imagination. Critique in terms of imagination has been presented by Wright Mills (1959) in terms of sociological imagination. Through such imagination, one can reveal that something being the case could be seen differently. This means that one reveals a fact as being, not a necessity but a contingency. I find that a pedagogical imagination constitutes part of a critical educational endeavor.

Let us assume that we want to establish democracy in the classroom. We want teacher and students to address each other on equal terms and for interaction to become dialogical. We might imagine (or conjecture) that this facilitates new learning processes and ensures a greater empowerment of the students. They could develop new and more profound competences in reading and writing the world, to use some formulations that have been explored by Eric Gutstein (2006). We could go further and expand our imagination by exploring not only the notion of democracy but also dialogue, empowerment, and related notions such as equality and social justice.

However, one can see all such ideas as illusory by highlighting constraints that configure the educational reality. We can remind ourselves of the power network that any classroom practice is a part of. Imaginations do not make changes possible; instead, imagination could be the carrier of illusions. Inspired by Nietzsche, one can see pedagogical imagination as just worshipping idols. The point of the careful genealogy that has been conducted with respect to mathematics education is precisely to reveal that beneath the decorative discourse of aims and possible improvements, one finds a disciplinary regime. Considering the decorative discourse, everything seems possible, while considering the disciplinary regime nothing appears possible. And certainly not to establish any genuine classroom democracy as mathematics education as a discipline is doomed to insert a prescription readiness into the students and to subject them to a meticulous labeling. The deconstruction of a classroom reality brings into the open a grid of power relations that establishes a range of necessities which turn make a pedagogical imagination illusory. This is one part of an argument that might ultimately bring imagination into discredit.

There is, however one more line of argument discrediting imagination, drawing directly on Nietzsche's radical postmodernity. What are we to think of the range of the very concepts that resource imaginations such as democracy, dialogue, empowerment, equality, social justice, and so on? According to the deconstruction of the educational reality, such notions might bring about illusions of what could be done in the classroom. According to Nietzsche, however, such ideas are themselves idols. Following the presentation in *Twilights of the Idols* (2003), they should be shattered with a hammer; idols need to be destroyed. Thus, the idols developed through Modernity represent religious ideas, albeit in philosophical disguise. And the ideas that resource pedagogical imaginations belong to the same family of idols. By celebrating such idols, one falsely condemns reality by providing suggestions for improvements. However, according to Nietzsche, there are no norms and standards that operate from any elevated position, and certainly not from any transcendental position. There are no ideas positioned outside the stream of life. Everything makes part of a chaotic power dynamics. Within a radical postmodern approach to education, there is no use for ideas that are seen as stimulating a pedagogical imagination; they just represent romanticism and a naïveté that falsely condemn the educational reality.

But could we get around this sharp corner, and move beyond postmodernity? Foucault (2000, p. 33) raises the question if one can talk about "powerless truth" and "truthless power." In other words, Could there be any truths that are not aligned with power? Could there be false powers? Or does power simply define truths? I do not clearly see Foucault's response to this question. I want, however, to answer clearly myself: Truth can be powerless, and power can be truthless. If we make this assumption, we are moving away from Nietzsche's radical postmodernity, but are we instead heading straight into romanticism?

We have to be careful here. Nietzsche knocked down any ideals as if they were idols. He feverishly attacked any form of transcendental idealism. To Nietzsche, any ideas are formed within the stream of life—something I fully agree with. So let us not try to revive any new form of transcendence. But does that put an end to any ethical principles and any conception about democracy, equality, or justice? My principal point is that ethical principles and idealized notions need not be revealed nor discovered. They can be constructed. *I want to propose a social constructivism with respect to ethics*. I see ethical standards and idealized notions as emerging through laborious social processes. They are human constructions, and as such, they are historical; they are temporary; they can be changed. They represent human visions, expectations, and hopes; and precisely visions, expectations, and hopes make up a part of the stream of life. For me, imaginations cannot be discharged, as is done with reference to Nietzsche, through the claim that they assume a transcendental reality. They do not. Imaginations make up part of life. They represent the claim that what is the case is not *necessarily* the case. They try to reveal facts as contingencies. To me, such revelations are important critical activities.

Let us now return to the claim that imaginations are illusions due to the very nature of educational realties. Schools as disciplinary institutions are no simple contingencies. They constitute part of the whole power network within which the educational system is caught, like a small insect in a spider's web. This power network includes so many restrictions that one cannot assume that, say, democratic principles could be established in the classroom. This could be obstructed by the traditions of the classroom, the students' expectations, the demands for test and control, the expectations of parents, the organizations of the textbooks, the general time pressure for the teachers, and so on. I agree that all such observations could very well bring us to the conclusion that establishing democracy in a particular classroom is (almost) impossible. Still, and this is my point, imaginations—also imaginations that cannot be implemented—are of critical importance. For me, it is an important critical step to explore the scope of contingencies for any particular social institution is positioned between being a necessity or an open contingency. This also applies to the education system. There is no *a priori* analysis that designates every feature of this disciplinary institution as being necessary. It is, like any other institution, crisscrossed by contingencies. Naturally, seeing a difference is very different from making a difference, but this does not make imaginations superfluous. Seeing a difference is an initial critical act; making a difference might be possible.

I have been concerned about some postmodern limitations of critical activities which reduce critique to genealogy. I agree that a genealogy is an extremely important critical device. I also agree that this form of deconstruction has gained tremendous force through postmodern investigations. My position, however, is that critique has more dimensions. Genealogy is important, but so are generative investigations, and so is imagination.

As Conclusion: Beyond Postmodernity

In this chapter, mathematics education has been critiqued though genealogical investigations. In this way, it has been revealed how a disciplining order has become incorporated into this form of education and how power dynamics are operating in and behind classroom practices. However, there is no need to stop the critical investigations by such deconstructions. We can address the generativity of mathematics education in different contexts and with respect to different groups of students. What possibilities could be constructed for them? This question could be addressed not only with reference to further disciplinary initiatives, but also in terms of students' perspectives and with reference to their life-worlds. It is important that any educational practice should become recognized as a contested practice, and this brings us directly to the notion of pedagogical imagination.

I suggest that a critique of a phenomenon, referring to mathematics or to mathematics education, must address its genealogy, its generativity, and that it should be contrasted with imaginations. The postmodern form of critique focused on genealogy, which gave an impressive depth to many critical studies, but it also inserted limitations. My point is to broaden the critical activities; I suggest that it is verey possible to move beyond some of the limitations that postmodernity has placed on critical enterprises.

NOTE

This paper was published in 2012 in *The Mathematics Enthusiast*, *9*(3), 233–252.

REFERENCES

Bauman, Z. (1989). *Modernity and the holocaust*. Cambridge, UK: Polity/Blackwell.

Bury, J. B. (1955). *The idea of progress: An inquiry into its origin and growth*. New York, NY: Dover. (Original work published 1932)

Clark, M. (1990). *Nietzsche on truth and philosophy*. Cambridge, UK: Cambridge University Press.

Deleuze, G. (1986). *Nietzsche and philosophy*. London, UK; New York, NY: Continuum. (Original work published 1962)

Derrida, J. (1974). *Of grammatology*. (G. C. Spivak, Trans.). Baltimore,MD: Johns Hopkins University Press. (Original work published 1967)

Foucault, M. (1991). *Discipline and punish: The birth of the prison*. London, UK: Penguin. (Original work published 1975)

Foucault, M. (2000). *Power*. New York, NY: New Press.

Grabiner, J. V. (1986). Is mathematical truth time-dependent? In T. Tymoczko (Ed.), *New directions in the philosophy of mathematics* (pp. 201–213). Boston, MA: Birkhäuser.

Gutstein, E. (2006). *Reading and writing the world with wathematics: Toward a pedagogy for social justice*. New York, NY; London, UK: Routledge.

Hales, S. D., & Welshon, R. (2000). *Nietzsche's perspectivism*. Urbana and Chicago: University of Illinois Press.

Menghini, M., Furinghetti, F., Giacardi, L., & Arzarello, F. (Eds.). (2008). *The first century of the International Commission of Mathematics instruction (1908–2008): Reflecting and shaping the world of mathematics education*. Rome, Italy: Instituto della Enciclopedia Italiana Fondata da Giovanni Treccani.

Nietzsche, F. (1968). *The will to power* (W. Kaufmann & R. J. Hollingdale, Trans.). New York, NY: Vintage. (Original work published 1901)

Nietzsche, F. (1974). *The gay science* (W. Kaufmann, Trans.). New York, NY: Vintage. (Original work published 1882)

Nietzsche, F. (1998a). *Beyond good and evil* (M. Faber, Trans. & Ed.). Oxford, UK: Oxford University Press. (Original work published 1886)

Nietzsche, F. (1998b). *On the genealogy of morals* (M. Clark & A.J. Swensen, Trans.). Indianapolis (Cambridge): Hackett Publishing Company.

Nietzsche, F. (2003). *Twilight of the idols and the anti-Christ* (R. J. Hollingdale, Trans.). London, UK: Penguin. (Original work published 1888)

Nietzsche, F. (2009). *Ecce homo*. São Paulo, Brazil: Editoria Schwarcz. (Original work published 1888)

Nietzsche, F. (2010). On truth and lie in a nonmoral sense. In (T. Carman, Trans. & Ed.). *On truth and untruth: Selected writings* (pp. 17–49). New York, NY: Harper Perennial/ Modernthought. (Original work published 1873)

Nisbet, R. A. (1980). *History of the idea of progress*. New York, NY: Basic.

Popkewitz, T. (2004). School subjects, the politics of knowledge, and projects of intellectuals in change. In P. Valero & R. Zevenbergen (Eds.), *Researching the socio-political dimensions of mathematics education: Issues of power in theory and methodology* (pp. 251–267). Dordrecht, The Netherlands: Kluwer Academic.

Popkewitz, T., & Brennan, M. (Eds). (1998). Foucault's challenge: *Discourse, knowledge and power in education*. New York, NY: Teachers College Press.

Rousseau, J.-J. (1984). *A discourse on inequality* (M. Cranston, Trans.). London, UK: Penguin. (Original work published 1755)

Said, E. W. (1979). *Orientalism*. New York, NY: Vintage.

Skovsmose, O. (2008). Mathematics education in a knowledge market. In E. de Freitas & K. Nolan (Eds.), *Opening the research text: Critical insights and in(ter)ventions into mathematics education* (pp. 159–174). New York, NY: Springer.

Skovsmose, O. (2009). *In doubt: About language, mathematics, knowledge and life*-world. Rotterdam, The Netherlands: Sense.

Wright Mills, C. (1959). *The sociological imagination*. New York, NY: Oxford University Press.

CHAPTER 22

MODERNITY, APORISM, AND MATHEMATICS EDUCATION

Was "modern mathematics" actually modern? Of course, in historical terms, it was in fact modern. However, a more satisfactory answer to that question requires some clarification of the notion of modern *mathematics* as well as of *modernity*. I shall try to characterize modern *mathematics* in two ways: first in terms of its intrinsic properties and second by its extrinsic properties. The notion of *modernity* I shall try to specify by only referring to one aspect: the assumption of a harmony between the development of scientific rationality and social progress in general.

Modern mathematics provides the foundation for the new math movement that emerged in the late 1950s, which certainly represented the modern assumption that mathematics ensures progress. However, this modern mathematics education was soon heavily criticized, with the result that very quickty, new trends in mathematics education were unveiled.

Were these critical reactions to modern mathematics education modern or postmodern? For instance, does constructivism—radical or social—represent modernity or postmodernity? I find that several of the present trends in mathematics education continue to embark on the modern perspective: mathematics ensures progress. To me, this represents a fundamental problem of mathematics education

Critique as Uncertainty, pages 317–331.

today, and in this paper, I want to identify the nature of this problem, but will not rush towards any conclusion. First I shall try to characterize modern mathematics.

Modern Mathematics

Modern mathematics grew out of developments in mathematics research in the 19th century and matured as an elaborated perspective on mathematics at the beginning of the 20th century. However, modern mathematics can be characterized from a philosophical perspective and in particular, I want to draw attention to the following three aspects: the development of formal axiomatics; an interpretation of mathematical notions which (apparently) does not presuppose the existence of any mathematical entities; and finally an interpretation of mathematical truth which (apparently) is without any referential assumptions. These three aspects represent an extreme purification of mathematics which is a trademark of *modern* mathematics.

In *Principia Mathematica I-III*, written by Alfred N. Whitehead and Bertrand Russell, the building up of mathematics starts with axioms, five in total, and proofs are constructed by applying only two simple rules of deduction. In this way, Russell and Whitehead wanted to demonstrate that mathematics as a whole can be built up from a foundation of logic. *Principia Mathematica* represents modern mathematics as it develops the symbolic power of formalism to such a degree that all empirical and intuitive elements of mathematics seem to be eliminated.[1]

However, *Principia Mathematica* also follows the classic Euclidean paradigm by assuming that mathematics can be demonstrated to consist of absolutely true statements. In order to demonstrate the truth of all mathematics, this axiomatic organization of mathematics is of great help. This axiomatic organization reduces the seemingly overwhelming task of assigning truth to all mathematical statements; theorems can be reduced to assigning truth to the axioms of the system. How, then, are we to assign truth to axioms? This project cannot be based on deduction. Instead, we have to rely on intuition. This was Aristotle's approach and is how the Euclidean paradigm has been dealing with the problem ever since. But how can we rely on intuition in such important matters? The simplicity of the axioms seems to be a precondition, and the axioms of *Principia Mathematica* seem to be so simple that their truth can be grasped by, in most cases the, not too reliable, human intuition. From then on, intuition has no role to play in mathematics. The theorems rely on logical deduction, and this deduction has the property that if "*A* implies *B*" is true, and "*A*" is true, then "*B*" is true also. So when intuition has provided the axioms with truth, then deduction will deliver truth to all theorems. So runs the ideal of the Euclidean paradigm, and *Principia Mathematica*

[1] That the project of logicism, the reduction of mathematics to logic, was not completed by *Principia Mathematica I-III* (some elements of intuition an empirical observations seem impossible to eliminate) is a well known. And soon the whole project of logicism was given up.

represents this paradigm. However, *Principia Mathematica* becomes the last main work in mathematics which assumes the Euclidean perspective.

In *Grundlagen der Geometrie,* David Hilbert (1968) suggests that Euclidean geometry, or any other mathematical theory, does not presuppose references to any entities. Axiomatic geometry contains notions like point, line, and plane, but these have nothing to do with the empirical interpretation of point, lines, and planes, nor does a Platonic world provide references for these terms. Hilbert shows how mathematical theories can be developed without any reference to a reality outside mathematics. Hilbert does not assume the possibility of assigning any truth to the axioms of geometry or to a mathematical theory in general. He simply skips from mathematics the task of assigning truth to mathematical axioms. In this respect, *Gundlagen der Geometrie* breaks with the Euclidean paradigm.

The axioms of a geometric theory (or of any other mathematical theory) express relationships between undefined notions; in this case between points, lines, and planes. These relationships are the only thing we know about these notions. Any particular quality of these notions is irrelevant to geometry. The development of the geometry, then, consists in developing the implications of the axioms, and here only logical *deduction* can be used. Any reference to intuitive properties of points, lines, and planes would be illegitimate. Thus, a mathematics textbook in geometry would not be in need of any diagrams or visual illustration, and if such illustrations were included, then they only serve an instructional purpose.

Hilbert provides a surprising and provocative answer to the question of mathematical existence, which provides a clue to modern mathematics. Consistency of a mathematical theory presupposes that it not possible to prove a statement, say *p*, and at the same time (by using a different deductive route) to prove the statement not-*p*. The completeness of a theory refers to the idea that of any two pairs of statements, the one being a negation of the other (*p* and not-*p*), only one of these statements can be proved within the theory. So while consistency presupposes that not too much can be proved, then completeness presupposes that not too little can be proved.

Hilbert's hope was to provide formalizations of mathematical theories that were both consistent and complete. Had this task been fulfilled, Hilbert would also have made an original contribution to the question of mathematical existence, as he suggests that the question of existence of mathematical entities can be reduced to the question of consistency of a mathematical theory. In particular, points, lines, and planes exist if the axiomatic system of geometry can be proved to be consistent.

In the most dramatic way, Kurt Gödel proved that Hilbert's hope must remain a dream: A formalism, rich enough to contain the theory of natural numbers, would be incomplete if it was the case it was consistent. But whether he was a dreamer or not, Hilbert transformed the basic philosophical questiosn related to mathematics to *meta*mathematical questions. Metamathematics became a theory about mathematical formalisms clarifying for instance questions of consistency and completeness, and it gave further inspiration for formalism as a philosophy of mathematics.

In *Outlines of a Formalist Philosophy of Mathematics*, Curry (1970) provides a further clarification of the notion of mathematical truth. A formalized mathematical theory presupposes a formal language, in which basic units are symbols. Symbols can be organized in sequences, and some of these sequences can be counted as formulas; some formulas are established as axioms. The rules of deduction identify when a formula is a consequence of other formulas. A proof can then be described as a sequence of formulas with the following property: Any formula in the sequence must either be an axiom or a consequence of some of the previous formulas in the sequence (according to stated rules of deduction). A theorem, them, is any formula which occurs as the last formula in a proof.

Based on this clarification, the definition of mathematical truth splits up into two parts. First, a formula can be true in the sense that it can be a theorem in formalized theory. Thus, truth of a mathematical formula becomes relative to a particular formalism. Truth of a mathematical statement (a formula) means provability of this statement within a certain formalized theory.

However, according to Curry, it makes sense also to discuss the truth of a mathematical theory, but this discussion is completely different from the investigation of the truth of mathematical statements. The possible truth of a mathematical theory is not a mathematical question but a genuine empirical question. A mathematical theory may be true if it is possible to apply the theory. Thus, Euclidean geometry may be true if we can find applications for this theory. If we are concerned with "local space," then Euclidean geometry may be sufficient.

A more advanced formalization of mathematical theories was developed by Bourbaki, a pseudonym for a group of mainly French mathematicians.[2] As the formalist philosophy was stated by Curry, is seems to make sense to investigate implications of any axiomatic systems. But Bourbaki eliminated this relativism by identifying some structures as mathematical mother structures, and he showed how different areas of mathematics can be described in terms of these mother structures.

These ideas came to play a significant role in mathematics education. In Europe, the main initiative for the new math movement can be found in the Royaumont seminar held in 1959, where the main presentation was made by Jean Dieudonné, who perhaps more that any other represents Bourbakism.[3]

In summary, the position is this: Modern mathematics represents a purification of mathematics by establishing mathematics as formal structures and by interpreting questions of mathematical truth and mathematical existences as having to do with properties of formal structures. In this way, modern mathematics represents an internalism. It reduces questions about mathematics to questions which can

[2] See Bourbaki (1949, 1950); and Dieudonné (1970).

[3] See OEEC (1961).

be dealt with mathematically. So, was modern mathematics modern? However, before facing this question, I shall make a further clarification of modern mathematics in terms of what it would mean to apply mathematics.

Mathematics Provides Descriptions: A Picture Theory of Mathematical Modeling

In general terms, mathematical modeling refers to establishing a relationship between a piece of a mathematical world and a piece of an empirical reality. Proponents of modern mathematics were not blind to the applicability of mathematics, but still they tried to maintain the purity of mathematics.

Curry has certainly eliminated the discussion of applications of mathematics from mathematics itself, as he has designated questions of possible relationships between mathematics and reality to be merely of an empirical nature. These relationships can, however, be clarified further in terms of the picture theory of language as presented by Ludwig Wittgenstein (1992) in *Tractatus*. According to Wittgenstein, the basic function of language is to provide a picture of reality. The language Wittgenstein has in mind (while writing the *Tractatus*) is the formal language presented in *Principia Mathematica*. So, when Wittgenstein speaks about language, he in fact speaks about the language—the formalized language—of logic and mathematics. The presentation in *Tractatus* of the relationship between mathematics and reality supports the following idea: Mathematics provides possibilities for establishing a more or less adequate picture of pieces of reality. Mathematics provides the raw material for making a picture, while the actual evaluation of the picture is an empirical question.

The picture metaphor indicates that mathematics does not influence the reality which is pictured. The picture represents, more or less accurately, something to be pictured. But what is pictured is not influenced by the picture. The picture metaphor supports the idea that mathematics represents a tool for representing reality and that mathematics does not influence this reality. The picture metaphor, therefore, implies that there are no particular ethical, political, or other complex issues related to the application of mathematics. Mathematics is thus a detached science, and this detachment ensures ethical neutrality; applications of mathematics mean pure picturing.

The picture theory can be taken a bit further. As Wittgenstein speaks of the language, and the language is the formalized mathematical language, then as a consequence mathematics comes to play a special role in science. Wittgenstein indicates the existence of limits of language. Certainly he pays much emotional attention to what is beyond these limits, but this does not influence the interpretation of what the language is providing. What can be said can be said clearly (in the formal language of *Principia Mathematica*) and that which can be said clearly belongs to the natural sciences. Therefore, it makes sense to consider mathematics as the language of science. Mathematics represents the grammar of scientific rationality.

Modern mathematics is, thus, enveloped in a set of ideas which provide the whole enterprise of doing mathematics with an unproblematic nature. The presentation of mathematics as formalized theories splits up the notion of mathematical truths into the provability of formulas and the applicability of theories. Questions about mathematics of a philosophic nature become internalized, while the whole relationships between mathematics and the external world is taken care of by the metaphor of the picture theory.

Modernity: Trust in Mathematical Rationality

In order to address the question, "Was modern mathematics modern?" it becomes important to consider the notion of modernity and to clarify how this concept can be specified with reference to different sociological perspectives. Thus, a clarification of notions like late-modernity, reflexive modernity, risk society, postmodernity not to forget the very notion of modernity itself, seems relevant.[4] However, in clarifying the notion of modernity, I shall restrict myself to refer to certain assumptions concerning the role of science.

According to the perspective that represents modernity, science is playing a solid and important role in social development. The development of science and social progress, including even political progress and the development towards welfare and democracy, unite smoothly within scientific rationality. Modernity does not question the rationality of science. Naturally, I am not making the claim that science during some period of modernity in fact represented a solid foundation for progress. Instead, I claim that modernity assumes that we are concerned with a harmonious relationship between social progress and scientific rationality. In what follows however, I have natural science and mathematics in mind.

Scientific rationality is expressed by scientific methodology. In science, this methodology has been presented as the experimental method, the inductive method, or the hypothetical-deductive method. Whatever the interpretation of scientific rationality has been, the assumption of modernity has been the same: A support of this rationality needs no particular justification. By means of scientific rationality, all kinds of superstition and dogmatism can be confronted. This form of rationality was cultivated by the Enlightenment. A remarkable cultivation of scientific rationality was performed by logical positivism, and I see logical positivism as a particular clear expression of the assumption of modernity. Logical positivism is not only "positive" paying a special attention to the given. It is also "logical" meaning that logic and mathematics are assigned particular importance in the construction of the unified system of science. This unified system represents one version of the scientific ideal of modernity. That scientific knowledge can be presented in a unified form has also been expressed by the Enlightenment in the production of Encyclopedias.

---

[4] See, for instance, Beck (1992, 1998); Beck, Giddens and Lash (1994); and Lyotard (1984).

Modern mathematics presents a purified expression of rationality, and therefore it finds itself protected by the assumption that this rationallity serves an ultimate good. Modernity makes a critique of mathematical thinking superfluous. Instead, the clarity of mathematics becomes a model for critical thinking: being without clearly expressed assumptions and devoted to strict argumentation.

Jürgen Habermas has been interpreted as representing modernity, and this assumption can be supported by his study *Knowledge and Human Interest*. It is well known that Habermas (1971) is making a strong critique of the social sciences. A main aim of this study is to establish social science outside the paradigm provided by logical positivism. Habermas makes his point by suggesting the existence of different knowledge constituting interests, in particular an emancipatory interest constitutes social science. Should social sciences develop with a knowledge constituting interest, as if it were belonging to the natural sciences, it would serve the interests not emancipation, but of domination and manipulation. However, in *Knowledge and Human Interest*, Habermas does not provide a basis for a critique of natural science. It is guided by a technical interest, and it must be guided by this interest in order to constitute knowledge as part of natural science. The technical interest is, so to say, proper for its own domain. Therefore, we do not find a resource for a critique of rationality as expressed by the natural science in *Knowledge and Human Interest*; and mathematics is not touched by any critique. In this way, the works of Habermas can be read as accepting (part of) the assumption of modernity that rationality in the domain of natural sciences needs no criticism. Habermas' criticism of the rationality of natural science consists of limiting the scope of this rationality to what he finds to be its proper domain.

Let me summarize: A basic assumption of modernity is that the rationality which is exercised by natural science and purified by mathematics is a resource for progress. Rationality has to be developed and sharpened in its many forms. Obstacles for progress have to be identified and criticized, and scientific rationality represents the standards of critique.

I see modern mathematics as modern, as it draws heavily on this assumption of modernity. Modern mathematics has turned metareflection "inwards," and this is consequential, as there is no point in submitting mathematics to an external critique. The inner growth of mathematics represents an exemplary form of a development of rationality.

Modern Mathematics Education: Trust in Mathematics

Since the Royamont seminar, modern mathematics education in the form of the new math movement spread worldwide and influenced a variety of curriculum reforms directed towards the question, How do we provide learning environments, including textbooks and curriculum structures, which can make sure that students learn a substantial part of modern mathematics? Modern mathematics education has tried to open a main road for students into modern mathematics.

This route became legitimized by theories of children's epistemic development. In particular, Jean Piaget supported the idea that the logical structuring of mathematics, as developed by Bourbaki, in fact fits the natural learning trajectories of the child. This surprising idea was realized in a seminar in 1952 where both Piaget and Dieudonné participated,[5] and it was only later this understanding developed into a broad support for the new math movement.

To Piaget, the important notion is reflective abstraction.[6] This represents the epistemic action of the child when, for instance, he or she considers a certain set of operations on objects and then, by identifying some regularities in these operations, takes a further step by recognizing some of the unifying principles in these operations. In this case, reflective abstraction deals with properties of operations on objects (and not with properties of the objects). Reflective abstraction represents a constructive element in epistemic development.

Reflective abstractions, as described by Piaget, represent individual faculties by means of which the child becomes able to establish a different perspective of certain operations. Reflective abstractions becomes a mode of constructing more abstract mathematical notions, and in this way the child finds his or her way into mathematics. But this route into mathematics bypasses all possible reflections on the social role of mathematics. Reflective abstraction is presented as a modus of construction of mathematical knowledge in its structural form. There are no resources for a critique of the mathematical knowledge as such in the conceptual framework as presented by Piaget. There seems to be no place in Piaget's works where the idea is presented that mathematical rationality may have problematic consequences and that the learning of mathematics may also help to provide a critical uncertainly about this rationality. Piaget shares with Bourbaki the trust that rationality is represented by the logical architecture of mathematics

Modern mathematics education represents modernity in the sense that it assumes (now with an educational interpretation) that mathematical thinking has to be developed and sharpened, that obstacles to mathematical learning has to identified and eliminated, but that the social role of mathematical thinking need not to be questioned. Mathematics education is a true source for social and technological development in general. The possibility that it may be essential for mathematics education to prepare students for a discussion of mathematics, including the social role of mathematics, is not part of the program of modern mathematics education. Modern mathematics education recapitulates the assumption of modernity: Mathematics is not an object of critique, it is instead a resource of rationality, and the task of education becomes to bring students into mathematics, not to open mathematics as object of critique.

The educational thinking of John Dewey is not concerned with mathematics in particular. However, Dewey, more that any other educator and philosopher,

[5] See Beth and Piaget (1968).

[6] See Beth and Piaget (1968); Piaget (1970); and Dubinsky (1991).

has expressed the assumption of modernity in education. Dewey includes in his educational thinking a grand assumption of harmony of different elements of progress, namely, scientific methodology, inquiry-based learning process, and an endeavor for democracy.[7]

When properly identified, the scientific method will contain several qualities. First, it represents scientific rationality, and here Dewey has, first of all, the natural sciences in mind. According to Dewey, science starts with observations and collection of data. This data becomes then condensed into guesses and hypotheses which in their turn can be further confirmed or falsified. More data has to be collected, and the process of scientific investigations continues. Science is first of all characterized by a never-ending process of investigations.

Secondly, this process of investigation represents the proper way of coming to know, even if the subject of this process is not the scientific community. According to Dewey, there exists a strong similarity between the scientific method and the way children come to learn. Therefore, paying attention to the scientific method also means being able to identify a proper approach to learning and teaching. Dewey does not find any principal difference between a scientific community making investigations and a classroom doing the same thing. Dewey thus supports the introduction of the experimental method in schools, and this idea has become the central pinnacle of progressive education.

Thirdly, Dewey finds an intrinsic resonance between a concern for democracy and education. The proper schooling prepares students for the democratic way of life, and this preparation can also be related to the scientific method. This method represents proper investigation and, in general terms, a way of approaching a question in a nondogmatic way. Naturally, Dewey does not claim that any form of actual scientific approach would have these qualities. His point is that the scientific method, when developed in a proper way, has qualities which reach far beyond science.

I find Dewey's grand assumption of harmony is the basic ingredient in the modern approach to education. Modern mathematics education expresses a similar assumption.

Mathematics Provides Changes: The Formatting Power of Mathematics

Could we make the simple connection that postmodernism represents a disbelief in the rationality of natural science as crystallized by mathematics? I do not think this claim makes much sense. However, it becomes important to consider what it would mean to give up the assumption of modernity: that there exists a harmonious relationship between social progress and scientific rationality.

In order to illustrate that it is essential to address the social role of mathematics by a critique, I shall refer to the special issue of *Zentralblatt für Didaktik*

[7] See Dewey (1963, 1966, 1974). See also Chapter 11.

der Mathematik: *Mathematics, Ethics and Peace*. Here, a wide range of concerns about the social role of mathematics is expressed. Mathematics is presented as a fundamental aspect of our culture. A critique of culture must also include a critique of technology, including mathematics; otherwise, a "critique of culture" will lose significance, especially in the context of our technologically based society. In the article, "Mathematics and Peace: Our Responsibilities," this far-reaching cultural perspective on mathematics is expressed by D'Ambrosio in the following way:

> It is undeniable that mathematics is well integrated into the technological, industrial, military, economy and political systems of the present "Westernised" world. Indeed mathematics has been relying on these systems for the material basis of its continuing progress. We may say that mathematics is intrinsic to today's culture. (D'Ambrosio, 1998, p. 67)[8]

If mathematics is intrinsic to today's culture, then the picture theory of mathematical modeling becomes misleading. In fact, the picture theory becomes revealed as a metaphor which hides the mathematical involvement in social affairs. The complexity of mathematical modeling is emphasized by the notion of the *formatting power of mathematics*, which has been developed further through the discussion of mathematics in action (see, for instance Chapters 14, 15, and 16).

This discussion then questions the picture theory of mathematical modeling but furthermore, questions the following claims: that mathematics can be associated with abstract axiomatics; that mathematical notions can be described without making references; and the discussion of mathematical truth can be split up into a discussion of provability and of applicability. When the picture theory of mathematical modeling is revealed as an illusion, then the other assumptions establishing the self-understanding of modern mathematics also breaks down.

Mathematics, being part of technological systems and therefore part of our culture, also turns into an object of critique. We cannot undertake a critique if culture is excluded from critique or indeed if any critical investigation of any essential aspect of our social environment is excluded from critique. The thesis of the formatting power of mathematics provokes a demand for a critique of mathematical rationality as it functions as elements in any kind of technological planning and construction processes.

Aporism: Uncertainty About Rationality

The Greek word *poria* means direction, and *a-poria* can be read as "without direction." With reference to persons, the word *aporia* means (among other things)

[8] See also Appelbaum (1995); Restivo, Bendegem, and Fisher (1993); Niss (1994); and Porter (1995).

"want of means" or "want of resources." [9] I shall interpret this as beginning without sufficient epistemological resources for pointing out a direction. Modernity assumes that it makes sense to talk about social progress being associated with science. And the very directedness of progress is pointed out by scientific rationality. This rationality represents a direction, *aporia*. This leads to the intellectual optimism which we find in modernism. (For a further elaboration of the notion of aporia, see Chapter 23).

Recently it has become apparent that science not only produces wonders but also horrors. And the thesis of the formatting power of mathematics emphasizes that mathematics operates in the midst of this production.

The alternative to the modernist assumption is not to claim that the direction of scientific rationality leads to catastrophes, and that we should find the new directions by searching in the opposite direction. I simply do not find that the notion of opposite direction makes sense. It might be claimed that a strong belief in scientific rationality has brought us into a risk society, but it does not make sense to assume that negating scientific rationality may place us at a safe distance from risk.

When we give up the assumption of a harmonic relationship between social progress and scientific rationality, we find ourselves in quite a different epistemic situation. We are without direction, and acknowledging this is the characteristic of *aporism*. We are in a situation without guidelines for progress. In particular, scientific rationality does not provide such a guideline. Aporism can also be characterized in terms of the (limited) power of reflection. If scientific rationality can be questioned, we could try to look for a platform of a critique of scientific methodology. We need a critique of rationality which has different sources than this rationality. And this represents the epistemic difficulty acknowledged by aporism. A critique of rationality is needed. Sources for such a critique are not identified. However, critique cannot be postponed as we are entering a risk society. Scientific rationality appears so powerful in terms of constructive forces; at the same time we seem to be in such a weak position for reflecting on this constructive force.

We now see why Habermas' interpretation of natural science in *Knowledge and Human Interest* represent modernism. According to aporism, it is important to question the very knowledge constituting interest of natural science. They are, however, accepted by Habermas only paying attention to what this interest may cause of problems outside natural science.

Aporism does not accept the assumption that harmony exists as a given between scientific rationality and progress. But aporism does not make the opposite claim, that disharmeony exists. Aporism simply claims that critique of scientific rationality, including mathematics, is essential. At the same time, aporism acknowledges that there is no platform to be found for such a critique of rationality.

---

[9] This notion has been suggested to me by Irineu Bicudo from the State University of São Paulo, Rio Claro, Brazil.

We are left in the same situation as were we swimming in the open sea trying to construct a ship.

It does not any longer make sense to assume a strong connection between scientific methodology, learning in school, and democratic development. The Dewey assumption can only be characterized as naïve. Aporism becomes not only a substitute for the modernist assumption of this great harmony, it becomes also a possible perspective from which to discuss education, and mathematics education in particular.[10]

Aporism and Critical Mathematics Education

As part of the ethnomathematical approach, the rationality of mathematics, as presented in certain forms of mathematics education, has been linked to the idea of colonizing. It has been asked if Western mathematics really is a colonizer.[11] My point is not that such claims are verified. Instead my point is that such uncertainties about the role of mathematics are essential to consider as part of mathematics education.

Some of the recent discussions of mathematical modeling also represent uncertainty about the implication of what can be done by mathematics. The point is not simply to present examples of mathematical modeling, and by adding up a sufficient number of such cases, then to claim that mathematics is useful. Certainly mathematics can be useful, and certainly this usefulness can be illustrated by mathematical modeling. But whenever we face mathematical modeling, we face all the uncertainty linked to an interpretation and evaluation of the formatting power of mathematics.[12]

In *Towards a Philosophy of Critical Mathematics Education*, I have made suggestions for educational practices which keep in mind a critical perspective on mathematics. I have tried to illustrate the possibility of bringing students into a situation which makes it possible for them to consider critical aspects of mathematical thinking. In one case, I discussed in that book that students were asked to distribute child benefits to families living in a microsociety (described by the students themselves). In order to prepare the distribution, the students had to formulate principles for how to distribute the money. These principles were then specified in terms of the income of the family, the number of children of the family, and whatever they found relevant to include in the principle for distribution. Next, the students were asked to formulate an algorithm for actually carrying out the distribution, and the students experienced the difficulty of turning the principles into these algorithms. Simplifications seem to be needed in order to provide

[10] I shall not try to clarify to what extent the postmodern condition can be related to aporism, although such a clarification might be very useful.

[11] See Bishop (1990); and Powell and Frankenstein (1997).

[12] See, for instance, Booss-Bavnbek (1991, 1995); Christiansen (1996); Dorling and Simpson (1999); Højrup and Booss-Bavnbek (1994); and Keitel (1993).

an efficient system of doing the calculation. When the results of the calculation were presented, some of the students realized that in making their calculations, they had come to ignore so of the ideas stated in their general principles. They had performed a technocratic simplification, so often associated with real applications of mathematics.

My point is not that by being presented with such examples, students will come to reflect on the formatting power of mathematics. My point is that it is a concern of a mathematics education, which does not accept the assumption of modernity and which acknowledges that a critique of mathematical rationally is relevant, to *provide opportunities* for the students to reflect on what mathematics might be doing in terms of wonders and horrors.

A general implication of aporism is that a critique of rationality, and also of scientific rationality, becomes an educational task. This is a task of any education which is to be called critical. In particular critical mathematics education does not subscribe to the Dewey assumption that mathematics education maintains an intrinsic resonance with a democratic concern.[13] Critical mathematics education, however, shares with Dewey the concern for developing education as part of a democratic endeavor, and this means struggling with the episemic uncertainty related to mathematics.

REFERENCES

Appelbaum, P. (1995). Popular culture, educational discourse and mathematics. New York, NY: SUNY Press.

Beck, U. (1992). Risk society: Towards a new modernity. London, UK: Sage (Original work published 1986).

Beck, U. (1998). Politics of risk society. In J. Franklin (Ed.), *The politics of risk society* (pp. 9–22). Cambridge, UK: Polity/Blackwell.

Beck, U., Giddens, A., & Lash, S. (1994). *Reflexive modernization: Politics, tradition and aesthetics in the modern social order.* Cambridge, UK: Polity.

Beth, E. W., & Piaget, J. (1968). *Mathematics, epistemology and psychology*. Amsterdam, The Netherlands: Reidel.

Bishop, A. J. (1990). Western mathematics: The secret weapon of cultural imperialism. *Race and Class, 32*(2), 51–65.

Booss-Bavnbek, B. (1991). Against ill-founded, irresponsible modelling. In M. Niss, W. Blum, & I. Huntley (Eds.), *Teaching of mathematical modelling and applications* (pp. 70–82). Chichester, UK: Ellis Horwood.

Booss-Bavnbek, B. (1995). Model, prediction, action: Mathematics induced changes in science, technology and public thinking. *Res Humana, 1*, 113–172.

Bourbaki, N. (1949). Foundations of mathematics for the working mathematician. *Journal of Symbolic Logic, 14*, 1–8.

Bourbaki, N. (1950). The architecture of mathematics. *The American Mathematical Monthly, 57*, 221–232.

[13] See Skovsmose and Valero (1999).

Christiansen, I. M. (1996). *Mathematical modelling in high school: From idea to practice* (R-96-2030). Aalborg, Denmark: Department of Mathematics and Computer Science, Aalborg University.

Curry, H. B. (1970). Outlines of a formalist philosophy of mathematics. Amsterdam, The Netherlands: North-Holland. (Original work published 1951)

D'Ambrosio, U. (1998). Mathematics and peace: Our responsibilities. *Zentralblatt für Didaktik der Mathematik, 98*(3), 67–73.

Dewey, J. (1963). *Experience and education.* New York, NY: Macmillan. (Original work published 1938).

Dewey, J. (1966). *Democracy and education.* New York, NY: Free Press (Original work published 1916).

Dewey, J. (1974). *On education* (R. D. Archambault, Ed.). Chicago, IL: University of Chicago Press.

Dieudonné, J. A. (1970). The work of Nicholas Bourbaki. *The American Mathematical Monthly, 77*, 134–145.

Dorling, D., & Simpson, S. (Eds.). (1999). *Statistics in society: The arithmetic of politics.* London, UK: Hodder Arnold.

Dubinsky, E. (1991). Reflective abstraction in advanced mathematical thinking. In D. Tall (Ed.), *Advanced mathematical thinking* (pp. 95–123). Dordrecht, The Netherlands: Kluwer Academic.

Habermas, J. (1971). *Knowledge and human interests.* Boston, MA: Beacon. (Original work published 1968)

Hilbert, D. (1968). *Grundlagen der Geometrie.* Stuttgart, Germany: B. G. Teubner. (Original work published 1899)

Højrup, J., & Booss-Bavnbek, B. (1994). On mathematics and war: An essay on the implications, past and present, of the military involvement of the mathematics sciences for their development and potentials. In J. Højrup (Ed.), *Measure, number, and weight: Studies in mathematics and culture* (pp. 225–278). Albany: State University of New York Press.

Keitel, C. (1993). Implicit mathematical models in social practice and explicit mathematics teaching by applications. In J. de Lange, I. Huntley, C. Keitel, & M. Niss (Eds.), *Innovation in maths education by modelling and applications* (pp. 19–30). Chichester, UK: Ellis Horwood.

Lyotard, J. F. (1984). *The postmodern condition: A report on knowledge* (G. Bennington & B. Massumi, Trans.). Manchester, UK: Manchester University Press. (Original work published 1979.)

Niss, M. (1994). Mathematics in society. In R. Biehler, R. W. Scholz, R. Sträßer, & B. Winkelmann (Eds.), *Didactics of mathematics as a scientific discipline* (pp. 367–378). Dordrecht, The Netherlands: Kluwer Academic.

OEEC. (1961). *New thinking in school mathematics.* Paris, France: OEEC.

Piaget, J. (1970). *Genetic epistemology.* New York, NY: Columbia University Press.

Porter, T. M. (1995). *Trust in numbers: The pursuit of objectivity in science and public life.* Princeton, NJ: Princeton University Press.

Powell, A. B., & Frankenstein, M. (Eds.). (1997). *Ethnomathematics: Challenging Eurocentrism in mathematics education.* Albany: State University of New York Press.

Restivo, S., van Bendegem, J. P.,& Fisher, R. (Eds.). (1993). *Math worlds: Philosophical and social studies of mathematics and mathematics education.* Albany: State University of New York Press.

Skovsmose, O., & Valero, P. (1999). *Breaking political neutrality: The critical engagement of mathematics education with democracy.* Aalborg, Denmark: Centre for Research in Learning Mathematics, Royal Danish School of Educational Studies, Roskilde University, Aalborg University.

Whitehead, A. N., & Russell, B. (1910/1913). *Principia mathematica I-III.* Cambridge, UK: Cambridge University Press.

Wittgenstein, L. (1992). *Tractatus logico-philosophicus* (C. K. Ogden, Trans.). London, UK: Routledge. (Original work published 1921)

CHAPTER 23

APORISM AND CRITICAL MATHEMATICS EDUCATION

A THOUGHT EXPERIMENT: Imagine yourself back in the Middle Ages joining a conference on religious education. In the dim light of the conference center, we are listening to many interesting lectures. One lecturer, who happens to come from France, makes an exegesis of the teaching of the Holy Trinity. He has investigated the sources of the Holy Writ and has arrived at new conclusions concerning the basic structure of the Trinity. He suggests that the curriculum in religious education, including the teaching of how to pray, should be changed in accordance with his findings. Instead of relating to a patchwork of information about the Trinity, the prayers can be categorized into simple structures, and a few basic prayers can serve as "mother structures" in the development of all sorts of advanced praying. To change the curriculum in accordance with this insight would mean that children become better equipped for further education, which naturally means further praying.

Another lecturer has found it possible to interpret the basic concepts expressing the structure of the Trinity in a way comprehensible to every child, independent of the stage of the child's intellectual development. This lecturer has obviously already been listening to the first one. Other lecturers announce that they

Critique as Uncertainty, pages 333–346.

have produced textbooks in accordance with these new ideas about how to teach the Holy Trinity.

However, critical opinions are voiced. Why not listen to the way children already produce their own simple prayers? Children are already preoccupied with praying! The only thing needed is to push the children smoothly in the direction of praying for the real Trinity. In this way, the teaching of the Holy Trinity can be based on children's own prayers with which they are familiar.

This suggestion is supported by scholars not living in the metropolis hosting the conference. They explain that in their country, they have observed many sorts of old and well-established prayers. Long before praying for the Holy Trinity was institutionalized, people have been praying, and in many respects, these old prayers anticipate praying for the Holy Trinity. The foreign scholars suggest that the religious curriculum should take into consideration these ethnoprayers. Using these as the basis, learning can be directed towards the paradigmatic prayers for the Holy Trinity. But why, in fact, prefer some prayers to others? Why stick to the old paradigm? All prayers might be equal.

During the happy hour of the conference, somebody raises the following questions: Why this concern for religious education? Why teach children to memorize prayers? What are the social and political functions of teaching everybody to pray? What is the purpose? Why not discuss the social function of religious education? These questions might show the voice of critical religious education.[1]

Critical Mathematics Education

What, then, might be the voice of critical mathematics education? I do not see critical mathematics education as a particular "topic" in the same sense that we talk about "elementary mathematics," "Cabri geometry," and "advanced calculus." Instead, I see critical mathematics education as an expression of some broader concerns about mathematics education. In our chapter, "Critical Mathematics Education" in the *International Handbook of Mathematics Education*, Lene Nielsen and I (1996) tried to enumerate some of these concerns:

- Education cannot be discussed only in terms of preparation for further education or for the labor market. Schooling also means preparation for citizenship and participation in social and political life. What does this mean to mathematics education?
- Could mathematics serve as a tool for identifying and analyzing critical features of social life?
- How could mathematics education consider the students' interests and competence for the development of knowledge and understanding?

[1] A similar thought experiment was presented in Skovsmose (1998).

- Mathematics education might provide "cultural filters" by, for instance, being a gatekeeper for the technological society. How could questions of equality, equity, and justice be reflected in the mathematics classroom?
- Mathematics might become a problematic tool for solving a wide range of problems, as mathematics itself is part of the technological society. Mathematics cannot only be a tool for critique; mathematics itself also has to be addressed by a critique, and in this sense become an "object of critique." What does this mean for mathematics education?
- Every classroom becomes a microsociety and may represent democracy *in spe* (or otherwise). What does this mean for the interactions between students and teacher in the mathematics classroom?

I do not think that responses to such concerns can be punctuated into particular claims about the organization of the mathematics classroom.[2] Critical mathematics education cannot be translated into principles for, say, curriculum development. Nevertheless, I find that educational practices can be discussed in terms of such concerns.

In what follows, I shall concentrate on the mathematical aspects of those concerns. My aim is to search for a working philosophy *of mathematics* for critical mathematics education. In doing so, I shall naturally be keeping in mind that critical mathematics education needs other working philosophies—of classroom interaction, of democracy, and so on.

Absolutism and Fallibilism

Absolutism refers to the basic assumption that it is possible, at least within certain areas of knowledge, to get access to absolute truths. The question is, however, how paths may be found that lead into this mesmerizing landscape. Which faculty of the mind could provide human beings with some knowledge which they could share with an omniscient God? The most tried-out route has been through purified reason.

Different areas of understanding have been proposed as candidates for absolute knowledge, most often mathematics. Absolutism has been supported by the idea that certain kinds of knowledge can be established *a priori*, independently of empirical evidence. *Apriorism* has been manifested in many forms in the philosophy of mathematics: For instance, in trying to build the architecture of mathematics on

[2] However, these concerns have certainly served as inspiration for much innovation in the classroom. For instance, the notions of "project work" and "thematizations" have been suggested: see, for example, Vithal, Christiansen and Skovsmose (1995). However, I find it important to distinguish between "inspiration" and "justification." I do not think that any conception of critical mathematics education can "justify" a particular practice. Critical mathematics education expresses a set of concerns which might become a challenge to both practice and educational theory and, in this sense, it might become a source of inspiration.

a foundation of logic, Gottlob Frege wanted to ensure that no empirical elements could destabilize the foundation of mathematics.

It is not surprising, then, that Imre Lakatos chose to outflank apriorism. He advocated *aposteriorism*: the idea that even mathematical knowledge is ultimately based on empirical evidence. In "A Renaissance of Empiricism in the Recent Philosophy of Mathematics," Lakatos (1978) refers to a great number of philosophers who, in one way or another, have provided support for aposteriorism. Human senses have cheated the human mind too often; and in classical philosophy, it is a well-accepted idea that those commodities which our senses bring to the stock of knowledge easily become tainted. So, if mathematics becomes pervaded by aposteriorism, then *fallibilism* will follow and certainty will be tarnished by the mold of doubt.

In Search of a Working Philosophy for Critical Mathematics Education

Fallibilism opens new doors in mathematics education. The constructivist idea, that knowledge cannot be imposed on students but must grow in a natural process of construction, deconstruction, rebuilding, repairing, is difficult to establish within an absolutist framework. Therefore, fallibilism has become a most welcome working philosophy of progressive mathematics education (see, for instance, Ernest, 1991, 1998).

However, can fallibilism be a working philosophy for critical mathematics education? My answer is "no." Why? Well, not because I find fallibilism to be a "wrong" philosophy of mathematics. In fact, I agree that many studies demonstrate that Lakatos' (1976) analysis reveals interesting aspects of mathematics (see, for instance, Koetsier, 1991). Nevertheless, the problem of fallibilism is its limited scope.

Lakatos' fallibilism is inspired by Karl Popper's (1963, 1965, 1972) critical rationalism and its internalist perspective on science. It is "internalist" in the sense that the focus is on the development of scientific concepts, and this development is defined only with reference to science itself. "The logic of scientific discovery" refers to a rational process taking place in what Popper calls the "third world"; the first world being that of experiences and the second that of physical objects. The third world is inhabited by theoretical constructions. While Popper's philosophy of critical rationalism provides an overall description of the development of the third world, Lakatos provides a description of a particular province of this world, that of mathematics.

This grand theorizing of scientific development provides a constructivist description of scientific progress which is linked with fallibilism; this I will label *rational constructivism*. This description does not represent actual history but the rational history of the "third world." This description is without any reference to the social context. The Popperian third world is one which seems to exist and to develop without interaction with social reality.

Rational constructivism represents a conception of epistemic progress. Although Popper gives up on the possibility of ever coming to grasp absolute truths, as all prevailing theories are only not-yet-refuted guesses, he assumes the possibility of getting closer and closer to the truth. Rational constructivism presupposes an epistemic optimism: Knowledge is progressing! Lakatos does not make any reference to the notion of mathematical truth, yet still his work shares a rational optimism: Science makes progress! (See also Lakatos, 1970, for a discussion of the methodology of scientific research programs.)

The assumption of rational constructivism is that, by being rational, scientific development represents a benign growth. The logic of scientific development does not present us with ethical problems: It represents the ultimate progress intrinsic to science itself.

As already mentioned, Lakatos' work has provided much inspiration for mathematics educators, but it has, nevertheless, a huge limitation as a working philosophy of mathematics for critical mathematics education. Nowhere does fallibilism, as developed with reference to rational constructivism, accommodate a broader and more fundamental interpretation of the social role of mathematics. This fallibilism is deeply ingrained in the internalist perspective on scientific development.

The optimism of rational constructivism, which means a belief in the notion of scientific progress, also legitimizes internalism. Therefore, the notion of critique, developed by critical constructivism, becomes intrinsic to scientific development, and is to promote the "logic of scientific discovery," not to establish an extrinsic critique of this development. Therefore, I find rational constructivism (including fallibilism) inadequate as a working philosophy of mathematics for critical mathematics education.

Aporism

The *paradox* which aporism acknowledges concerns the role of mathematics in society.[3] On the one hand, mathematics, the "queen of science," seems to be "pure" by only providing new extensions of human rationality. This must represent a worthwhile enterprise in itself. On the other, this extended rationality seems to play an essential role in social and technological development, manifested in the technonature which has come to contain devastating new risk structures. How can it be that the pure queen of science seems to play a contradictory role in social affairs?

By *aporism*, I refer to a certain perspective on mathematics which pays special attention to this paradox by grappling with the questions, What is mathematics doing to society? How can we conceptualize the social role of mathematics? These two questions will be discussed in what follows.

[3] See also Chapter 22.

Mathematics: A Formatting Power?

The thesis of the formatting power of mathematics states that, in many cases, social actions and decisions are prescribed by means of mathematics.

What social circumstances provide the basis for a certain development in mathematics? This question is closely linked with the concern about the social role of mathematics: What impact does mathematics have on society? From an analytic point of view, it makes sense to separate these two questions, although, naturally, they are closely linked. The thesis of the formatting power concentrates on what mathematics might be doing to society, leaving aside discussion of the social construction of mathematics.

It has to be acknowledged that this thesis is more complex than is indicated by its first simple formulation. It is not mathematics as such which is claimed to be doing anything; neither is it targetting mathematicians as a group. Rather, the thesis is about the interplay between mathematical techniques (systems produced in the process of mathematizing) and resources for technological development.

It makes sense to claim that modem warfare could not exist in any of its present forms without mathematics as a constituent part (see Tymozcko (1994), Højrup & Booss-Bavnbek, 1994, and Emmer, 1998). Mathematics is integrated in business and economic management in such a way that it does not make sense to try to separate this management from its mathematical expression. Mathematics is an integral part of all aspects of technology. These are just different manifestations of the thesis of the formatting power of mathematics. But is the thesis in fact true?

In *A Mathematician's Apology*, G. H. Hardy (1967) discusses the usefulness of mathematics. His general conclusion is

> If useful knowledge is . . . knowledge which is likely, now or in the comparatively near future, to contribute to the material comfort of mankind, so that mere intellectual satisfaction is irrelevant, then the great bulk of higher mathematics is useless. (p. 135)

Could mathematics, nevertheless, do any harm? "So a real mathematician has his conscience clear; there is nothing to be set against any value his work may have; mathematics is . . . a 'harmless and innocent' occupation" (pp. 140–141). In the final pages of his *Apology*, Hardy writes this about his own particular work in mathematics: "I have never done anything 'useful.' No discovery of mine has made, or is likely to make, directly or indirectly, for good or ill, the least difference to the amenity of the world" (p. 150). The thesis of the formatting power of mathematics contradicts this position directly: The Hardian claim of innocence is not tenable.

To emphasize the complexity of the role of mathematics within society, we can take a look at the notion of modeling. This notion is often used in the discussion of mathematics and its application. Modeling has become part of a well-elaborated "metaphysics" about mathematics. The idea is simply that "reality" is "pictured" by means of mathematics. A model is constituted by a mathematical vocabulary and connections between certain parameters are established in terms of mathematical equations. The different elements in the mathematical model refer to entities in reality, and manipulations of the model provide information about reality. The model, naturally, has to be validated, and this concerns the degree to which the model actually "pictures" reality. If this is an adequate interpretation of mathematical modeling, then Hardian innocence, most certainly, can be extrapolated to applied mathematics.

However, I do not think that a picture theory of mathematics-in-use provides an adequate understanding of what mathematics is doing to society. Instead of a picture theory of applications, we have to understand the deeper involvement of mathematics in social affairs.

By means of the notions of *modulation* and *constitution*, I have tried to express the idea that when part of reality becomes modeled and remodeled, then this process also influences reality itself. Mathematics applied, for instance, in business does not consist of "pictures" of a reality which exists prior to and independent of the modeling process. Mathematical models of advertising, marketing, investments, and so on become part of the economic reality themselves. They serve as a basis for decision making and for economic transactions. In this way, mathematics has become part of the economic reality. This not only applies to business but to economic policymaking in general, and not only to economy, but to categories like time, space, communication, transport, war. Modeling becomes a continuous process of the modulation of reality, and the thesis of the formatting power of mathematics can now be formulated as fundamental categories of techno-nature are continually modulated and eventually constituted by mathematics. By talking about technonature, I refer to the fact that we no longer live in nature but in an environment reconstructed by means of technology.

The thesis might be true. It might be false. Well, it *might* be true, and this is the possibility which is considered by aporism.[4] The concept of the "formatting power of mathematics" is at present best treated as a metaphor. In order to develop further the thesis that mathematics has formatting power, the notions of "formatting," "power," and "mathematics" are in need of a good deal of further investigation. As a first inspiration for this challenging analysis, I want to refer to a study by Keiko Yasukawa (1998).

In her paper, "Looking at Mathematics as Technology: Implications for Numeracy," Yasukawa examines ways in which mathematics works in society. She emphasizes that mathematics must be interpreted as a technology.

---

[4] If this is the case, then basic theoretical conceptions of social development and social nature appear to be inadequate. How could social theorizing, which proposes fundamental ideas for understanding the nature and the development of society, ignore the social role of mathematics? In this sense, the thesis of the formatting power of mathematics establishes a major challenge to social theorizing.

> I would suggest that mathematics itself can be viewed as a technology, although it is clearly not associated with an artefact such as computers are with information technology, solar panels are with power generation technology, and ticket machines are with transport technology. (1998, p. 352)

The technology of mathematics can then be characterized also as an "invisible technology":

> Mathematics models, as experienced by the community at large, is an invisible technology. It is invisible not only in the actual numbers that the models produce as "solutions," but invisible in its derivation. . . . Risks, pay rises, new tax rates, cut-off scores for university entrance are all prescriptions which mathematical models generate. . . . What the general public sees is typically the answer only. Rarely do they even see the original question that drove the modelling process. Yet, they become co-opted as consumers of these model-generated solutions. (1998, p. 353)

asukawa also finds useful the characterization used by Porter (1995) of mathematics as a technology of distance, in the sense that mathematics may provide distance between social groups. She writes:

> The authority of mathematics, based on the perceived objectivity of what "truths" it can convey, is at one level a critical part of explaining how mathematics creates a distance (and in some cases gulfs) between the "haves" and "have nots" of mathematical knowledge. (1998, p. 355)

So, a first step into the analysis of the formatting power of mathematics could concern the invisibility of the technology of mathematics and the distancing which this technology may produce (see also Keitel, 1993). Such a discussion would lead directly to the notion of power and to a clarification of the nature of the power associated with mathematics.

Critique of Pure Reason?

The title of Immanuel Kant's work (1933), *Critique of Pure Reason* (*Kritik der reinen Vernunft*), refers to a fundamental epistemic dilemma. No doubt, reason (pure or not) needs to be criticized. In classical epistemology, the task took the following form: The foundation of knowledge has to be prepared by means of a fundamental critique. Rationalists, positivists, and many others have embarked on such a once-and-for-all critique. But from what source did they carry out such a critique? By means of pure reason itself? This problem of self-reference is nicely illustrated by the title *Critique of Pure Reason.*

In the tradition of critical theory, a substantial amount of critical argument has been leveled at *instrumental reason*, which is certainly distinct from pure reason. Instrumental reason is related to Max Weber's concept of *Zweckrationalität* (goal rationality), which refers to a mode which would develop civilization in bureaucratic forms. In order to discuss aspects and expressions of instrumental reason, critical theory has tried to incorporate a network of topics. Two questions, however, remain: How to carry out a critique of the even more complex form of "reason," which is produced in the process of mathematical development, and which expresses itself in mathematical formatting. How to provide a critique of the rationality of rational constructivism.

Let me try to express this problem in a terminology which has been suggested by Ulrich Beck (1992, 1994). In this way, I want to relate the discussion of aporism to a recent discussion in sociology. He puts forward the possibility "that the transition from one social epoch to another could take place unintended and unpolitical, by-passing all the forums for political decisions, the lines of conflict and the partisan controversies" (1994, p. 3). Beck has also introduced the notion of *risk society*: "This concept designates a developmental phase of modem society in which the social, political, economic and individual risks increasingly tend to escape the institutions for monitoring and protection in industrial society" (p. 5). Beck suggests *reflexivity* as an overall description of the automatic feedback system which has come to govern social development. A risk society is produced by a forceful "feedback" of industrial society:

> One can virtually say that the constellations of risk society are produced because the certitude of industrial society . . . dominate[s] the thought and action of people and institutions in industrial society. Risk society is not an option that one can choose or reject in the course of political disputes. It arises in the continuity of autonomised processes which are blind and deaf to their own effects and threats. Cumulatively and latently, the latter produces threats which call into question and eventually destroy the foundations of industrial society. (pp. 5–6)

It is produced because of the predominant "certitude" of the industrial society, and Beck summarizes: "Let us call the autonomous, undesired and unseen, transitions from industrial to risk society reflexivity" (p. 6). Reflexivity, therefore, refers to self-confrontations. Industrial society will contain certain products of society itself, and the accumulation of these products redirects society into new forms. The thesis of the formatting power of mathematics can now be stated in a different way: "Some basic elements of social reflexivity are modulated and eventually constituted by mathematics." In other words, mathematics takes part in the process leading us into risk society. This opens the possibility of a most interesting study: How could "formalization" become a vehicle for risk production?

Every society might try to develop an understanding of its social dynamics. We need *reflection* to grasp reflexivity. Beck again: "In the risk society, the recognition of the unpredictability of the threats provoked by techno-industrial development necessitates self-reflection on the foundations of social cohesion and the examination of prevailing conventions and foundations of 'rationality.'" (1994, p. 8)

The challenging problem is that (epistemic) reflection might be much weaker than (social) reflexivity. We might find ourselves trapped in a situation without sufficient means for grasping the social reality in which we are situated. In particular, we might be unable to grasp the role of mathematics in social affairs. It might be impossible to carry out a critique of reason, in case this reason has expressed itself as mathematical formatting. Aporism acknowledges this possibility as a basic condition of humankind.

Naturally, a weakness of reflection would not constitute any problem, in the case that Hardy's analysis is adequate. If not, we face a big problem: We are not even able adequately to reflect upon and to understand what we ourselves have constructed, namely, technological society. How is it possible that reason seems to express itself much stronger in social reflexivity than in epistemic reflection? The production of reason gets beyond the reach of reflection, and we are left with the devastating doubt: Have we become cheated by reason itself?

Supported by Michel Foucault, we have come to realize the possibility that the whole rational enterprise brings humankind into a calamitous situation. This is the problem of the Enlightenment. The rational development of society, of technology, and of the economy (supported by the most advanced elements of science) might bring us to disaster itself. The presumptions of the Enlightenment do not allow mathematical rationality to be identified as a critical epistemic challenge. Critical rationalism shares the same presumptions: Scientific development means progress! And, certainly, mathematics does not cause any harm!

It is a well-known fact that it was for disobedience (human knowledge) that we were driven from the Garden of Eden. However, the further development of the rational enterprise might bring us into a much more heated situation.

Aporism as a Working Philosophy of Mathematics

Classical philosophies of mathematics do not face the possibility that a fundamental reflection of what is produced by mathematics is of social and ethical importance. Instead, they relate the discussion of mathematics to questions like What is mathematical truth? What is mathematical reality? What is mathematical development?

Aporism suggests a different agenda in the philosophy of mathematics. Aporism tries to emphasize issues which are important to a "politics of mathematics."[5]

Mathematics is considered a cultural system, and not only in the sense that it may be engraved in a certain exotic culture which we can explore by, for in-

[5] Hersh (1998) set up some "criteria" for a philosophy of mathematics. In this way, he establishes a critical stance in the philosophy of mathematics, breaking away from the tradition that the philosophy of mathematics first of all has to do with the foundations of mathematics. Of the 13 criteria Hersh does mention, he emphasizes the following 3 as essential: (1) The philosophy of mathematics should recognize "the scope and variety of mathematics," it must fit "into general epistemology and philosophy of science," and it must be "compatible with mathematical practice—research, application, teaching, history, calculations, and mathematical intuition" (see p. 33). Considering this broad picture of mathematics, aporism represents a smaller element. Nevertheless, I see aporism as essential in order to relate an understanding of mathematics to a broader understanding of society. See also the criteria set up by Ernest (1998).

stance, ethnographic means. Mathematics, including research mathematics and mathematics education, is seen as a fundamental aspect of our culture. In this way, the thesis of the formatting power of mathematics can be read as a thesis about conditions for the understanding of culture.

Thus, the thesis is indicative of our "critique of culture"—critique of art, architecture, literature, and so on. A critique of culture must also include a critique of technology, including mathematics; otherwise the term a "critique of culture" will lose significance, especially in the context of our technologically based society.

In summary, aporism emphasizes two basic elements of uncertainty about mathematics. The first concerns the possible roles of mathematics in social affairs; the second concerns our ability to come to identify and criticize these roles. These two elements are important for the development of critical mathematics education. What is more, aporism might be relevant outside the scope of mathematics education. If the claims of aporism are valid, they will be highly relevant to any projects in social theorising.

What Then Does Aporism Mean for Mathematics Education?

I have suggested aporism as a working philosophy of mathematics for critical mathematics education. But what does this mean? As emphasized in the introduction, I do not see critical mathematics education as a particular topic. Instead, I have made an attempt to describe it in terms of concerns, referring to mathematics education in general. Aporism might help to accentuate some of them.

Aporism will challenge any working philosophy of mathematics education which does not provide a critique of scientific knowledge, including mathematical knowledge. A general implication of aporism is that a critique of knowing, and also of scientific knowing, becomes an educational task. Such a critique cannot be dismissed merely as a philosophical task or one for scientific methodology alone. The claim of any critical education is that all forms of knowing must be addressed by a critique as part of the educational process.

This has an implication for constructivism which has developed as a most popular working philosophy of mathematics education. I do not claim that constructivism is false. Constructivism has helped to "humanize" the mathematics classroom by embracing some of the assumptions of fallibilism which, first and foremost, concern the nature of mathematical knowledge. However, fallibilism has not made any great inroads into a critical investigation of the role of mathematics in social affairs. Fallibilism has been developed in accordance with rational constructivism and constructivism as a perspective on education and learning, and has repeated the internalist perspective of knowing. Nowhere in *Radical Constructivism in Mathematics Education*, edited by Ernst von Glasersfeld (1991), or in his own book, *Radical Constructivism: A Way of Knowing and Learning* (1995), do we find considerations related to "a politics of knowing." Radical constructivism shares with rational constructivism the assumption that mathematics, as a scientific discipline, is in no need of an "extrinsic" critique of rationality.

But what about social constructivism? First, a remark about Lev Vygotsky. It is clear that Vygotsky (1978, 1986) emphasized the importance of the social aspects of learning processes. At the same time, Vygotsky seems to have been influenced by the scientific paradigm which claims that science, once organized in a "proper" way, does not need any further critical assistance from other faculties, in particular not from education; in fact, no critical education developed from within a Marxist or positivist paradigm. Certainly, positivism suggests a critique of science to eliminate any kind of metaphysics from this field. Marxism has also suggested a critique of science to eliminate any "bourgeois" elements which might have corrupted science. However, these suggestions have not been transformed into suggestions for a critical educational practice. The task of education, then, becomes the enculturation of students into the scientific area by supporting them in their constructions of knowledge and in challenging of students' knowledge from the position of the socially accepted form of scientific knowledge. The basic aim of education is merely to provide a scaffolding, not a critique.

In *Social Constructivism as a Philosophy of Mathematics*, Paul Ernest (1998) suggests Lakatos and Wittgenstein's philosophies of mathematics as the "two main sources for the social constructivist philosophy of mathematics" (p. 134). Both philosophies, however, ignore the social functions of mathematics, although Wittgenstein certainly acknowledges the sociel formation of mathematics. Therefore, a considerable amount of reworking seems necessary. Ernest formulates a set of adequacy criteria for a philosophy of mathematics (pp. 56–57). Such a philosophy must be able to account for questions related to the following fields: mathematical knowledge, mathematical theories, the objects of mathematics, the applications of mathematics, mathematical practice, and the learning of mathematics. He confronts social constructivism with this set of criteria and manages to show that the adequacy criteria have in fact been met.

However, in his discussion, Ernest does not raise questions constituting a critique of mathematics. I do not see in Ernest's evaluation of social constructivism as a philosophy of mathematics a concern for topics and questions essential to aporism. In his final chapter, Ernest discusses the important subject of values and social responsibility but adds the following remark: "Treating these issues exceeds the primary objective of social constructivism in accounting for mathematics and mathematical knowledge naturalistically" (1998, p. 268). As a consequence, I find social constructivism, in this formulation at least, inadequate.

In its recent formulation, constructivism, radical or social, does not provide a framework which acknowledges the problematic nature of scientific rationality. In this sense, constructivism is still situated within the paradigm of scientific optimism related to the intemalism which characterizes rational constructivism.

Aporism attempts to consider mathematics in its social, cultural, and political complexities and to express uncertainties about the possible role of this science as an educational concern. In this way, aporism might assist in articulating the voice of critical mathematics education.

NOTE

This paper was published in 2000 in *For the Learning of Mathematics, 20*(1), 2–8.

REFERENCES

Beck, U. (1992). *Risk society: Towards a new modernity* (M. Ritter, Trans.). London, UK: Sage. (Original work published 1986)

Beck, U. (1994). The reinvention of politics: Towards a theory of reflexive modernization. In U. Beck, A. Giddens, & S. Lash, *Reflexive modernization: Politics, tradition and aesthetics in the modern social order* (pp. 1–55). Cambridge, UK: Polity.

Emmer, M. (1998). The mathematics of war. *Zentralblatt für Didaktik der Mathematik, 98*(3), 74–77.

Ernest, P. (1991). *The philosophy of mathematics education.* Brighton, UK: Falmer.

Ernest, P. (1998). *Social constructivism as a philosophy of mathematics.* Albany: State University of New York Press.

Hardy, G. H. (1967). *A mathematician's apology.* Cambridge, UK: Cambridge University Press. (Original work published 1940)

Hersh, R. (1998). *What is mathematics, really?* London, UK: Vintage.

Højrup, J., & Booss-Bavnbek, B. (1994). On mathematics and war: An essay on the implications, past and present, of the military involvement of the mathematics sciences for their development and potentials. In J. Højrup (Ed.), *Measure, number, and weight: Studies in mathematics and culture* (pp. 225–278). Albany: State University of New York Press.

Kant, I. (1933). *Critique of pure reason* (N. K. Smith, Trans.). London, UK: Macmillan. (Original work published 1781)

Keitel, C. (1993). Implicit mathematical models in social practice and explicit mathematics teaching by applications. In J. de Lange, I. Huntley, C. Keitel, & M. Niss (Eds.), *Innovation in maths education by modelling and applications* (pp. 19–30). Chichester, UK: Ellis Horwood.

Koetsier, T. (1991). *Lakatos' philosophy of mathematics: A historical approach.* Amsterdam, The Netherlands: North-Holland.

Lakatos, I. (1970). Falsification and the methodology of scientific research programmes. In I. Lakatos, & A. Musgrave (Eds.), *Criticism and the growth of knowledge* (pp. 91–196). Cambridge, UK: Cambridge University Press.

Lakatos, I. (1976). *Proofs and refutations: The logic of mathematical discovery* (J. Worrall & E. Zahar, Eds.). Cambridge, UK: Cambridge University Press.

Lakatos, I. (1978). A renaissance of empiricism in the recent philosophy of mathematics. In J. Worrall & G. Currie (Eds.), *Mathematics, science and technology: Philosophical papers* (Vol. 2, pp. 24–42). Cambridge, UK: Cambridge University Press.

Popper, K. R. (1963). *Conjectures and refutations.* London, UK: Routledge and Kegan Paul.

Popper, K. R. (1965). *The logic of scientific discovery.* New York, NY: Harper and Row. (Original work published 1934)

Popper, K. R. (1972). *Objective knowledge.* Oxford, UK: Oxford University Press.

Porter, T. M. (1995). *Trust in numbers: The pursuit of objectivity in science and public life.* Princeton, NJ: Princeton University Press.

Skovsmose, O. (1998). Critical mathematics education: Some philosophical remarks. In C. Alsini, J. M. Álvarez, B. Hodgson, C. Laborde, & A. Pérez (Eds.), *8th International Congress on Mathematical Education: Selected Lectures* (pp. 413–425). Sevilla, Spain: S.A.E.M. 'THALES'.

Skovsmose, O., & Nielsen, L. (1996). Critical mathematics education. In A. Bishop, M. A. Clements, C. Keitel, J. Kilpatrick, & C. Laborde (Eds.), *International handbook of mathematics education* (pp. 1257–1288). Dordrecht, The Netherlands: Kluwer Academic.

Tymozcko, T. (1994). Humanistic and utilitarian aspects of mathematics. In D. F. Robitaille, D. H. Wheeler, & C. Kieran (Eds.), *Selected lectures for the 7th International Congress on Mathematical Education* (pp. 327–339). Sainte-Foy, Canada: Les Presses de l'Universite Laval.

Vithal, R., Christiansen, I. M.,& Skovsmose, O. (1995). Project work in university mathematics education—A Danish experience: Aalborg University. *Educational Studies in Mathematics*, *29*(2), 199–223.

von Glasersfeld, E. (Ed.) (1991). *Radical constructivism in mathematics education*. Dordrecht, The Netherlands: Kluwer Academic.

von Glasersfeld, E. (1995). *Radical constructivism: A way of knowing and learning*. London, UK: Falmer.

Vygotsky, L. (1978). *Mind in society: The development of higher psychological processes* (M. Cole, V. John-Steiner, S. Scribner, & E. Souberman Eds.). Cambridge, MA: Harvard University Press.

Vygotsky, L. (1986). *Thought and language* (A. Kozulin, Trans.). Cambridge, MA: MIT Press. (Original work published 1934)

Yasukawa, K. (1998). Looking at mathematics as technology: Implications for numeracy. In P. Gates (Ed.), *Proceedings of the First International Mathematics Education and Society Conference* (pp. 351–359). Nottingham, UK: Centre for the Study of Mathematics Education, University of Nottingham.

CHAPTER 24

MATHEMATICS EDUCATION VERSUS CRITICAL EDUCATION

Several times it has been emphasized that mathematics education, to be understood as research into the mathematics educational process, bears a complicated relationship to other subjects of research. In what follows, my use of the concept mathematics education should in some cases be identified with the German concept *Didaktik der Mathematik*, in other cases, it simply refers to the educational process in mathematics.

The development of mathematics education as a scientific discipline, which was initiated in the late 1960s, has led to the following questions being asked: (1) What is the object(s) of the discipline? (2) What scientific method(s) should be used? (3) What relations does the new field have to other more established scientific disciplines? Anna Zofia Krygowska has given a very precise description of the problems:

> Didactics of Mathematics is developing as a typical "border" discipline. Every independent discipline is characterised by the specificity of its problems, of its language, and of its research methods. In its first phase of development the border subject has a rather vague status. In particular, its methods of research can be quite inhomogeneous. On the one hand mathematics education develops at the border of mathemat-

Critique as Uncertainty, pages 347–362.

ics, its philosophy, and its history; on the other hand at the border of pedagogy and psychology. (Krygowska, 1971, p. 118)

In the following, I shall discuss a point in relation to the above question (3); I shall look at the connections between mathematics education and general educational theories. In my opinion, these connections appear to be rather selective. An important branch of general education, interwoven with an important school of epistemology and philosophy, is hardly mentioned and not conceptualized in mathematics education (ME). I am thinking of critical education (CE), critical pedagogy, and ideas of critical theory.[1] And my basic postulates are as follows:

It is necessary to increase the interaction between ME and CE, if ME is not to degenerate into one of the most important ways of socializing students (to be understood as students or pupils) into the technological society and at the same time destroying the possibilities of developing a critical attitude towards precisely this technological society.

It is important for CE to interact with subjects from the technical sciences, and among these ME, if CE is not to be taken over by the technological development and fade away into an unimportant and uncritical educational theory.

In the following, I shall concentrate on postulate (A), but postulate (B) should be kept in mind.[2]

Critical Education

CE has several sources of inspiration. There exists a strong association with Karl Marx's understanding of humanism and of society, especially as expounded by the Frankfurt School[3] (or Critical Theory). The original main figures of this school were Theodor W. Adorno, Max Horkheimer, and Herbert Marcuse. They worked at the Institut für Sozialforschung in Frankfurt am Main; the institute was founded in 1923, and in 1923–1929 Carl Grünberg was the director of the institute. He was succeeded by Horkheimer, who in his inaugural address (1931) put forward three themes which should characterize the School of Frankfurt:

The first [theme] suggests the necessity of re-specifying "the great philosophical questions" in an interdisciplinary research programme. The second theme, more implicit but made clearer in later essays, is a call for a rejection of orthodox Marxism and its substitution by a reconstructed understanding of Marx's project. The third emphasizes the necessity for social theory to explicate the set of interconnections (mediations) that make possible the reproduction and transformation of society, economy, culture and consciousness. (Held, 1980, p. 33)

[1] The neglect of CE is obvious, for instance, in Hans-Georg Steiner's introduction in Steiner (1978).
[2] In Skovsmose (1980, 1981a, 1981b, 1984), I have discussed relationships between ME and CE.
[3] For a further exposition of Critical Theory, see, for instance, Jay (1973) and Held (1980).

When the Nazis came to power in Germany, the staff of the institute continued their work outside Germany, especially in the United States. After the war, Adorno and Horkheimer returned, but Marcuse remained in the United States. Later, Jürgen Habermas became the dominant figure of the school.[4]

Another, but much less important source of inspiration for CE is to be found in the *Geisteswissenschaftliche Pädagogik*. In a fundamental way, this educational theory is inspired by hermeneutics, as worked out by Wilhelm Dilthey.[5] Other important persons in the *Geisteswissenschaftliche Pädagogik* are Eduard Spranger, Theodor Litt, and Herman Nohl. This pedagogy dominated the educational discussions in Germany between the two World Wars; and again after World War II. Later, Wolfgang Klafki, belonging to this school of education, also shows inspiration from the Frankfurt School.

Some of the main persons in the (first) more theoretical development of CE are Herwig Blankertz, Wolfgang Lempert, Klaus Mollenhauer, and also Wolfgang Klafki. Basically, they try to develop pedagogy as a praxiological research discipline as a reaction to the empirical-positivistic tradition in pedagogy; for example, Blankertz tries to relate pedagogy to Habermas' theory of interests. Similar problems are discussed by Klafki (1971).

Also Oskar Negt (1964) reveals a close relation to critical theory in his theoretical basis, but he gives CE a more independent and original foundation, as he takes into consideration a broad spectrum of political, economic, and psychological aspects. Negt does not limit the development of CE primarily to philosophy, which has been a characteristic tendency in the first phase of CE. In Scandinavia, Negt's version of CE has been discussed in great detail and developed for instance by Knud Illeris (1974, 1981).[6]

Main Terms in Critical Education

It is, of course, impossible to sum up the ideas of CE in a few statements. However, I shall make an attempt, although it may create misunderstandings.

For CE, the relation between teacher and students plays an important role. Several types of relationship are possible, but CE emphasizes that an important principle is that the partners are equal. Paulo Freire has discussed the teacher-students relationship in connection with what he calls "emancipatory pedagogy":

> Through dialogue, the teacher-of-the-students and the students-of-the-teacher cease to exist and a new term emerges: teacher-student with students-teachers. The teacher

[4] For an exposition and examination of the philosophy of Habermas see McCarthy (1978). Also Habermas relates the Critical Theory to the hermeneutics of Dilthey; see the discussion in Habermas (1968).

[5] See, for instance, Nohl (1970).

[6] Identifying these alternatives, primarily I am concerned with ME in Europe since the mid-1960s. I believe that a quite different conceptual framework has to be introduced if trends and problems related to the countries of the so-called Third World are to be discussed.

> is no longer merely the-one-who-teaches, but one who is himself taught in dialogue with students, who in their turn while being taught also teach. They become jointly responsible for a process in which all grow. (Freire, 1972, p. 53)

The ideas concerning the dialogue and the student-teacher relationship are developed from the general point of view that education must belong to a process of democratization. If a democratic attitude is to be developed through education, education as a social relationship should not contain fundamentally undemocratic features. It is not acceptable that the teacher (alone) has the decisive and prescribing role. Instead the educational process must be understood as a dialogue.

Summarily, we can specify the first key term in CE as involvement of the students in the control of the educational process. Or, in other words, in CE the students (and the teachers) are attributed a *critical competence.*

This competence is attributed mainly to the students for two reasons. First, for *factual* reasons, that the students, although their experience is faulty, fragmentary, and such, also have a general experience which in a dialogue with the teacher enables them to identify relevant subjects for the educational process; relevant both in relation to the immediate interests of the students but also in relation to the general perspective of the educational process. Secondly, for *principle* reasons, that if education should develop a critical competence, such a competence cannot be imposed upon the students, but must be developed from an already existing capability.

An educational process involves persons (students, teachers) but of course also a subject matter (the curriculum). The next basic key term of CE is critical consideration of contents and other subject-matter aspects. Formulated otherwise, in CE, both the students and the teacher must establish a *critical distance* to the content of the education. In German, this key term is *Fachkritik*, translated into English, "curriculum critique" perhaps could be used.

In a curriculum critique, we try to put the seemingly objective and value-free principles for the structuring of the curriculum into a new perspective as we try to reveal such principles as value loaded. Questions in a curriculum critique concern the following:

- The applicability of the subject matter: Who use it? Where is it used? Which types of qualifications are developed in ME?
- The interests behind the subject: Which knowledge-constituting interests are connected with the subject?
- The assumptions behind the subject: Which questions and which problems have generated the concepts and the results in mathematics? Which contexts have promoted and controlled the development?
- The functions of the subject: Which possible social functions could the subject have? This question does not primarily concern the possible applications, but the implicit function of a ME with regard to attitudes towards

technological questions, the student's attitudes towards his own capacities and so on.

- The limitations of the subject: In which areas and in relation to which questions is the subject without any importance?

The last key term of CE concerns conditions outside the educational process. It could be formulated as problem orientation of the teaching-learning process. The essential thing is that the educational process is related to problems existing outside the educational universe. Moreover, several criteria can be used to select these problems. The two fundamental criteria are the following. The subjective one: The problem must be conceived as relevant from the perspective of the students, it must be possible to encircle and define the problem in close connection to the experience and conceptual framework of the students. And the objective one: The problem must have a close relation to objective existing social problems.

This last key term about problem orientation gives perspective to the previous terms concerning critical competence and critical distance. The problem orientation implies that the dimension of *critical engagement* must be involved in education.

Alternatives in Mathematics Education

Mathematics education includes several very different, even inconsistent, ideas about mathematics and education. To make the discussion as precise as possible, I shall distinguish between three alternatives in ME: (1) structuralism, (2) pragmatism, and (3) process orientation.[7] Furthermore, I shall relate the above alternatives to the key terms of CE in order to show the extent to which ME and CE contradict each other.

Structuralism is characterized by the following assumptions: The essence of mathematics can be determined by crystallizing fundamental concepts through logical analysis of existing mathematical theories; these fundamental concepts can be conveyed to the learner by means of suitable concretizations in accordance with the epistemological potentials of the child. The structural point of view is then characterized by an idea about mathematics (to be associated with the name of Nicolas Bourbaki), an idea about educational communication and transformation (Bruner, 1960), and an idea about epistemology (Piaget, 1970).

The assumptions of structuralism now exemplify an almost complete negation of CE, a fact which does not seem to have affected the proponents of structuralism in any way. I shall illustrate this by a remark of Jean Dieudonné:

> If the people responsible for drawing up school curricula may be persudaded to consult professional mathematicians in order to understand the relevance of their decisions to science as it is practiced in the university and beyond, we may yet wit-

[7] A basis for an applied oriented ME is found in Pollak (1979).

> ness one day some sensible teaching of mathematics from kindergarten to graduate school. (Dieudonné, 1973, p. 19)

Here we have an explicit negation of the idea about critical competence, and implicitly the ideas about critical distance and critical engagement are also negated.

Structuralism bears a close relationship to the tradition in pedagogy: "Teach the disciplines!" This could be interpreted both as a principle governing the selection of school subjects as well as a principle governing the presentation of the subject matter. Basic to this tradition is the idea that the students' knowledge has to be built up in accordance with structures and contents identified independently of the students.

In accordance with the *pragmatic* trend in ME, the essence of mathematics is to be found in its applications and thus, in a sense, outside mathematics. In the process of education, it is therefore highly important to illustrate the various ways in which mathematics can be useful. This pragmatic trend can be understood in a broad sense, and many arguments have been put forward to support an application-oriented ME.[8] This broad pragmatic trend can be interpreted as a reaction against the structuralism of the 1960s.

However, the pragmatic trend can also be interpreted in a much more specific way. In the article "Mathematics, Applicable versus Pure-and-Applied," Christopher Ormell emphasizes that the work of the Sixth form Mathematics Project at Reading is based on a specific Working Philosophy of mathematics which he relates to the philosophy of Charles Sanders Peirce who has noted, "Mathematics is the study of what is true of hypothetical states of things" (Peirce, 1960, p. 193). Ormell stresses the relevance of this working philosophy about mathematics as follows:

> What has altered the situation today is the emergence of the Peircean point of view; which provides for the first time a definite, coherent account of the role of mathematics in science and affairs. It has become clear that the unique contribution which mathematics makes to society operates not via its relevance to the actual, but via its relevance to the not yet and would be actual: via the light it sheds on the hidden and semi-hidden implications of candidate innovations, hypotheses, proposals, programmes and projects of all kinds. The "relevance" of mathematics is not to a crass utilitarianism, but to the possibilities of new patterns of explanation in science and new patterns of organization in social and technical affairs. (Ormell, 1972a, pp. 129–130)

The basic principles of Ormell's interpretation of mathematics are as follows: First, mathematics is the science of hypothetical situations. Secondly, the central

[8] The introduction to "Mathematics Applicable" is *Mathematics Applicable: Teaching Mathematics Applicable, Introductory Guide*, School Council Sixth Form Mathematics Project, Heinemann Educational Books for the School Council, London 1979.

thing in ME is as simply as possible to illustrate how mathematics can be used to handle and to investigate hypothetical situations.

Ormell seems to argue in favor of the first proposition on two levels—not clearly separated; and it is not evident whether Ormell finds that the two levels ought to be separated. At one time he says that mathematics could be interpreted as if it *in fact* is a science of possible situations. In "Mathematics, Applicable versus Pure-and-Applied," Ormell (1972a) develops this point of view by, among other things, looking at the history of mathematics. He finds that even the development of Greek geometry fits into the pattern of the Peircean view. In "Mathematics, Science of Possibility," Ormell (1972b) gives further reasons for the descriptive part of the argument. At other times, he says that mathematics ought to be interpreted as a science of possible situations, if we are to develop an adequate interpretation of mathematics. This is the normative part of the argument in favor of the Peircean view.

Secondly, it was stated that the crucial point of ME is to explain how mathematics can become a science of hypothetical situations; and the textbooks *Mathematics Applicable* illustrate how this can be done in elementary ways.[9]

If we now turn to the key terms of CE, it is obvious that *Mathematics Applicable* does not consider the passage about critical competence. Although the adoption of the Peircean approach implies that mathematics must be seen primarily as a model-building activity, and not as a body of results, in *Mathematics Applicable* it is difficult to find a platform for an authentic teacher-student dialogue. The textbooks contain examples of mathematical modeling, and the process of education must, at least according to the exposition, move from one thoroughly prepared example to another thoroughly prepared example. What the students have to do is to calculate, to find the solutions, and to solve already well-defined problems.

With regard to the key term, critical consideration of contents and other subject matter aspects, the situation is somewhat different. An important element in the curriculum critique concerns the question, Which (type of) problems and which (type of) questions have generated the technical terms? To some extent, this aspect could be compatible with the pragmatic trend. But the element of "propaganda for the usefulness of mathematics," which is a rather characteristic feature of *Mathematics Applicable*, is quite foreign to curriculum critique.

The pragmatic trend is problem-oriented. However, in CE, it is essential that the problems have to do with fundamental social situations and conflicts, and it is important that the students simultaneously can recognize the problems as "their own problems." Compare the objective and the subjective criteria of problem identification in CE. Problems must not belong to "play realities" without any significance except as an illustration of mathematics as a science of hypothetical situations. In general, therefore, there exists a large gap between CE and the pragmatic trend in ME.

[9] For a general exposition of the "genetic" trend in ME, see Schubring (1978).

When looking at the general theory of education, we can easily relate the pragmatism in ME to the pragmatic educational philosophy, as for instance worked out by John Dewey (1938). It is, however, not possible to equate the two types of pragmatism. Dewey emphasizes the experimental activity of the students to a much higher degree than the textbook system *Mathematics Applicable* does. Furthermore, it is difficult to use any ready-made educational program in the framework of the pedagogy of Dewey. However, in a metaperspective, we find the CE criticism of the Dewey pragmatism to be parallel to the above-sketched criticism of the Ormell pragmatism.

The third trend in ME which should be mentioned is *process orientation.* According to this viewpoint, the essence of mathematics is neither connected to particular concepts nor to the applicability (the usefulness) of mathematics as such, but to the processes of thought which have led to the mathematical insight. And it is stressed that the main concern of mathematics education is to give students opportunities for making reinventions of their own.[10]

One of the most important proponents of this view is Hans Freudenthal. In the preface of *Mathematics as an Educational Task*, Freudenthal writes, "My educational interpretation of mathematics betrays the influence of L. E. J. Brouwer's view on mathematics (though not on education)" (Freudenthal, 1973, p. ix.). If Brouwer's intuitionistic interpretation of mathematics should be summarized in few words, it could be that mathematics is a human construction. This constructivism implies that the mathematical concepts and truths are to be constructed, not to be discovered. Freudenthal uses the formulation, "Mathematics is a human activity." Here he stresses, first, that the essential thing is the activity, the process of thought, which led up to the mathematical concepts. Secondly, that this activity is a general human activity, not an exclusive one only for specific talented persons.

From this point of view, the fundamental thing in education is to make the students able to create mathematics. Freudenthal writes,

> Science at its summit has always been creative invention, and today it is even so at levels lower than that of the masters. The learning process has to include phases of directed invention, that is, of invention not in the objective but in the subjective sense, seen from the perspective of the student. It is believed that knowledge and ability acquired by re-invention are better understood and more easily preserved than if acquired in a less active way. (Freudenthal, 1973, p. 118)

In many ways the work of IOWO (Instuut voor de Ontwikkeling van het Wiskunde Onderwijs) has illustrated possible practical implications of such general phrases.[11]

This process orientation has something in common with CE. What I especially have in mind has to do with critical competence. In the work of IOWO, a crucial

[10] See Brouwer (1975).

[11] See Five Years IOWO (1976).

point is the refusal to use all sorts of prestructured and ready-made material as developed in connection with structuralism. Instead, IOWO tries to create situations which lead to mathematization, which means development of mathematical ideas and concepts. A process-oriented ME must reject the idea of concretization, to be understood as an elementarization of (abstract) mathematical concepts, in favor of a mathematization. But there exists an important asymmetry between these two sorts of activities. To concretize is to give more abstract terms a concrete interpretation and in this way to make them more comprehensible. The activity of concretization is reserved for the planners of the curriculum (compare with the remark of Dieudonné), and as such this activity is removed from the educational process. To mathematize means, in principle, to formulate, to criticize, and to develop ways of understanding; and hence mathematization must take a leading role as an integrated part of the educational process: both the students and the teacher must be involved in the control of this process. However, it is important to note that the control in question has certain limits, because the students have to criticize inside a preestablished situation. So, the process orientation (in the program of IOWO) builds upon a limited concept of critical competence and not upon the broader concept in CE, which ascribes critical competence to the students.

In *Mathematics as an Educational Task*, Freudenthal (1973) rejects the widespread use of "play realities" in education. Instead, he recommends mathematics fraught with relations:

> When speaking about mathematics fraught with relations, I stressed the relations with a lived-through reality rather than with a dead mock reality that has been invented with the only purpose of serving as an example of application. This is what often happens even in arithmetic teaching. I do not repudiate play-realities. At a low level games may be useful means of motivation. But it is dangerous to rely too much on games. Ephemeral games are no substitutes for lived-through reality. The rules of games that are not daily exercised are as easily forgotten as mathematics or even faster. The lived-through reality should be the backbone which joins mathematical experiences together. However motivating and charming games may be, they can never fill this place. (pp. 78–79)

But according to CE, several of the examples developed by IOWO look like play realities, and the concept of problem orientation in the IOWO context is not identical with the corresponding concept of CE. It is as if only a part of the key terms of CE is to be found in the process orientation of ME as exemplified by Freudenthal and IOWO.

The process orientation of ME has a rather definite relationship with general educational traditions. Like the Reform Movement, or the Progressivism, which was introduced at the beginning of the last century by Ellen Key, Maria Montessori, and others, this process orientation puts the child into the center of the educational process. The curriculum content should not be imposed upon the child, instead the potentials of the child should be developed in a rich and stimulating milieu. The process orientation of IOWO also has roots in the tradition of Alexander I. Wittenberg and Martin Wagenschein. Wittenberg (1963) connects his philosophy of ME with *Geisteswissenschaftliche Pädagogik* and Wagenschein (1970) stresses the principles of exemplarity. Both relate their ideas to a genetic principle, which has to do with process orientation. The tradition of Wittenberg and Wagenschein is also related to CE, and the same sort of limitations, as mentioned above, are found again. For instance, it is interesting to note the remarks of Heinz-Dieter Hermann about Wittenberg and Wagenschein concerning ME in the Glocksee project.[12]

What is the Problem?

Obviously, the conclusion of the previous section is that there does not exist any integration—nor even any close relationship—between ME and CE. The strongest contradiction is found between structuralism and CE, but neither pragmatism nor process orientation has much in common with CE. But what is the problem? Why should any such integration exist? Is it important to establish one? Through some theses which are to be formulated in the following, I shall argue in favor of the importance of this integration. Compare my postulates A and B.

The first thesis has to do with the relationship between technology and society, and I shall underline an idea, already formulated by the French philosopher of technology, Jacques Ellul, in *The Technological Society*. According to Ellul (1964), technology has replaced nature as the environment of man. Technology must be conceived as a closed circle around man. In Ellul's words:

> Technique has progressively mastered all the elements of civilisation. . . . Man himself is overpowered by technique and becomes its object. The technique which takes man for its object thus becomes the centre of society: this extraordinary event . . . is often designated of technical civilization. The terminology is exact and we must fully grasp its importance. Technical civilization means that our civilization is constructed by technique (makes a part of civilization only what belongs to technique), for technique (in that everything in this civilization must serve a technical end), and is exclusively technique (in that it excludes whatever is not technique or reduces to technical form). (pp. 127–128)

Apart from finding that technology completely closes around man, Ellul finds that the evolution of technology is fully determined by technology itself, and that society and the way of living is fully determined by this evolution; therefore Ellul's view on technology is deterministic—and pessimistic too. However, I do not intend to discuss the different kinds of determination between technology and society, only to stress the postulate about the total integration of society into technology and vice versa. Let me emphasize it in the following: *Thesis of technology*

[12] Detailed information about the Glocksee project is found in the Danish journal *Kontext*, No. 35, 1978.

(*Ellul's thesis*): "Technology is the dominating feature of civilization, and man is completely immersed in technology." This thesis deals with power and power relations, because via technology it is possible to establish and/or to increase power relations. A corollary of the above thesis is that man is located in a civilization with power relations determined by and integrated in a technological structure.[13]

The next thesis deals with education, especially the curriculum. It is my opinion, and a general one too, that power relations can be embedded into a specific curriculum organization.[14] I shall give a single formulation:

> The original introduction of pre-packed material was stimulated by a specific network of political, cultural, and economic forces, originally in the 1950s and 1960s in the United States. The views of academics that teachers were unsophisticated in major curriculum areas "necessitated" the creation of what was called teacher-proof material. The cold war climate (created and stimulated by the State in large part) led to a focus on the efficient production of scientists and technicians as well as a relatively stable work-force; this, the "guaranteeing" of this production through the school curriculum became of increasing import. (Apple, 1982, p. 150)

In this formulation, we have interwoven, at least implicitly, the following two propositions: (1) The contents of the curriculum are determined, not primary by factual causes having to do with the logical structure of the curriculum, but with economic and political forces having to do with the power relations of the society; and (2) the curriculum can function as an extension of the existing social relations. It must be noted that (1) does not imply that for instance the specific contents in the structuralistic curriculum are not derived from logical analysis of mathematics, but (1) implies that the structuralistic reform is made possible because a structuralistic curriculum seems to be in accordance with dominating economic and political interests: The interests in a stable workforce with technical skills without regard to political, humanistic, and moral reflections.

Summarizing the above, we state the following thesis: *Thesis of curriculum*: "The fundamental structuring principles of the curriculum are derived from or are in accordance with the dominant power relations of the society." After formulating the two general theses, we shall now look more closely at education in mathematics. In this connection, I find two different aspects important. First, how does ME serve to introduce a particular way of thinking, acting, evaluating and so on in a technological society? To what extent does ME in the educational system work as an instrument for technological interests? These questions naturally deal with the problems about the hidden curriculum in ME.[15]

---

[13] For more details about philosophy of technology, see, for instance, Rapp (1981).

[14] See, for instance, the article "An Approach to the Study of Curricula as Socially Organized Knowledge" by M. F. D. Young in Young (1971, pp. 19–46). Also, B. Bernstein's conceptions are of special importance in this relation (Berstein, 1971/1975).

[15] An important analysis and discussion of the hidden curriculum is found in Broady (1981).

The next aspect which I find important concerns the students and the relationship among the students. In an educational system, the structure of knowledge (in a narrow sense) of the students is developed and transformed, but also the structure of social relations among the students and the students' experience of it are changed. Several investigations indicate that ME plays a part in reorganizing the world around the students. They take up an attitude towards the technological society. They learn that some people can manage technological problems, and that some people cannot. And consequently, the "incapable" students learn to become servile to technological questions and to be servile to those who can manage such questions.

I shall state my idea about ME as follows:

> *Thesis about mathematics education*: *In the educational system*, ME serves as the most significant introduction to the technological society. It is an introduction which both endows (a part of) the students with relevant technical skills, and at the same time endows (all) the students with a "functional" attitude towards the technological society. ("Functional" is seen from the perspective of the dominating power structures.)

The basic axiom in CE is that education must not serve as a passive reproduction of existing social relations and power relations. This axiom makes sense in the talk about critical competence, critical distance, and critical engagement. Education has to play an active role in identifying and combating social disparities. Naturally, education does not play any important role in social and technological changes; such changes do not follow educational enterprises; but education must strive to play an active role parallel to other critical social forces.

In education, theory as well as practice, CE is the most important movement which tries to negate the thesis of curriculum. The intention of CE is to unmask the dominating structuring principles of the curriculum as historical and accidental. In the light of Ellul's thesis, the most important school subjects must be "technologies" (in a broad sense), with respect to ME this is underlined by the thesis about mathematics education. So, these were the general ideas on which I have based my postulate (A): that it is necessary to increase the interaction between ME and CE if ME is not to degenerate into one of the most important ways of socializing students into the technological society.

Any Solutions?

There are examples of CE carried out in practice; in Denmark, both in the primary and in the secondary schools, but also at university level, particularly at the University Centre of Roskilde and at the University Centre of Aalborg. Two different strategies are used when a CE practice is developed: *thematization* or *project organization*. While thematization is widely used in the primary school and (perhaps to less extent) in the secondary school, the project orientation is found at university level, particularly at the abovementioned university centers.

It is possible to use the strategy *thematization* even if the educational structure of the school system is rather traditional. By integrating different school subjects and by cooperation between the teachers, it could be possible to eliminate the demarcations between the school subjects and to put the timetable out of function, to make room for long and continuous periods of work. The strategy *project organization* is more demanding, in that it (as developed at university level) requires not only a special organization of study program, but also a special arrangement of the university building. When elaborating projects, the students have to work together in groups, and to make their work much like "real" research work, it is important that every project group has a room of their own at the university, where to work, to collect material, to receive advice, and so on. Much experience in connection with these strategies has been gained through the last 10 years, and evidently, much thematized and project organized education is developed without a close relation to CE. But from the perspective of CE, these two strategies are still considered to be "reasonable" (if not necessary), but they are in no way sufficient as strategies. Problem orientation has to play an important role.

CE has had almost all its influence on school subjects in the humanities and the social sciences. As previously indicated, there does not exist much influence on the technical subjects, with exceptions naturally, for instance, at the abovementioned university centers. However, this does not imply that recommendations do not exist for such an integration of CE. We have an example in the article "Plädoyer für einen problemorientierten Mathematikunterricht in emanzipatorischer Absicht" by Volk (1975).

These and other such recommendations I could summarize in some criteria for the selection of the problem in a problem-oriented ME as follows:

- It should be possible for the students to realize that the problem is of importance. That is, the problem must contain subjective relevance for the students. It must have to do with situations connected to the experience of the students.
- The problem must have to do with important processes in society.
- Somehow and to some extent the engagement of the students in the problem situation and in the process of problem solving should prepare a basis for (later) political and social engagement.

These criteria constitute an explicit specification of the intentions behind CE, but such types of criteria do not play any important role in the discussion of the contents in ME. The criteria normally found in ME are in some way related to mathematics itself: to the logic of mathematical structures as in structuralism, to the applicability of mathematics as in pragmatism, or to the mathematical way of thinking as in the process orientation.

It seems to be more and more difficult to realize a CE the closer the concerned section of the educational system is to technology. This has to do with Ellul's thesis and the thesis of the curriculum. To preserve the functionalism of the educational system, it is necessary, particularly for the technological subjects (and for mathematics), to maintain a demarcation between the factual and the normative, including all ethical questions, questions about relevance and so on. This demarcation is intimately connected with the postulate of the social neutrality of the criteria for selection of curriculum content. This postulate is asserted in structuralism in the most explicit way and negated in CE.

The whole idea of CE is to interpret the curriculum and education as a normative structure; and if the conceptualization of the (from a sociological point of view) most important normative curriculum structures is to be effected in practice, CE has to be explicitly integrated into the technical sciences and into ME. CE cannot do without such an integration. That is the background for my postulate (B): that it is important for CE to interact with the technical sciences and mathematics to prevent CE fading away into an unimportant and uncritical educational theory.

However, as described, CE has not influenced ME in any important respect. Moreover, most of the advocates of CE who have developed its special aspects do not show any interest in ME. (Exceptions like Volk are involved in ME.) The CE ignorance of ME is built into the conceptual framework of CE, and in my opinion, this ignorance is a catastrophe for CE, because it will destroy its critical power.

One of the explanations of this CE ignorance is found in the philosophy of technology of Critical Theory, which concept of technology is based on the assumption that technology can be identified with the natural sciences. In Jürgen Habermas' philosophy the knowledge-constituting interest of the natural sciences is technical and quite different from the emancipatory interest of the social sciences; so obviously, an emancipatory enterprise as worked out in CE can ignore technology, including mathematics.

This conception of technology is, however, strongly criticized. It does not establish a basis for analyzing the function of technology in society. The philosophy of technology in Critical Theory cannot be used if the intention is to influence the technological structures themselves.

Therefore, one of the main jobs for a CE is to develop a more adequate philosophy of technology so that CE can manage and interpret technical education, so CE and ME can become integrated, so ME can become a critical education, and so CE again can become critical.

NOTE

This paper was published in 1985 in *Educational Studies in Mathematics*, *16*(4), 337–354. The original title of the publication was "Mathematical Education versus Critical Education," however I prefer the term "mathematics eduation."

REFERENCES

Apple, M. W. (1982). *Education and power*. London, UK: Routledge & Kegan Paul.

Bernstein, B. (1971/1975). *Class, codes and control* (Vols. 1–3). London, UK: Routledge & Kegan Paul.

Broady, D. (1981). *Den dolda läroplanen* [The hidden curriculum]. Stockholm, Sweden: Symposion Bokförlag.

Brouwer, L. E. J. (1975). Consciousness, philosophy, and mathematics. In A. Heyting (Ed.), *L. E. J. Brouwer: Collected works, Volume I, Philosophy and foundations of mahematics* (pp. 480–494). Amsterdam, The Netherlands; Oxford, UK; New York, NY: North-Holland/American Elsevier. (Original work published 1948)

Bruner, J. (1960). *The process of education.* Cambridge, MA: Harvard University Press.

Dewey, J. (1938). *Experience and education.* Columbus, OH: Kappa Delta Pi.

Dieudonné, J. A. (1973). Should we teach "modern" mathematics? *American Scientists, 61*, 16–19.

Ellul, J. (1964). *The technological society.* New York, NY: Free Press.

IOWO. (1976). Five years IOWO. *Educational Studies in Mathematics, 7*, 189–368.

Freire, P. (1972). *Pedagogy of the oppressed.* Harmondsworth, UK: Penguin.

Freudenthal, H. (1973). *Mathematics as an educational task.* Dordrecht, The Netherlands: D. Reidel.

Habermas, J. (1968). *Erkenntnis und Interesse.* Frankfurt, Germany: Suhrkamp.

Held, D. (1980). *Introduction to critical theory.* London, UK: Hutchinson.

Illeris, K. (1974). *Problemorientering og deltagerstyring* [Problem orientation and participant direction]. Copenhagen, Denmark: Munksgaard.

Illeris, K. (1981). *Modkvalificeringens pædagogik* [Pedagoy of resistance]. Copenhagen, Denmark: Unge Paedagoger.

Jay, M. (1973). *The dialectical imagination.* London, UK: Heinemann.

Klafki, W. (1971). Erziehungswissenschaft als kritischkonstruktive theorie: Hermeneutik—Empirie—Ideologiekritik [Educational research as criticla constructive theory: Hermeneutics—Empiricism—Critique of ideology]. *Zeitschrift für Pädagogik, 17*, 351–385.

Krygowska, A. Z. (1971). Mathematik: Didaktische Forschung an der Pädagogischen Hochschule Krakau [Mathematics: Educational research at the educational highschool krakau]. *Beiträge zum Mathematikunterricht*, 117–125.

Lempert, W. (1971). *Leistungsprinzip und Emanzipation* [Performance and emancipation]. Frankfurt, Germany: Suhrkamp.

McCarthy, T. (1978). *The critical theory of Jürgen Habermas.* London, UK: Hutchinson.

Mollenhauer, K. (1973). *Erziehung und Emanzipation* [Education and emancipation]. Munich, Germany: Juventa Verlag.

Negt, O. (1964). *Soziologische Phantasie und exemplarisches Lernen* [Sociological imagination and exemplary learning]. Frankfurt am Main, Germany: Europäische Verlagsanstalt.

Nohl, H. (1970). *Die pädagogische Bewegung in Deutschland und ihre Theorie* [The educational movement in Germany and its theory]. Frankfurt, Germany: Verlag G. Schulte-Buhnke.

Ormell, C. P. (1972a). Mathematics, applicable versus pure-and-applied. *International Journal of Mathematical Education in Science and Technology, 3*, 125–131.

Ormell, C. P. (1972b). Mathematics, science of possibility. *International Journal of Mathematical Education in Science and Technology, 3*, 329–341.

Peirce, C. S. (1960). *Collected papers, Vol. IV: The simplest mathematics*. Cambridge, MA: Belknap Press of Harvard University Press.

Piaget, J. (1970). *Genetic epistemology*. New York, NY; London, UK: Columbia University Press.

Pollak, H. O. (1979). The interaction between mathematics and other school subjects. In ICMI (Ed.), *New trends in mathematics teaching IV* (pp. 232–248). Paris, France: UNESCO.

Rapp, F. (1981). *Analytical philosophy of technology*. Dordrecht, The Netherlands: D. Reidel.

Schubring, G. (1978). *Das Genetische Prinzip in der Mathematik-Didaktik* [The genetic principle in mathematics education]. Stuttgart, Germany: Klett-Cotta.

Skovsmose, O. (1980). *Forandringer i matematikundervisningen* [Changes in mathematics education]. Copenhagen, Denmark: Gyldendal.

Skovsmose, O. (1981a). *Matematikundervisning og kritisk padagogik* [Mathematics education and critical education]. Copenhagen, Denmark: Gyldendal.

Skovsmose, O. (1981b). *Alternativer i matematikundervisningen* [Alternatives in mathematics education]. Copenhagen, Denmark: Gyldendal.

Skovsmose, O. (1984). *Kritik, undervisning og matematik* [Critique, education, and mathematics]. Copenhagen, Denmark: Lærerforeningernes Materialeudvalg.

Steiner, H.-G. (Ed.). (1978). *Didaktik der Mathematik* [Mathematics education]. Darmstadt, Germany: Wissenschaftliche Buchgesellschaft.

Volk, D. (1975). Plädoyer für einen problemorientierten Mathematikunterricht im emanzipatorischer Absicht [Plea for a problem-oriented mathematics education with an emancipatory interest]. In M. Ewers (Ed.), *Naturwissenschaftliche Didaktik zwischen Kritik und Konstruction* [Natural science education between critique and construction] (pp. 203–234). Weinheim, Germany: Belz Verlag.

Wagenschein, M. (1965, 1970). *Ursprüngliches Verstehen und exaktes Denken, I-II*. [Original understanding and exact thinking, I–II] Stuttgart, Germany: Ernst Klett Verlag.

Wittenberg, A. I. (1963). *Bildung und Mathematik* [Education and mathematics]. Stuttgart, Germany: Ernst Klett Verlag.

Young, M. F. D. (Ed.). (1971). *Knowledge and control*. London, UK: Collier Macmillan.

NAME INDEX

Critique as Uncertainty, pages 363–367.

SUBJECT INDEX

Critique as Uncertainty,, pages 369–374.

www.ingramcontent.com/pod-product-compliance
Lightning Source LLC
LaVergne TN
LVHW052223200926
848481LV00001B/197
* 9 7 8 1 6 2 3 9 6 7 5 3 6 *